<u>a 7</u>
6, 7, 11, 12
due Wed.

Introduction to Dynamic System Analysis

Introduction to Dynamic System Analysis

NORMAN H. BEACHLEY
HOWARD L. HARRISON

UNIVERSITY OF WISCONSIN–MADISON

Harper & Row, Publishers

New York, Hagerstown, San Francisco, London

Sponsoring Editor: Charlie Dresser
Project Editor: David Nickol
Designer: Robert Sugar
Production Supervisor: Marion A. Palen
Compositor: Science Typographers, Inc.
Printer and binder: The Murray Printing Company
Art Studio: J&R Technical Services, Inc.

Introduction to Dynamic System Analysis

Copyright © 1978 by Norman H. Beachley and Howard L. Harrison

All rights reserved. Printed in the United States of America. No part of this book may be used or reproduced in any manner whatsoever without written permission except in the case of brief quotations embodied in critical articles and reviews. For information address Harper & Row, Publishers, Inc., 10 East 53rd Street, New York, N.Y. 10022.

Library of Congress Cataloging in Publication Data

Beachley, Norman H   1933-
    Introduction to dynamic system analysis.

    Bibliography: p.
    Includes index.
    1. Vibration.   2. Feedback control systems.
3. System analysis.   4. Dynamics.   I. Harrison,
Howard L., joint author.   II. Title.
TA355.B38   620'.72   78-16608
ISBN 0-06-040557-0

Contents

Preface

Chapter **1**	**Introduction**	*1*
Chapter **2**	**Writing Differential Equations**	*4*

    2-1   Introduction *4*
    2-2   Mechanical Systems *4*
    2-3   Hydraulic Systems *18*
    2-4   Electric Systems *21*
    2-5   A Thermal System *25*
    2-6   An Ecological System *26*
    2-7   A Political-Military System *29*
    2-8   Conclusion *30*
          Problems *30*

Chapter **3**	**Solving Differential Equations**	*40*

    3-1   Introduction *40*
    3-2   Solution of Homogeneous Equations *43*
    3-3   Consideration of Initial Conditions *47*
    3-4   Solution of Nonhomogeneous Equations *49*

- 3-5 Transient and Steady-State Responses *52*
- 3-6 Combining Simultaneous Equations *53*
- 3-7 Conclusion *53*
  Problems *54*

Chapter 4 Linear Equations for Modeling Nonlinear Systems — 57

- 4-1 Introduction *57*
- 4-2 Examples of Nonlinearities *58*
- 4-3 Principles of Linearization *62*
- 4-4 Conclusion *73*
  Problems *73*

Chapter 5 Introduction to Vibrations — 80

- 5-1 Introduction *80*
- 5-2 Basic Principles and Definitions *82*
- 5-3 Units *88*
- 5-4 Conclusion *90*
  Problems *90*

Chapter 6 Free Vibration: Systems with a Single Degree of Freedom — 96

- 6-1 Introduction *96*
- 6-2 Undamped Free Vibration *97*
- 6-3 Damped Free Vibration *104*
- 6-4 Determination of Damping Ratio from Experimental Data *110*
- 6-5 Conclusion *112*
  Problems *113*

Chapter 7 Forced Vibration: Systems with a Single Degree of Freedom — 121

- 7-1 Introduction *121*
- 7-2 Examples of Mechanical Vibrating Systems with Sinusoidal Inputs *122*
- 7-3 Frequency Response *125*

CONTENTS

- 7-4 Vibration Caused by Rotating Unbalance *137*
- 7-5 Displacement Input Acting Through a Dashpot and Spring in Parallel *141*
- 7-6 Transmissibility *146*
- 7-7 Systems with Simultaneous Inputs at Two or More Frequencies *149*
- 7-8 Conclusion *153*
  Problems 156

Chapter 8 More Complex Single-Degree-of-Freedom Systems — 166

- 8-1 Introduction *166*
- 8-2 Determining the Degrees of Freedom of a Mechanical System *166*
- 8-3 Method of Analysis *169*
- 8-4 Equivalent Inertia, Damping, and Spring Rate *176*
- 8-5 Conclusion *178*
  Problems 179

Chapter 9 Vibrating Systems with More Than One Degree of Freedom — 186

- 9-1 Introduction *186*
- 9-2 Writing the Equations of Motion *186*
- 9-3 Two-Mass System Without Damping *191*
- 9-4 Two-Mass System with Damping *198*
- 9-5 Vibrating Systems with More Than Two Degrees of Freedom *202*
- 9-6 Conclusion *202*
  Problems 203

Chapter 10 Distributed Parameter Systems — 209

- 10-1 Introduction *209*
- 10-2 Rigorous Analysis of a Distributed Parameter System *211*
- 10-3 Beams with Other Boundary Conditions *218*

10-4 Conclusion *219*
Problems 219

Chapter 11 Critical Speeds of Rotors — 221

11-1 Introduction *221*
11-2 Analysis of a Simple Lumped Parameter Rotor *222*
11-3 Effect of Compliance in Bearings and Bearing Mounts *226*
11-4 Other Critical Speed Considerations *229*
11-5 Similar Rotor Instability Phenomena *234*
11-6 Conclusion *235*
Problems 236

Chapter 12 Balance of Rotors — 242

12-1 Introduction *242*
12-2 Static Balance *243*
12-3 Dynamic Balance *245*
12-4 Conclusion *251*
Problems 252

Chapter 13 The Feedback Control System — 256

13-1 Introduction *256*
13-2 Home Heating Application *257*
13-3 Feedback Control Systems *258*
13-4 Conclusion *259*

Chapter 14 System Response and Stability — 261

14-1 Introduction *261*
14-2 Response of a Liquid-Level System *262*
14-3 First-Order System Response to a Step Input *264*
14-4 First-Order System Response to a Sinusoidal Input *266*

14-5 Second-Order System Response *270*
14-6 Response of Higher-Order Systems to Simple Sinusoidal Inputs *274*
14-7 The Concept of System Stability *275*
14-8 Conclusion *277*
  Problems 277

Chapter 15 Control Actions 284

15-1 Introduction *284*
15-2 Proportional Control *285*
15-3 Integral Control *288*
15-4 Proportional-plus-Integral Control *291*
15-5 "Bang-Bang" Control *293*
15-6 Other Control Actions *297*
15-7 Conclusion *297*
  Problems 298

Chapter 16 Block Diagrams 303

16-1 Introduction *303*
16-2 The Transfer Function *304*
16-3 The Block Diagram *307*
16-4 Block Diagram Algebra *308*
16-5 Liquid-Level Integral Control System Example *314*
16-6 Systems with Two or More Inputs *318*
16-7 Feedback Compensation *321*
16-8 Conclusion *323*
  Problems 324

Chapter 17 State-Variable Formulation and Computer Solutions 327

17-1 Introduction *327*
17-2 State-Variable Formulation *328*
17-3 State-Space Trajectories *334*
17-4 Analog Computer Solutions *336*
17-5 Digital Computer Solutions *339*
17-6 Conclusion *344*
  Problems 344

Chapter 18 Experimental Determination of System Dynamic Characteristics — 348

18-1 Introduction *348*
18-2 Test Equipment and Instrumentation *350*
18-3 Applying System Inputs *359*
18-4 Recording and Interpreting System Response *362*
18-5 Conclusion *371*
Problems 372

Appendix A Deflection of Beams — 376

Appendix B Alternative Mathematical Expressions for Certain Harmonic Functions — 380

Appendix C Solution of Equations — 383

Appendix D Steady-State Solutions by Rotating-Vector and Complex-Number Techniques — 386

Appendix E Rayleigh's Energy Method — 394

Appendix F Decibel Conversion Table — 397

Appendix G Routh's Criterion — 398

Bibliography *401*

Index *403*

Preface

This book is designed to be used for an introductory undergraduate course in dynamic system analysis. Although there is heavy emphasis on mechanical vibrations and feedback control systems, a wide variety of other dynamic systems is introduced, with examples, to illustrate the broad application of the theory developed. The material in the book is based primarily on linear theory. In keeping with this orientation, techniques for adequately representing nonlinear systems by linear equations are presented and discussed in some detail.

One of the principal goals of this book is to demonstrate the unity and broad applicability of dynamic system theory. An understanding of the material presented will give a student the proper background for a more detailed study of any of the more specific fields of dynamic systems, such as vibrations, control systems, and electric circuit analysis.

The underlying response characteristics of a dynamic system can be characterized by the form of the differential equation describing it rather than by the laws that were used in deriving the equation. This concept, and the unity of dynamic system theory in general, is often missed by the student who studies vibrations, control systems, and other dynamic system topics in separate courses rather than as a unified discipline. In addition, the separation of the topics often leads to alternate sets of terminology; it may take the student a long time (if he ever accomplishes it) to realize, for example, that "forced vibration" means the same as "system response to a sinusoidal input," that "free vibration" is virtually the same phenomenon as "step response," and so forth.

It is felt by the authors that dynamic system theory is a valuable tool with widespread application that is sometimes ignored by many researchers in fields outside of engineering. A better understanding of many biological, ecological, social, and political problems, and more accurate predictions of the results of different actions are possible, in many cases, by the application of dynamic system theory. Use of the theory can be valuable even when the laws governing a system, and the values of system parameters, are difficult to define precisely. The organization of the book allows its ready use by other than engineering students if they have an adequate background in mathematics and physics. A knowledge of calculus is essential. Previous study of differential equations is recommended, although it is not essential because the book briefly covers all the differential equation theory necessary for the material presented.

When used in a mechanical engineering curriculum, this book will allow a good introduction to the more important aspects of vibration theory and an introduction to the fundamental principles of feedback control system theory, while at the same time pointing out the broad applicability of generalized dynamic system theory to a wide range of other seemingly unrelated topics. This approach is particularly valuable for the colleges where the crowded engineering curriculum has room for only a single "vibrations" course.

Sufficient material is included for a one-semester, three-credit course. Although there is more material on mechanical vibrations than any other topic, the material has been arranged so that it is not necessary to cover all sections of the vibrations chapters in order to use and understand the remainder of the material in the book. By omitting some of the latter sections in the chapters on vibration, omitting entirely the chapters on distributed parameter systems, critical speeds of rotors, and balance of rotors (Chapters 10, 11, and 12), and possibly the material on linearization (Chapter 4), the remainder of the material in the book could be covered in much less than a semester's work. Used in this manner, the book can serve as a compact presentation of the basic theory required for specialized study in other areas of dynamic system analysis.

Although there have been many people who have given us useful inputs, we would like to acknowledge specifically the help of Dr. William H. Park of Pennsylvania State University, who went over the original manuscript in great detail, making many suggestions that were incorporated in the final draft.

Introduction to Dynamic System Analysis

Introduction

This book is devoted to the study of dynamic systems. We should therefore start out with some discussion of what is actually meant by the term *dynamic system*. The word *system* is defined in Webster's dictionary as "an assemblage of objects united by some form of regular interaction or interdependence." For *dynamic system* we add the restriction that the response of the system will vary with time when it is disturbed or acted upon by some external excitation. This dynamic behavior is typically defined by differential equations, with time as the independent variable. Dynamic systems meeting our definition can be found in nature as well as in man-made machines and devices.

Probably the most obvious example of a dynamic system is a vibrating system consisting of mass, spring, and friction elements. Here the dynamic action is quite obvious because the movement of the parts (i.e., the system response) can usually be seen or felt. A disturbance such as a blow from a hammer or a sudden movement of one of the components will cause a predictable vibration that eventually dissipates. A steadily applied periodic force will cause a vibration that continues with time, with characteristics determined by both the system parameters and the characteristics of the input. Although most people will envision a man-made machine whenever the term vibration is mentioned, there are many objects occurring in nature that are clearly vibrating systems. The vibration characteristics of a tree limb or leaf are not basically different from those of certain machine elements.

Other systems, though quite different in physical form, have dynamic characteristics that are mathematically the same as those of mechanical vibrating systems. For example, electric circuits composed of resistive, capacitive, and inductive elements will oscillate (vibrate) under the proper type of excitation, and the differential equations describing such systems are the same as those describing the analogous mechanical vibrating systems. The dynamic action in this case cannot be seen directly; it is not the movement of an object, but rather the rise and fall of voltages and currents. A hydraulic system has dynamic properties characterized by pressure changes and flow rate variations. Thermal systems may have dynamic variations in temperature and the flow rate of heat energy.

A feedback control system is another good example of a dynamic system. Because such a system is frequently used to control the position or velocity of some physical object, its dynamic action will often have the same physical appearance as that of a mechanical vibrating system.

There are many other systems whose dynamic properties are not quite so obvious as those discussed above. However, these properties, when mathematically analyzed, are found to be defined by differential equations of the same form. For example, the flow of traffic on a congested turnpike has been found to have dynamic characteristics. An analysis based on the fairly predictable characteristics of human beings behind the wheel will result in differential equations that can be used to predict the effects of various types of traffic disturbances. For example, a dog running across the highway might cause oscillations of the traffic flow rate that persist for a long period of time. [6].[1]

Ecological systems have dynamic characteristics. The population of a species of insect or vertebrate in a given region can vary markedly from year to year because of factors such as the number of predators (and interactions between the predators and prey), disease, parasites, weather conditions, and the food supply (which may be decimated by an overabundance of the animals). Many populations have been found to vary cyclically, with a period that is remarkably consistent; for example, a population peak has been found to occur every 9 to 10 years for mink, every 4 years for field mice, every 2 years for pink salmon, and so on [1].

---

[1]Numbers in brackets correspond to reference numbers in the Bibliography.

INTRODUCTION

The time variation of accumulation of pesticides, such as DDT, in various animals provides another example of an ecological dynamic system —one that can be defined and studied through the use of differential equations [8].

When, in our study of various types of dynamic systems, we find the same form of differential equation occurring for widely different physical systems, it will pay to take special note of the fact. The characteristics of one such system, obtained through detailed analysis or experimental studies, can often be applied directly to gain a better understanding of analogous systems. Sets of performance curves prepared for any particular system can readily be applied to all others defined by the same form of differential equation. Such curves are particularly useful and easy to use when they are prepared in dimensionless form.

The type of information that one wants to know about a dynamic system is essentially the same regardless of the physical details of the system. It is important to know:

1. How the system responds with time for any particular type of disturbance.
2. How long it will take for the dynamic action to dissipate if the disturbance is applied only briefly and then removed.
3. Whether the system is stable or if its oscillations will increase in magnitude with time even after the disturbance has been removed.
4. What modifications can be made to the system to improve its dynamic characteristics with regard to some specific application.

Although one of the goals of this book is to teach the fundamentals of dynamic system analysis as a *unified discipline*, the order of presentation is chosen so that the first systems studied are those whose dynamic actions are easy to visualize as actual motions of an object. Mechanical mass-spring-damper systems are emphasized in the first part of the book. When other dynamic systems are studied later, many of the principles and definitions learned from the study of vibrating systems can be applied directly.

In attempting to achieve a unified approach to the study of dynamic systems, we use the same nomenclature and terminology wherever possible for phenomena that are mathematically the same even though their physical manifestations may be different. One of the goals of this book is to develop in the student that proper analytical approach and insight that will allow him to apply the information learned to all types of dynamic systems, including those not specifically covered here.

Writing Differential Equations

2-1 Introduction

Differential equations are used to define the behavior of dynamic systems. The purpose of this chapter is to illustrate the way in which differential equations are obtained. The concept is simple: Differential equations are written from an understanding of the basic laws underlying the system being considered. Newton's second law of motion ($\Sigma F = Ma$) is an example of such a law. A number of different types of dynamic systems will be considered here.

2-2 Mechanical Systems

Our study begins by considering a number of spring-mass-damper systems encountered in vibration work. The fundamental law used is

$$\Sigma F = Ma \tag{2-1}$$

2-2 MECHANICAL SYSTEMS

in which F is force, M is mass, and a is acceleration.[1] With the traditional U.S. units, force may be expressed in pounds and mass in pound second^2 per inch (lb · s^2/in.); the corresponding units of acceleration are inches per second^2 (in./s^2). In SI units, force is expressed in newtons (N), mass in kilograms (kg), and acceleration in meters per second^2 (m/s^2). Of course, any other consistent units are valid.

SPRING-MASS SYSTEMS (DAMPING ASSUMED ZERO)

Figure 2-1a shows a mass M resting on a frictionless surface and connected to a rigid wall through a spring with rate k (N/m or lb/in.). The displacement of the mass is y. This is called a *translational* system, since the motion of the mass is translation. The term *linear system*, often used for systems of this type, is reserved in this book for systems described by linear differential equations (to be covered in Chapters 3 and 4). The problem is to write an equation (the differential equation for the system) that, when solved, will yield an expression for the displacement of the mass as a function of time.

To solve the problem, the displacement of the mass will be considered zero with the system at equilibrium; that is, the spring is neither in tension nor in compression and therefore exerts no force on the mass. Furthermore, displacements in the direction of the arrow (to the right in this case) will be taken as positive displacements, those to the left as negative. Thus, a displacement y can be zero, positive, or negative. Once a sign convention has been adopted, it must not be changed during the course of the problem. Also, and this is very important, the sign convention applies not only to displacement but to force, velocity, and acceleration as well. To illustrate, a velocity to the right is a positive velocity, whereas a force acting on the mass and urging it to the left is a negative force.

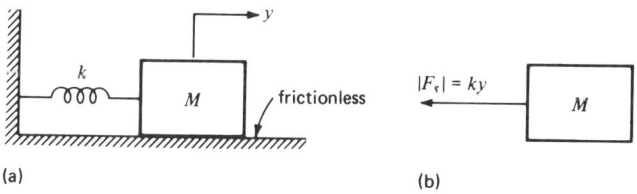

FIG 2-1. (a) Spring-mass system.
(b) Free-body diagram with mass displaced to the right (positive direction).

---

[1] A more general form of Newton's second law is $\Sigma F \propto Ma$. The equality of Eq. 2-1 is correct only when compatible units are used. A detailed discussion of units is given in Section 5-3.

Acceleration is, of course, the second derivative of displacement with respect to time. For this problem, Eq. 2-1 can therefore be written as

$$M\frac{d^2y}{dt^2} = \Sigma F$$

The only force acting on the mass is the spring force, which has a magnitude ky (spring rate multiplied by displacement); however, the correct sign must be associated with this magnitude. Let us assume that the mass is displaced in the positive direction (to the right); the spring force F_s will act to restore the mass to the equilibrium position as shown in Fig. 2-1b. Following the sign convention established, this force is negative. Therefore,

$$F_s = -ky$$

It is instructive to study this expression and to be convinced that it will yield the correct spring force for any possible displacement. If y is zero, F_s is zero. This is correct because the displacement was initially taken to be zero with the spring relaxed. If y is negative (that is, the mass is displaced to the left), F_s is positive. This is correct because the spring will be placed in compression and will act to move the mass in the positive direction. Thus, the expression for the spring force is correct for any possible displacement of the mass. Because the spring force is the only force acting on the mass, the differential equation for the system can be written as

$$M\frac{d^2y}{dt^2} = \Sigma F = F_s = -ky$$

or

$$M\frac{d^2y}{dt^2} + ky = 0 \qquad (2\text{-}2)$$

The behavior of any physical system is independent of the sign convention adopted. To prove this, the reader may wish to rework the problem with the sign convention reversed. The same differential equation will result.

The weight of the mass was not a factor in the system just analyzed. However, weight is a factor in the spring-mass system shown in Fig. 2-2a. Let us consider the displacement y to be zero, that is, choose a reference point, with the system at rest with the spring extended by some amount δ sufficient to support the weight of the mass. This, as will be shown, is a convenient reference point. The arrow indicates that the upward direction will be considered positive.

Figure 2-2b is a free-body diagram of the mass displaced upward from the equilibrium position. Two forces exist, the weight Mg (where g is the acceleration of gravity) acting downward and the spring force F_s acting upward. The spring force is equal to the spring rate k multiplied by the

2-2 MECHANICAL SYSTEMS

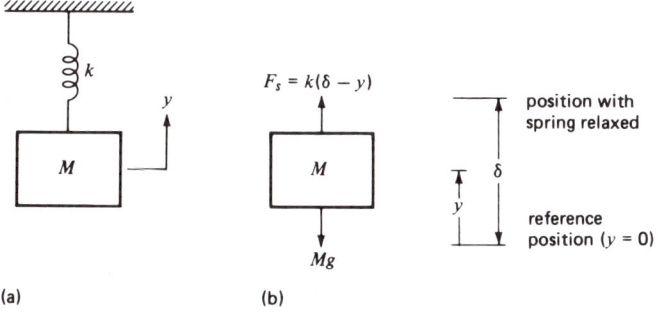

FIG 2-2. System in which weight is a factor.

extension of the spring, which is $(\delta - y)$. From Eq. 2-1,

$$M\frac{d^2y}{dt^2} = \Sigma F = k(\delta - y) - Mg = k\delta - ky - Mg$$

But $\delta = Mg/k$ = static deflection. Therefore,

$$M\frac{d^2y}{dt^2} = k\left(\frac{Mg}{k}\right) - ky - Mg = -ky$$

and

$$M\frac{d^2y}{dt^2} + ky = 0 \tag{2-3}$$

Two points should be noted. First, the sign convention adopted initially was followed in introducing the forces into the first equation: An upward force is positive; a downward force, negative. Second, by choosing the reference at the equilibrium position, the weight of the mass had no effect on the differential equation, which turned out to be the same as that of the previous system in which weight was not a factor. Hereafter the reference will always be chosen at the equilibrium position so that the weight of the mass will not enter the problem.

COMBINATIONS OF SPRINGS

A mass will often be acted upon by the combined effects of two or more springs. There are two basically different ways in which springs may be combined: in *parallel* and in *series*. As a general statement, *springs in parallel share a common deflection, whereas springs in series share a common force.*

In Fig. 2-3a a pair of springs is used in parallel to connect the mass M to the wall. Because the total spring force on the mass is

$$F_s = -k_1 y - k_2 y \qquad (2\text{-}4)$$

the two springs could be replaced by a single spring having the equivalent rate

$$k = k_1 + k_2 \qquad (2\text{-}5)$$

Although it may not be readily apparent, mounting the two springs as shown in Fig. 2-3b will give exactly the same result. A displacement y will cause the spring on the left to be stretched while the one on the right is compressed. Since the two resultant spring forces act in the same direction and the net spring force on the mass is the sum of the two, the springs in this configuration are acting in parallel. Again we may write for the resultant net spring rate

$$k = k_1 + k_2$$

It does not matter, for the system of Fig. 2-3b, whether or not the springs have an initial compression (or an initial tension) so long as we define $y = 0$ as the position of static equilibrium. Assume, for example, that it was necessary to compress the spring on the left by the amount δ_1 from its free length, and the spring on the right by the amount δ_2 in order to fit them into the space available. Under static equilibrium ($y = 0$), the two spring forces must be equal, so we may write

$$k_1 \delta_1 = k_2 \delta_2$$

If the mass is then given the displacement y, the sum of the two spring forces (with our chosen sign convention) is

$$F_s = -k_1(y - \delta_1) - k_2(y + \delta_2)$$

Combining the above two equations gives the result

$$F_s = -(k_1 + k_2) y$$

which is the same as Eq. 2-4. It is apparent, therefore, that with fixed spring rates, the initial compression does not affect the method of analysis.

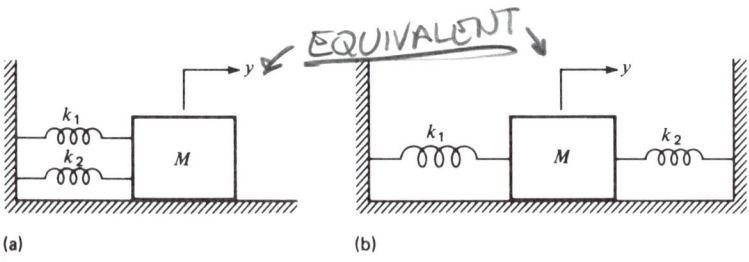

(a) (b)

***

FIG 2-3. Two examples of springs acting in parallel.

2-2 MECHANICAL SYSTEMS

The two springs in Fig. 2-4a, mounted end to end, are said to be in series. For this case it is also convenient to have an expression for the spring rate of a single spring that could replace the pair with no change in system characteristics. To obtain this equivalent spring rate, we first determine the total deflection y of the pair of springs when stretched by the tensile force F_s. The equivalent spring rate is then given by the fundamental equation $k = F_s/y$. Since both springs will have the tensile force F_s acting at each end (Fig. 2-4b) and since the total deflection is equal to the sum of the deflections of the two individual springs, we may write

$$y = y_1 + y_2$$

or

$$\frac{F_s}{k} = \frac{F_s}{k_1} + \frac{F_s}{k_2}$$

which may be simplified to

$$\frac{1}{k} = \frac{1}{k_1} + \frac{1}{k_2} \tag{2-6}$$

Solving for k,

$$k = \frac{k_1 k_2}{k_1 + k_2} \tag{2-7}$$

Spring rates for three or more springs acting in series may be similarly obtained. For three springs we find

$$k = \frac{k_1 k_2 k_3}{k_1 k_2 + k_2 k_3 + k_3 k_1} \tag{2-8}$$

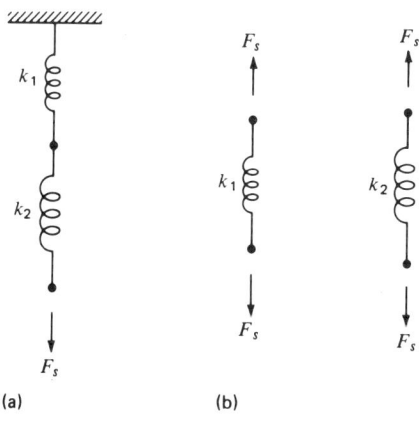

FIG 2-4. (a) Two springs acting in series.
(b) Free-body diagrams for the two springs.

For the general case of n springs in series, the equation

$$\frac{1}{k} = \frac{1}{k_1} + \frac{1}{k_2} + \cdots + \frac{1}{k_n} \tag{2-9}$$

derived in the same manner as Eq. 2-6, should be used.

BEAM SPRINGS

Up to this point we have not discussed types of springs, the illustrations having been drawn as if helical (coil) springs were used. Although helical springs are probably the most familiar type, there are many other types of springs that may be present in mechanical vibrating systems. As a matter of fact, any object that has compliance (i.e., anything that is not perfectly rigid) can be considered as a spring. The "springs" involved in many mechanical vibrating problems are structural members that were meant to be rigid, but in reality they have compliance that is quite significant.

One category of spring that warrants some discussion is the beam spring. Any beam—cantilever, simply supported, or whatever—can act as a spring. Spring rates of beam springs are readily calculated if force-deflection information is available. The basic definition of spring rate,

$$k = \frac{F_s}{y} \tag{2-10}$$

is used, with the deflection y determined for a given value of F_s. The important thing to remember when performing this type of calculation is that *the deflection y must be measured at the same point at which the load F_s is applied*. Appendix A lists a number of beam deflection equations that can be used for determining the spring constants of beams.

Example 2-1. Derive the differential equation of motion for a mass M attached to the end of a cantilever beam of length l having a uniform cross section with area moment of inertia I_a (Fig. 2-5). The mass of the beam can be considered negligible in comparison to M. It is also assumed that displacements are small and that the eccentric loading from gravity is negligible.

Solution: From Appendix A we find the equation for the deflection of a cantilever beam due to a force F_s applied at the end to be

$$y = \frac{F_s l^3}{3EI_a}$$

Substituting this into the basic equation defining a spring constant (Eq. 2-10), we obtain

$$k = \frac{F_s}{y} = \frac{3EI_a}{l^3}$$

2-2 MECHANICAL SYSTEMS

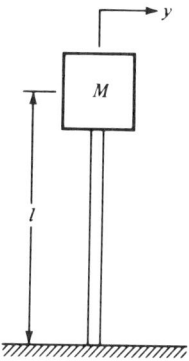

FIG 2-5. Spring-mass system with a cantilever beam serving as the spring.

Since the cantilever spring force is the only force acting on the mass in the y direction, application of Newton's second law gives

$$M \frac{d^2y}{dt^2} = -ky$$

which then becomes, for this system,

$$M \frac{d^2y}{dt^2} + \frac{3EI_a}{l^3} y = 0 \qquad (Ans.)$$

VISCOUS FRICTION AND THE DASHPOT

The nature of *viscous friction* (damping) is illustrated in Fig. 2-6. The curve shows that a device exhibiting viscous friction will offer a restraining force proportional to velocity

$$F_f = cv$$

where F_f is the friction force (N or lb), v is the velocity (m/s or in./s), and

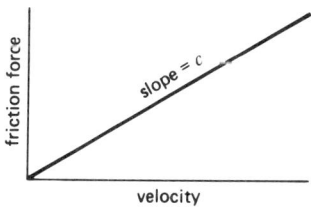

FIG 2-6. Nature of viscous friction.

c is the damping coefficient (N · s/m or lb · s/in.) and is the slope of the line shown in the figure.

By contrast, *Coulomb friction*, named in honor of the French physicist Charles Augustin de Coulomb, produces a constant friction force independent of velocity. A system exhibiting Coulomb friction is described by a differential equation more difficult to solve than that for a system with viscous friction. Therefore, in many cases, as a first approximation, Coulomb friction is assumed to be negligible.

The dashpot is a frequently used example of a device providing viscous friction, or damping. A schematic drawing of a dashpot is shown in Fig. 2-7a. The device consists of a piston inside a fluid-filled cylinder. The piston rod extends from the cylinder through a suitable seal. Relative motion between the piston rod and the cylinder is resisted by the fluid because the fluid must move from one side of the piston to the other through orifices provided in or around the piston. The greater the relative velocity, the greater must be the flow rate of fluid past the piston, and a pressure difference must exist across the piston to cause the fluid flow. It is this pressure difference that produces the restraining force. If the rate of fluid flow through the orifices is small, the flow is laminar (that is, the flow rate is proportional to the pressure drop) and the force is proportional to the relative velocity (viscous damping). If, however, the rate of fluid flow is high, the flow is turbulent (that is, the flow rate is proportional to the square root of the pressure drop), and the force is proportional to the square of the velocity. In both cases, other friction effects, such as that existing at the rod seal, have been neglected. (The reader should verify the statements made by analyzing a dashpot under the two given flow conditions).

In the discussion that follows, the assumption of an ideal (viscous friction) dashpot is made. A dashpot is usually represented as shown in

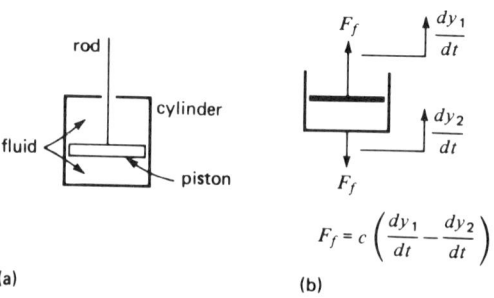

FIG 2-7. The dashpot.

2-2 MECHANICAL SYSTEMS

Fig. 2-7b. The equation defining the behavior of such a device is

$$F_f = c\left(\frac{dy_1}{dt} - \frac{dy_2}{dt}\right) \quad (2\text{-}11)$$

where dy_1/dt and dy_2/dt are the velocities of the piston rod and cylinder, respectively. It should be recognized that the dashpot symbol is frequently used to denote the assumption of viscous friction, rather than to indicate the presence of an actual dashpot.

SPRING-MASS-DAMPER SYSTEMS

Figure 2-8a shows a system that includes a dashpot; Figure 2-8b is a free-body diagram in which a positive (upward) displacement and a positive velocity are assumed. We should recall that, if the reference ($y = 0$) is chosen at the equilibrium position, the mass can be considered weightless. Two forces then exist on the mass. The spring force F_s, which equals ky, acts downward because of the "compression" in the spring. The dashpot (or damper) force from Eq. 2-11 is $c\,dy/dt$ (one end of the dashpot being fixed). It also acts downward in resisting the upward (positive) velocity. Therefore, both forces are negative in sign. From Eq. 2-1,

$$M\frac{d^2y}{dt^2} = \Sigma F = -ky - c\frac{dy}{dt}$$

or

$$M\frac{d^2y}{dt^2} + c\frac{dy}{dt} + ky = 0 \quad (2\text{-}12)$$

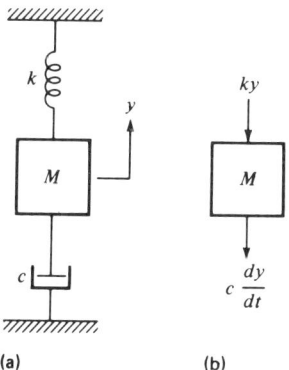

FIG 2-8. (a) Spring-mass-damper system.
(b) Free-body diagram assuming positive displacement and velocity.

Equation 2-12 is the differential equation of motion for the system. In the first equation the dashpot-force term $-c\, dy/dt$ can be examined to ensure that it is valid for all possible velocities. If dy/dt is zero, the force is zero; this is correct for a device exhibiting viscous friction. If dy/dt is negative, the force is positive. Looking at the system, this is correct because, if the mass has a downward velocity, the dashpot will provide an upward restraining force. Thus, the dashpot-force term is valid for any velocity even though it was obtained by assuming a positive velocity.

A variation of the preceding system is shown in Fig. 2-9a; here a provision is made for an input displacement x at the top of the spring. Figure 2-9b shows a free-body diagram of the mass responding to a positive input at x. It is assumed that x is displaced upward and that the mass is responding with a positive displacement y (less than x) and a positive velocity. With this assumption, the spring force is acting upward with a magnitude $k(x-y)$, and the dashpot force is acting downward with a magnitude $c\, dy/dt$. Therefore,

$$M\frac{d^2y}{dt^2} = k(x-y) - c\frac{dy}{dt}$$

or

$$M\frac{d^2y}{dt^2} + c\frac{dy}{dt} + ky = kx \qquad (2\text{-}13)$$

which is the differential equation for the system, in which y is the response to the input x.

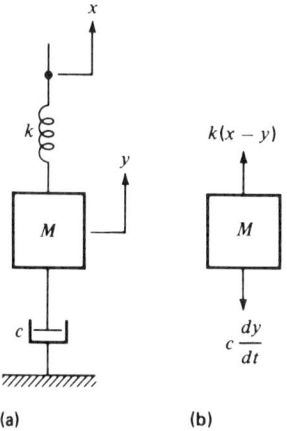

FIG 2-9. (a) System with an input.
(b) Free-body diagram of mass responding to a positive displacement of x.

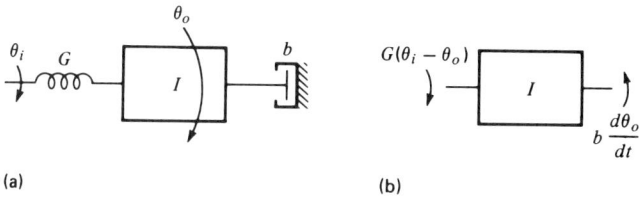

FIG 2-10. (a) Rotational system.
(b) Free-body diagram of mass responding to a positive displacement at θ_i.

ROTATIONAL SYSTEMS

The principles already presented for translational systems can easily be extended to set up the differential equation for the rotating system shown in Fig. 2-10a. The rotating mass I (kg · m² or lb · in.·s²), with an angular position θ_o (radians), is driven through a torsional spring with a spring rate G (N · m/rad or lb · in./rad). Rotation is resisted by a rotary dashpot with a damping coefficient b (N · m · s/rad or lb · in.·s/rad). The input displacement is θ_i. The arrows indicate the direction considered positive.

The fundamental law to be applied is the rotational equivalent of Newton's second law of motion,

$$\Sigma T = I\alpha \qquad (2\text{-}14)$$

where T is the torque (N · m or lb · in.) and α is the angular acceleration (rad/s²).[2] Figure 2-10b shows a free-body diagram of the mass, assuming that θ_i is displaced in the positive direction and that the mass is responding with a positive displacement θ_o (less than θ_i) and a positive velocity. With this assumption, the spring torque is acting in the positive direction with a magnitude $G(\theta_i - \theta_o)$, and the dashpot torque is acting in the negative direction with a magnitude $b\, d\theta_o/dt$. Therefore, from Eq. 2-14,

$$I\alpha = I\frac{d^2\theta_o}{dt^2} = \Sigma T = G(\theta_i - \theta_o) - b\frac{d\theta_o}{dt}$$

or

$$I\frac{d^2\theta_o}{dt^2} + b\frac{d\theta_o}{dt} + G\theta_o = G\theta_i \qquad (2\text{-}15)$$

which is the differential equation for the system. Note the similarity between this equation and Eq. 2-13.

---

[2]Equation 2-14 is readily derived from $\Sigma F = Ma$; see any basic physics text.

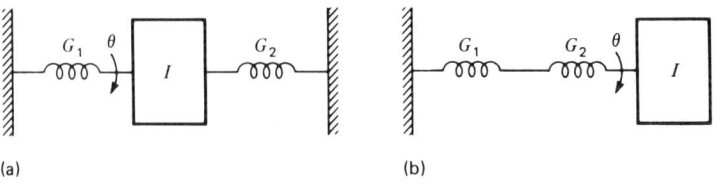

FIG 2-11. (a) Torsional springs acting in parallel.
(b) Torsional springs acting in series.

Torsional springs acting in series or in parallel are handled in exactly the same manner as translational springs. For two torsional springs in parallel (Fig. 2-11a) we can substitute the equivalent spring constant

$$G = G_1 + G_2 \tag{2-16}$$

If the two springs act in series (Fig. 2-11b), the equivalent spring constant is

$$G = \frac{G_1 G_2}{G_1 + G_2} \tag{2-17}$$

Torsional springs may take a number of different forms, such as helical springs, clock springs, and torsion rods. Any machine element or structure that incurs a significant twist for a given applied torque must be considered a torsional spring for purposes of vibration analysis.

THE SIMPLE PENDULUM

The simple pendulum, Fig. 2-12a, consists of a *point mass* suspended from a pivot by a massless rod. There are two alternative ways in which an equation for this system can be developed. It is very instructive to go through both methods, comparing the results and showing that they are indeed equivalent.

With M a point mass, its movement along the arc of a circle is equivalent to curvilinear translation.[3] The pendulum can therefore be analyzed as a simple translational system. The variable y is defined as the distance along the circular arc from the point of static equilibrium. F_s is defined as an equivalent spring force due to gravity that acts on the mass in a direction tangent to the path of motion. Using the trigonometric

---

[3] Curvilinear translation is defined kinematically as movement of a body along some path that is not a straight line that takes place without any rotation of the body. Rotation of the pendulum mass does occur, but rotation of a point mass has no dynamic significance, since $I = 0$. The simple pendulum can therefore be analyzed as if the mass moves with curvilinear translation.

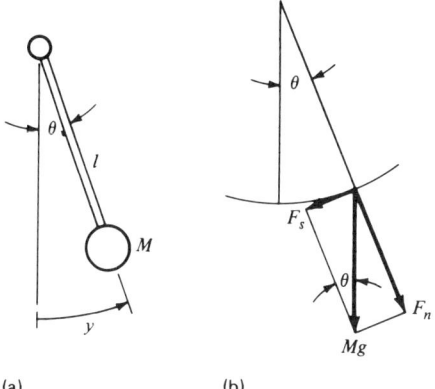

FIG 2-12. (a) A simple pendulum.
(b) Gravity force vectors: F_s = force component along path of M, F_n = force component normal to path (canceled by tension in rod).

relationships shown in Fig. 2-12,

$$F_s = Mg \sin \theta$$

Defining k as the effective spring rate due to gravity, and noting that

$$y = l\theta$$

we may write

$$k = \frac{F_s}{y} = \frac{Mg \sin \theta}{l\theta} \qquad (2\text{-}18)$$

The equation of motion can now be derived from Newton's second law to be

$$M \frac{d^2y}{dt^2} + \left(\frac{Mg \sin \theta}{l\theta} \right) y = 0 \qquad (2\text{-}19)$$

Equation 2-19 is exact, but it has been found convenient to use the approximation

$$\sin \theta \approx \theta \qquad (2\text{-}20)$$

which results in the simplified equation[4]

$$M \frac{d^2y}{dt^2} + \frac{Mg}{l} y = 0 \qquad (2\text{-}21)$$

---

[4] Equation 2-21 is a *linear equation*, which is easier to work with than Eq. 2-19, a *nonlinear* one. These definitions and their implications are covered in Chapter 3. Chapter 4 is devoted to the important topic of *linearization*.

Equation 2-21 is very accurate if the pendulum oscillates with small amplitude.

The simple pendulum can also be analyzed as a rotational system that oscillates about the point of suspension. The moment of inertia about this point is

$$I = Ml^2 \tag{2-22}$$

The force of gravity acting on the mass causes a torque on the pendulum

$$T = Mgl \sin \theta$$

which gives the effective torsional spring constant

$$G = \frac{T}{\theta} = \frac{Mgl \sin \theta}{\theta} \tag{2-23}$$

The resultant exact equation of motion is

$$Ml^2 \frac{d^2\theta}{dt^2} + \left(\frac{Mgl \sin \theta}{\theta} \right)\theta = 0 \tag{2-24}$$

Using the approximation $\sin \theta \approx \theta$ produces the simplified equation

$$Ml \frac{d^2\theta}{dt^2} + Mg\theta = 0 \tag{2-25}$$

To show that Eqs. 2-21 and 2-25 are equivalent, we use the trigonometric relationship

$$y = l\theta$$

and the derivatives

$$\frac{dy}{dt} = l \frac{d\theta}{dt}$$

$$\frac{d^2y}{dt^2} = l \frac{d^2\theta}{dt^2}$$

Substitution of the above into Eq. 2-21 shows that it is identical to Eq. 2-25.

2-3 Hydraulic Systems

Two distinct types of hydraulic systems will be discussed in this section. The first has to do with the liquid level in a tank, an illustration closely related to control problems encountered in the chemical processing industry [5]. The second type of hydraulic system is related to fluid power applications, which are common throughout industry and the military [12]. In this case the fluid is oil.

The fundamental relationship to be used for both types of systems is

$$q = Av \tag{2-26}$$

2-3 HYDRAULIC SYSTEMS

in which q is the volumetric flow rate (m³/s or similar units), A is the cross-sectional area (m²), and v is the velocity (m/s). The equation states that the flow rate is equal to the cross-sectional area multiplied by the average velocity normal to the area.

LIQUID-LEVEL SYSTEM

Figure 2-13 shows a tank with an inflow q_i. The liquid level h (for head) provides a potential that causes an outflow q_o through the outlet valve. It will be assumed that the outflow q_o is directly proportional to the head h, with the proportionality constant being G_v. The cross-sectional area A is constant throughout the height of the tank.

With Eq. 2-26, the net flow rate to the tank equals the area A multiplied by the velocity of the liquid level, or dh/dt. Therefore, based on the *conservation of mass*,

$$q_i - q_o = A \frac{dh}{dt} \qquad (2\text{-}27)$$

Note that if a rise in level is considered positive, the equation is correct in sign for any inflow-outflow condition. For example, if the inflow exceeds the outflow, the level will surely rise; in Eq. 2-27, substitution of the flow values will give a positive dh/dt.

It is assumed that the outflow q_o is directly proportional to the head h. This relationship can be expressed as (Bernoulli)

$$q_o = G_v h \qquad q_o = \sqrt{2gh}\, A_o \qquad (2\text{-}28)$$

Combining Eqs. 2-27 and 2-28 yields

$$A \frac{dh}{dt} + G_v h = q_i \qquad (2\text{-}29)$$

Equation 2-29 is the system differential equation, in which h is the response to an input q_i.

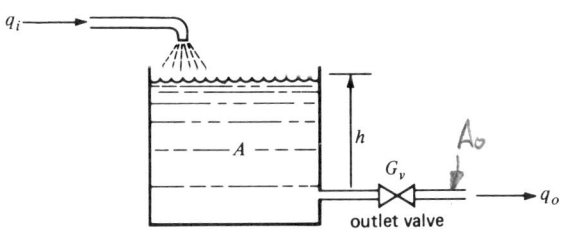

FIG 2-13. Liquid-level system.

FLUID POWER SYSTEM

Figure 2-14 is a schematic drawing of a valve-controlled hydraulic actuator, or cylinder. Oil under pressure is supplied to the pressure port of the valve. Two other valve ports are connected to the cylinder. With the valve spool centered, as shown, the cylinder ports are blocked, and the piston rod remains stationary. Should the spool be displaced to the right, fluid will enter the left end of the cylinder, causing piston rod motion to the right. Oil from the right end of the cylinder is expelled through the valve to the source of fluid. A leftward motion of the piston rod is obtained by displacing the valve spool to the left from the centered position.

Valve spool displacement x, the input, is considered to be zero with the spool centered. Piston rod displacement y, the response, is considered to be zero at the equilibrium position before an input x is applied. Displacements to the right are positive.

With the assumption of a given load on the piston, and therefore a given pressure drop across the valve spool, the flow rate of oil through a valve can be expressed as

$$q = K_v \Delta \tag{2-30}$$

in which q is the volumetric flow rate, K_v is the valve constant, and Δ is the valve opening. For the valve shown in Fig. 2-14, the valve opening is x, and Eq. 2-30 becomes

$$q = K_v x \tag{2-31}$$

The cylinder equation is based on Eq. 2-26; that is,

$$q = Av = A\frac{dy}{dt} \tag{2-32}$$

From conservation considerations, the flow through the valve equals

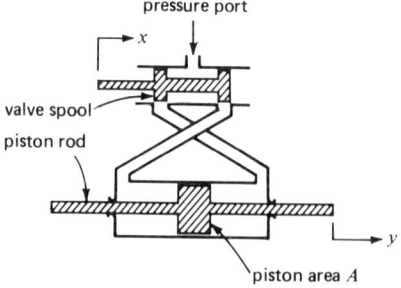

FIG 2-14. Valve-controlled hydraulic actuator.

2-4 ELECTRIC SYSTEMS

the flow to the cylinder. Thus, with Eqs. 2-31 and 2-32,

$$q = K_v x = A \frac{dy}{dt}$$

or

$$A \frac{dy}{dt} = K_v x \tag{2-33}$$

Equation 2-33 is the differential equation for the system as approximated.

2-4 Electric Systems

Electric systems form a very important group of dynamic systems. Systems composed of resistors, capacitors, inductors, and other electric components are represented by differential equations with time as the independent variable.

CIRCUITS

There are two fundamental principles, *Kirchhoff's first and second laws*, that are used for obtaining the differential equations that describe electric circuits. Kirchhoff's first law states that *the algebraic sum of the voltage differences around a closed loop is zero*,

$$\Sigma e = 0 \tag{2-34}$$

Kirchhoff's second law states that *the algebraic sum of the currents entering and leaving a node is zero*,

$$\Sigma i = 0 \tag{2-35}$$

Although either of the two laws may be applied to any given circuit, less labor is often involved if one is chosen over the other.

A summary of the equations for the common electric elements—resistance, capacitance, and inductance—is given in Table 2-1. Note that there are two alternative governing equations for each element. (These could be considered as different forms of the same equation.) The equations for voltage drop across the elements are used when applying Kirchhoff's first law to a circuit, while those for current through the elements are used with Kirchhoff's second law.

LOOP ANALYSIS

The circuit shown in Fig. 2-15a is composed of a resistance and inductance in series with an impressed voltage e. Application of Kirchhoff's first law (Eq. 2-34) to the closed loop, in combination with the equations for voltage

TABLE 2-1
Electric Elements

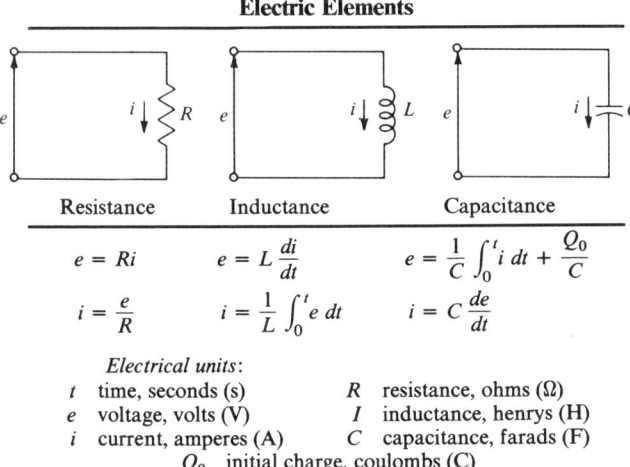

Resistance	Inductance	Capacitance
$e = Ri$	$e = L\dfrac{di}{dt}$	$e = \dfrac{1}{C}\int_0^t i\, dt + \dfrac{Q_0}{C}$
$i = \dfrac{e}{R}$	$i = \dfrac{1}{L}\int_0^t e\, dt$	$i = C\dfrac{de}{dt}$

Electrical units:
- t time, seconds (s)
- e voltage, volts (V)
- i current, amperes (A)
- Q_0 initial charge, coulombs (C)
- R resistance, ohms (Ω)
- I inductance, henrys (H)
- C capacitance, farads (F)

drop from Table 2-1, yields

$$e - Ri - L\frac{di}{dt} = 0$$

or

$$L\frac{di}{dt} + Ri = e \tag{2-36}$$

The latter equation is the differential equation that describes the circuit.

The circuit shown in Fig. 2-15b is handled in a similar manner. Assuming the capacitor initially discharged (i.e., $Q_0 = 0$), the result is

$$L\frac{di}{dt} + Ri + \frac{1}{C}\int_0^t i\, dt = e \tag{2-37}$$

This equation can be written in an alternative form by recalling that current is the time rate of flow of charge (i.e., $i = dq/dt$), where q is

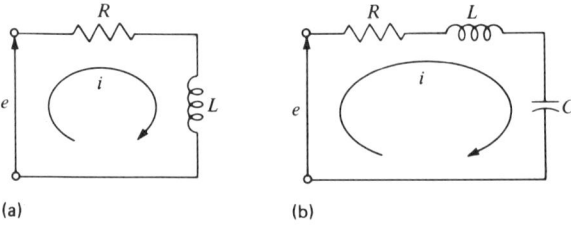

(a) (b)

FIG 2-15. Electric circuits.

2-4 ELECTRIC SYSTEMS

charge expressed in coulombs. Substitution of this relationship into Eq. 2-37 yields

$$L\frac{d^2q}{dt^2} + R\frac{dq}{dt} + \frac{q}{C} = e \tag{2-38}$$

NODE ANALYSIS

The circuit shown in Fig. 2-16 has two loops. If Kirchhoff's first law is to be used, it must be applied twice—once to each of the loops. A simpler solution is obtained in this case by applying Kirchhoff's second law to the single node, or point of unknown voltage e_1. Use of the equations for current through the elements in combination with Eq. 2-35 yields

$$-C\frac{de_1}{dt} + \frac{1}{R}(e - e_1) - \frac{1}{L}\int_0^t e_1\, dt = 0$$

or

$$C\frac{de_1}{dt} + \frac{1}{R}e_1 + \frac{1}{L}\int_0^t e_1\, dt = \frac{1}{R}e \tag{2-39}$$

To convert to a pure differential equation, a single differentiation is performed,

$$C\frac{d^2e_1}{dt^2} + \frac{1}{R}\frac{de_1}{dt} + \frac{1}{L}e_1 = \frac{1}{R}\frac{de}{dt} \tag{2-40}$$

Note that the dependent variable is *voltage* in the circuit equation developed by application of Kirchhoff's second law, whereas the dependent variable is *current* when Kirchhoff's first law is used.

DC MOTOR WITH LOAD

A schematic drawing of a dc motor with a load consisting of inertia and damping is given in Fig. 2-17. The elements R and L in the loop denote the resistance and inductance of the motor armature. A desirable system equation would be one which relates the output speed ω (rad/s) with the input voltage e to the motor armature. Note that the motor field is

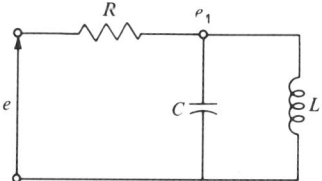

FIG 2-16. An electric circuit with a single node.

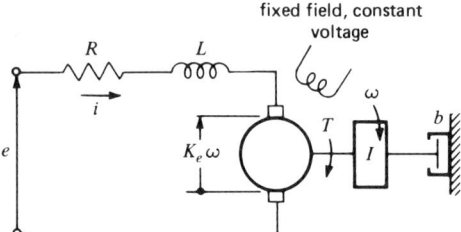

FIG 2-17. A dc motor with load.

supplied from another voltage source, which is independent of the armature voltage e. Three equations are required to define the behavior of this system.

The first equation is based on the algebraic sum of the voltages around the armature loop. In addition to the voltage drops across the armature resistance R and inductance L, the voltage induced by the armature winding cutting the lines of field flux must be considered. Recall that a voltage, proportional to velocity, is induced in a conductor that cuts lines of magnetic flux—provided, of course, that the flux field is constant, as is true in this case. Thus, the induced voltage in the armature winding is proportional to rotating speed and can be expressed as $K_e\omega$ where K_e is the voltage constant for the motor. This induced voltage opposes the impressed voltage e. Application of Kirchhoff's first law (Eq. 2-34) to the closed loop yields

$$e = Ri + L\frac{di}{dt} + K_e\omega \tag{2-41}$$

which is the first of the required equations.

The second equation relates the output torque T to the armature current i. Recall that a magnetic field surrounds a conductor carrying a current and that the intensity of the field is proportional to the current. Furthermore, if the current-carrying conductor is suitably located in a magnetic field, the interaction between the fields results in a force on the conductor. These phenomena are utilized in a dc motor (with a fixed field) to provide an output torque proportional to the armature current, or

$$T = K_t i \tag{2-42}$$

where K_t is the torque constant of the motor.

The final equation relates the output speed and the torque input to the load. This equation is obtained by applying Newton's second law to the rotating mass, which yields

$$T = I\frac{d\omega}{dt} + b\omega \tag{2-43}$$

2-5 A Thermal System

The three equations can, by methods to be presented later, be combined to yield a single equation defining the behavior of the overall system; that is, $\omega = f(e, t)$.

2-5 A Thermal System

Many thermal systems fit in the category of dynamic systems, with temperatures, heat flow rates, and so on that vary with time in response to a system input. One such thermal system will be considered in this section —a home heating system (Fig. 2-18).

We assume that the temperature outside the house stays constant at T_o, and that the inside temperature of the house, and of all its contents changes uniformly (i.e., the furniture, walls, air, and so forth are at the same temperature at any instant of time).

From the required *conservation of energy*, we may write

$$q_{in} - q_{out} = q_{stored} \tag{2-44}$$

where q_{in} is the heat flow rate, watts or Btu/h, from the furnace into the house, q_{out} is the flow rate of the heat being lost from the house (through the walls, by air leakage, and so on), and q_{stored} is the rate at which heat is stored within the house and its contents as their temperature T rises.[5]

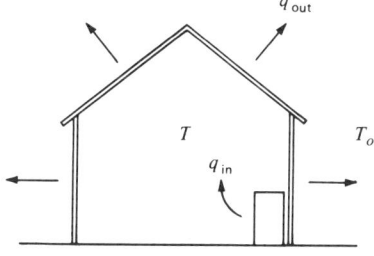

FIG 2-18. A house and its heating system as a dynamic thermal system.

The rate at which heat is produced by the furnace is essentially constant whenever it is on, so that

$$q_{in}(t) = Q_{in} \tag{2-45}$$

The rate at which heat is lost to the outside air can be considered to be directly proportional to the temperature difference between the house and

---

[5]Note the similarity to the analysis of the liquid-level system of Section 2-3.

the outside air,

$$q_{\text{out}} = H(T - T_o) \tag{2-46}$$

where H is an overall coefficient of heat transfer, in watts per kelvin (W/K) or Btu/(h · °F), determined either experimentally or from detailed analysis of the house design.

The rate at which heat is stored within the house and its contents is

$$q_{\text{stored}} = \sum_{i=1}^{n} C_{p_i} M_i \frac{dT}{dt} \tag{2-47}$$

where C_{p_i} and M_i are the specific heat and mass, respectively, of the n parts of the house and its contents. By defining an overall "heat capacitance" C, in joules per kelvin (J/K) or Btu/°F,

$$C = \sum_{i=1}^{n} C_{p_i} M_i \tag{2-48}$$

Eq. 2-47 becomes

$$q_{\text{stored}} = C \frac{dT}{dt} \tag{2-49}$$

Substituting Eqs. 2-45, 2-46, and 2-49 into Eq. 2-44 produces the differential equation of the thermal system,

$$Q_{\text{in}} - H(T - T_o) = C \frac{dT}{dt}$$

which may be manipulated to the more desirable form

$$C \frac{dT}{dt} + HT = Q_{\text{in}} + HT_o \tag{2-50}$$

Equation 2-50 is valid whenever the furnace is running. When the furnace is off, the same equation applies (but with $Q_{\text{in}} = 0$).

2-6 An Ecological System

The systems previously presented have been of an engineering nature. A study related to certain dynamic characteristics of an ecological system is now introduced to illustrate the fact that the approach to obtaining defining equations is the same despite the type of system involved [8].

A schematic representation of an ecosystem is shown in Fig. 2-19. The idea is that living matter can conveniently be thought of as existing in layers. Each layer, or level, is called a *trophic level*. The shape of the diagram brings out the fact that there is far greater plant biomass than that of, say, the top carnivore.

2-6 AN ECOLOGICAL SYSTEM

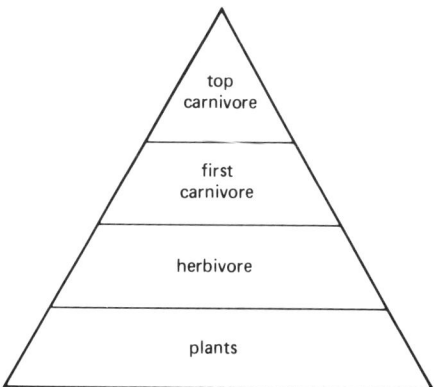

FIG 2-19. Schematic representation of an ecosystem.

Fig. 2-19 depicts a *food chain*; that is, the herbivores feed on the plant life and are fed upon by the first carnivores. The top carnivores feed primarily on the first carnivores.

We will assume that a pesticide is introduced into the ecosystem. The assumed characteristics of the pesticide are that it is highly persistent, is accumulated by living organisms, and is neither metabolized nor excreted. The point of interest will be the dynamic nature of the pesticide concentration in one of the trophic levels.

The system diagram of Fig. 2-20 will be used to obtain the required

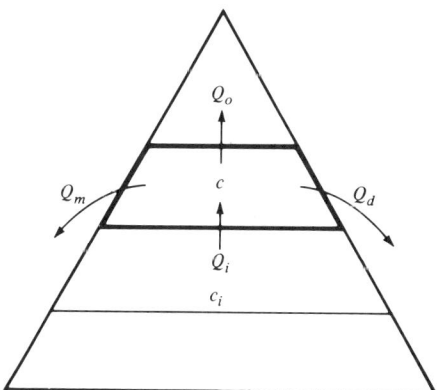

FIG 2-20. System diagram for the study of pesticide concentration in one trophic level.

defining equation. We will assume an established food chain with negligible population changes within the trophic levels. On this basis, the rate Q_i (mass per unit time) at which living matter (including pesticide) is eaten by level members is constant. Loss rates of mass from the level include the rate Q_o that corresponds to members being eaten, the rate Q_d that accounts for natural deaths, and the rate Q_m that results from metabolism and, ultimately, excretion. These rates are also constant. The law of the *conservation of mass* requires that

$$Q_i = Q_o + Q_d + Q_m \tag{2-51}$$

The flow rate q_i (mass per unit time) of pesticide into the trophic level can be expressed as

$$q_i = Q_i c_i \tag{2-52}$$

where c_i is the concentration in the level immediately below. The pesticide leaves the level under study through natural death and by predation. Thus,

$$q_o = Q_o c \tag{2-53}$$

$$q_d = Q_d c \tag{2-54}$$

where c is the pesticide concentration of the level.

In a time interval δt, the change δm of the mass of pesticide in the level is

$$\delta m = m_{in} - m_{out}$$

$$\delta m = q_i \, \delta t - q_o \, \delta t - q_d \, \delta t$$

or, in the limit,

$$\frac{dm}{dt} = q_i - q_o - q_d \tag{2-55}$$

With Eqs. 2-52, 2-53, and 2-54,

$$\frac{dm}{dt} = Q_i c_i - (Q_o + Q_d) c \tag{2-56}$$

The concentration of the pesticide in the level can be expressed as

$$c = \frac{m}{M} \tag{2-57}$$

where M is the constant mass of the members of the level. Upon the differentiation of Eq. 2-57,

$$M \frac{dc}{dt} = \frac{dm}{dt} \tag{2-58}$$

Combination of Eqs. 2-56 and 2-58 yields

$$M \frac{dc}{dt} + (Q_o + Q_d) c = Q_i c_i \tag{2-59}$$

Equation 2-59 is the required defining equation that relates the pesticide concentration of members in a level to that of the level immediately below.

Although a number of assumptions concerning the natures of the ecosystem and the pesticide were made for the purpose of simplification, the presentation does illustrate the basic approach involved in modeling dynamic systems.

2-7 A Political-Military System

The manner in which governments react to the actions of a real or potential enemy can be predicted to some extent from a knowledge of human nature and a study of past history. The armament buildups of two antagonistic nations can be considered as a dynamic system, changing with time in a manner that could be predicted if all the governing relationships and contributing factors were precisely known.

A pair of equations has been developed by Richardson to model an "armament race" between two nations [17],

$$\frac{dw_1}{dt} = a_1 w_2 - b_1 w_1 + c_1$$
$$\frac{dw_2}{dt} = a_2 w_1 - b_2 w_2 + c_2 \qquad (2\text{-}60)$$

where

w_1 and w_2 are a quantitative measure of the military weapons of nations 1 and 2, respectively

a_1 and a_2 represent the nations' "defense coefficients"

b_1 and b_2 represent the "fatigue and expense" coefficients associated with the weapons

c_1 and c_2 represent "attitude factors" covering the range from cooperation to antagonism

Equations 2-60 are somewhat different from the other dynamic equations developed in this chapter since they constitute a pair of interdependent differential equations that must be solved simultaneously.

The equations developed above obviously represent a simplification of a very complex system, but the relationships given do show the powerful forces that govern to a large extent the nations' actions. Solutions to Eqs. 2-60 for various sets of assumed coefficients will give some insight into the way that international events are influenced and provide useful input to those concerned with improving international relationships.

2-8 Conclusion

The purpose of this chapter is to introduce the concept of obtaining differential equations that represent dynamic systems. As illustrated, these defining equations are written from a knowledge of the basic laws underlying the system being considered. The idea is simple, but its application can be difficult in many cases. Not only must the system be thoroughly understood, but judgment and ingenuity are required to formulate equations that are reasonably representative of system behavior and yet are simple enough to be readily solved.

The chapter has discussed only a limited number of the many types of dynamic systems that exist. However, despite the diversity of actual systems, the defining equations are typically differential equations with time as the independent variable—a fact that permits a unification of dynamic characteristics for seemingly unrelated systems.

Problems

2-1. Obtain the differential equations of motion for the following systems.

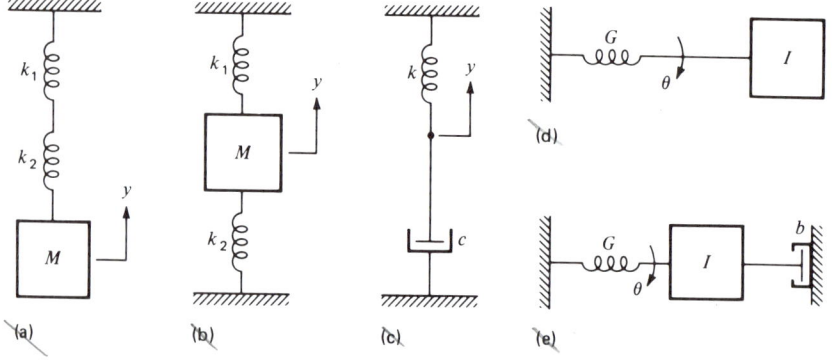

PROBLEMS

2-2. Obtain the appropriate differential equation for each of the following: In parts (a) through (e), x is input and y is response; in part (f), force F is input and y is response; in part (g), T is input and ω is response; in part (h), q_i is input and q_o is response.

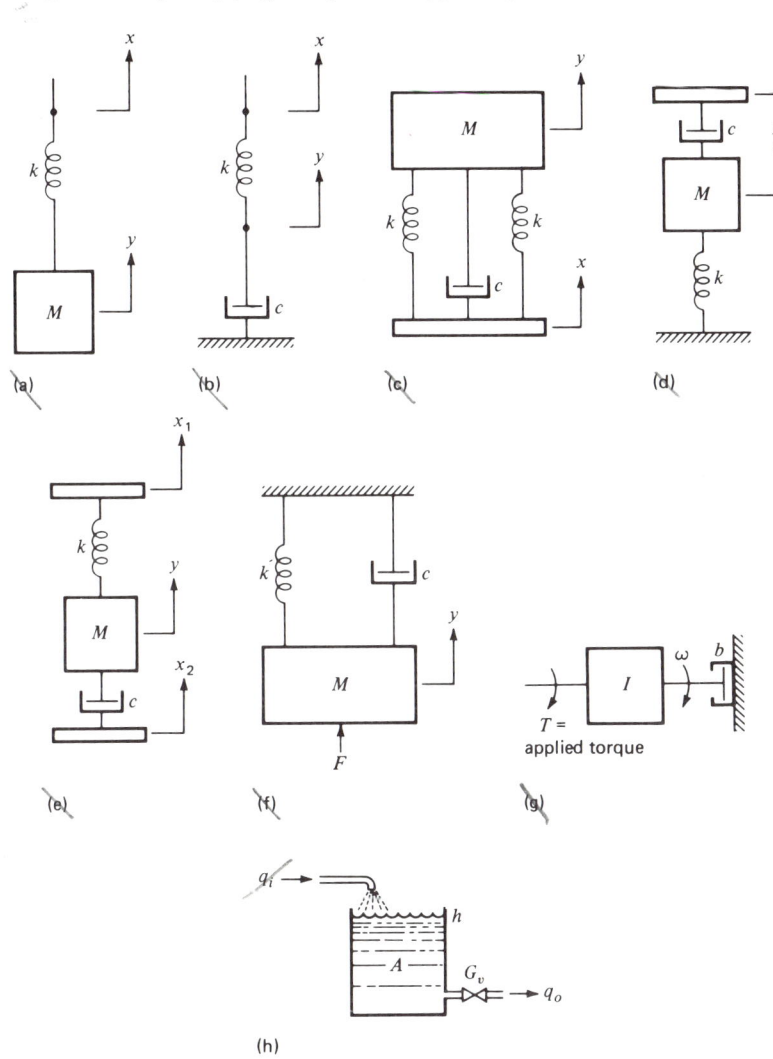

2-3. Obtain the differential equation of motion for each of the following systems.

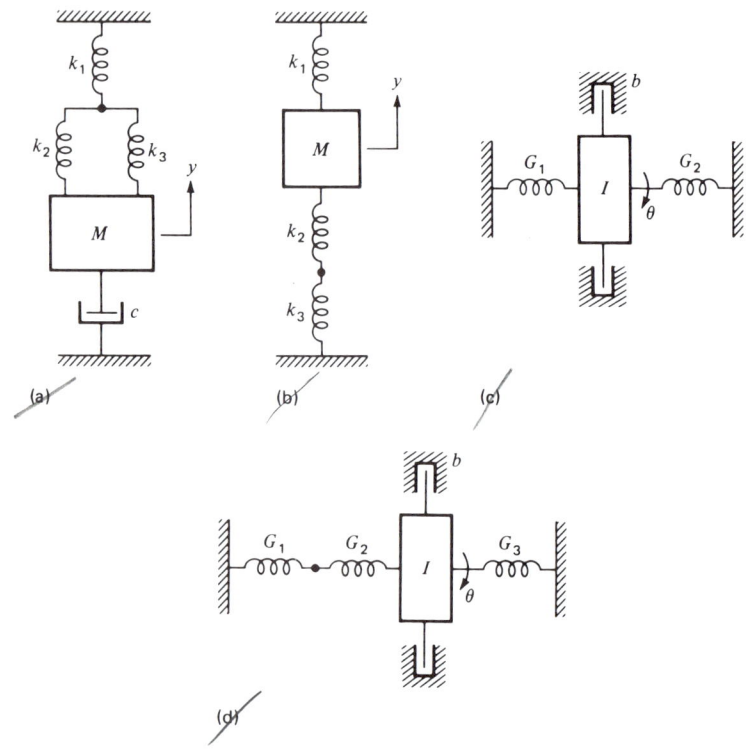

2-4. Obtain the differential equations of motion for the following systems, which contain both beam springs and coil springs. The beams all have a uniform cross section over their total length, with area moment of inertia I_a. Assume the mass of each beam to be negligible.

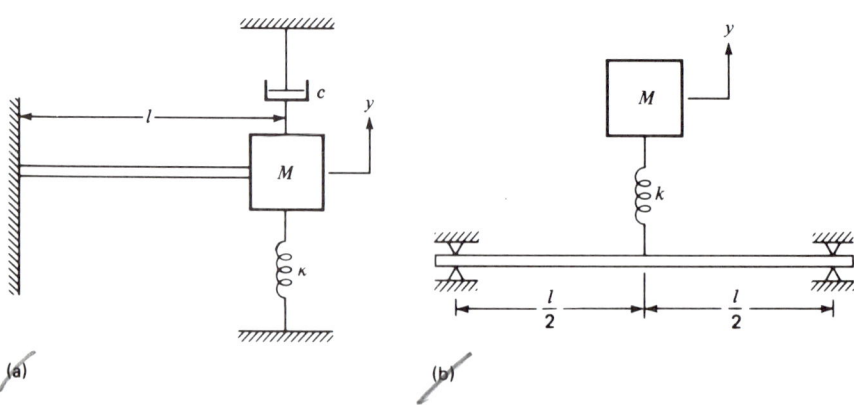

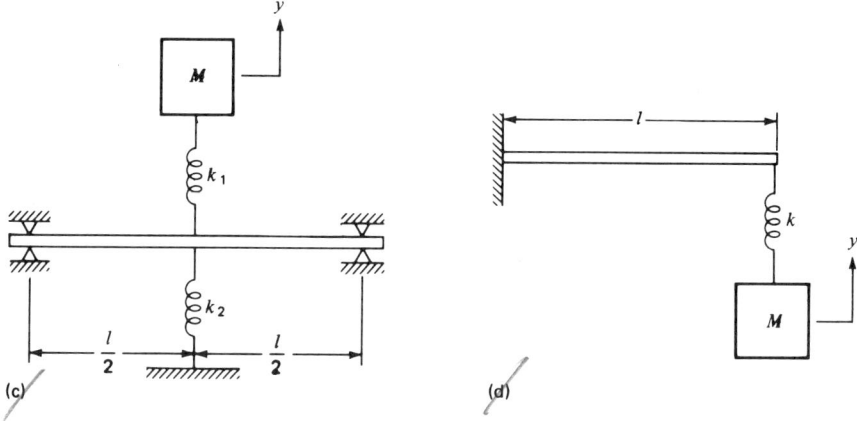

2-5. Obtain the differential equation of motion for each of the following pulley systems. Assume that the pulleys and cables have negligible inertia and that the systems are operated in a manner that causes tension always to exist in each cable.

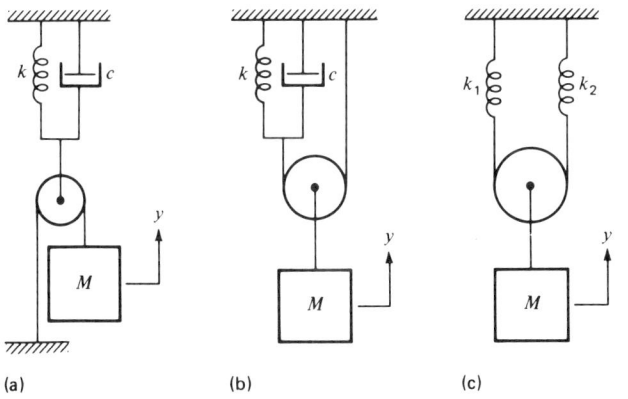

2-6. Determine the effective spring constant ($k = F/y$) of the system shown in the figure.

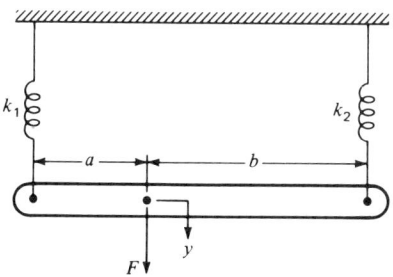

2-7. Obtain the differential equations of motion for the following systems. (Note that the helical torsional springs are represented by cross-sectional views).

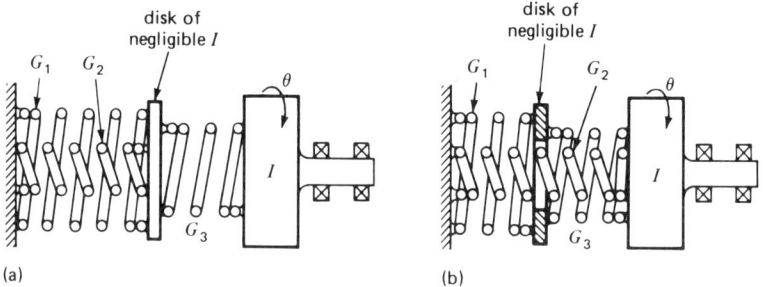

(a)   (b)

2-8. Obtain the differential equations of motion for the following systems, which have cylindrical torsion rods for springs. The equation relating torque and angular displacement of a solid cylindrical rod of diameter d and length l is

$$\theta = \frac{32l}{\pi d^4 G_m} T$$

where G_m is the shear modulus of elasticity of the material.

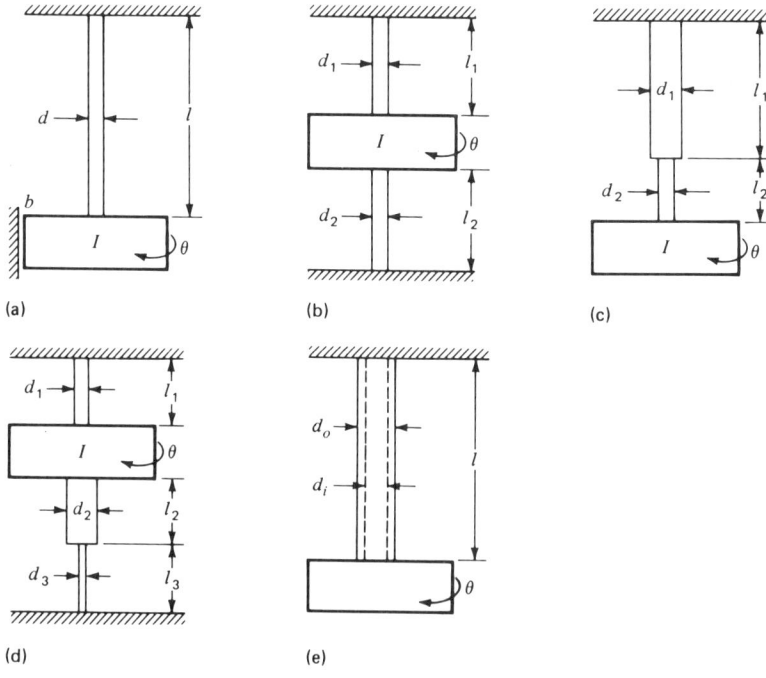

PROBLEMS

2-9. Obtain the equation of motion for vertical movement of the floating cylinder. Assume negligible damping. (M = mass, A = cross-sectional area, and γ = density of liquid.)

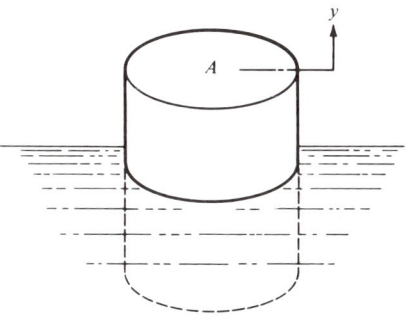

2-10. Obtain the differential equation that describes the motion of the liquid (of density γ) in the U-tube manometer. Both ends of the tube are open.

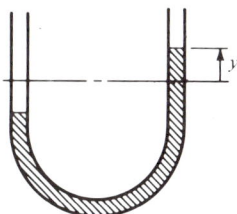

2-11. Assuming rigid beams of negligible mass, obtain the differential equations of motion of the following systems.

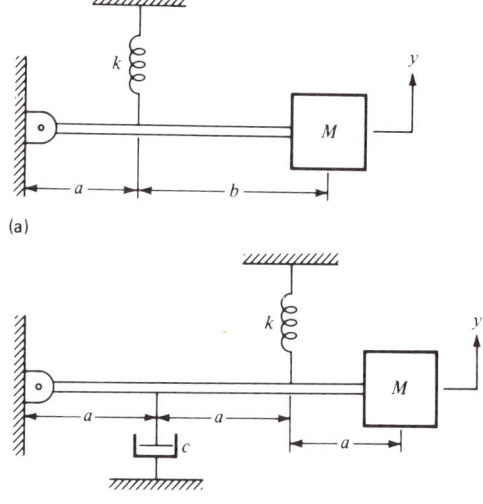

(a)

(b)

2-12. Obtain the differential equation for vertical motion of the mass suspended by a steel wire as shown in the figure. Start with the definition of Young's modulus to obtain the spring constant of the wire.

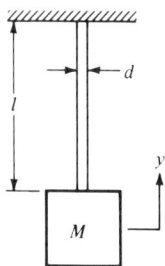

2-13. Obtain the differential equation, in which x is the input and y is the output, for the valve-controlled hydraulic actuator shown in the accompanying illustration. In this particular system the piston rod is fixed and the cylinder is free to move. The valve body is rigidly attached to the cylinder and moves with it. Displacement of the valve spool opens the valve, causing cylinder and valve body displacements that tend to close the valve. Therefore, the valve opening is $x - y$ rather than x.

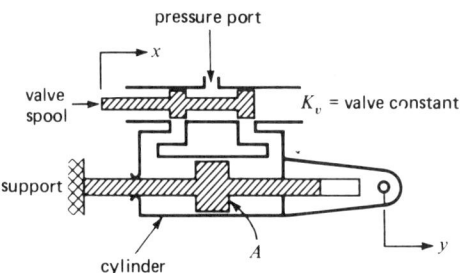

2-14. Obtain the differential equation for the valve-controlled hydraulic actuator shown in the figure. Note that the valve opening is a function of both the input x and the output y.

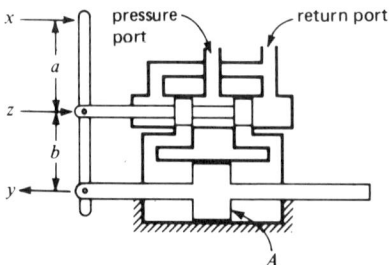

2-15. The viscosity μ of a given oil is defined by the equation

$$\frac{F}{A} = \mu \frac{dx/dt}{y}$$

based on relative motion between two flat plates separated by a thin film of oil as illustrated in part (a) of the figure. Starting with the above equation, obtain the damping coefficient (translational or torsional, as appropriate) for each of the dampers illustrated in parts (b) through (f) of the figure. The oil has a viscosity of 5.0×10^{-6} lb · s/in.² Assume in each case that pumping losses are negligible, and that damping is produced only where a very thin film of oil is sheared. (All dimensions are in inches.)

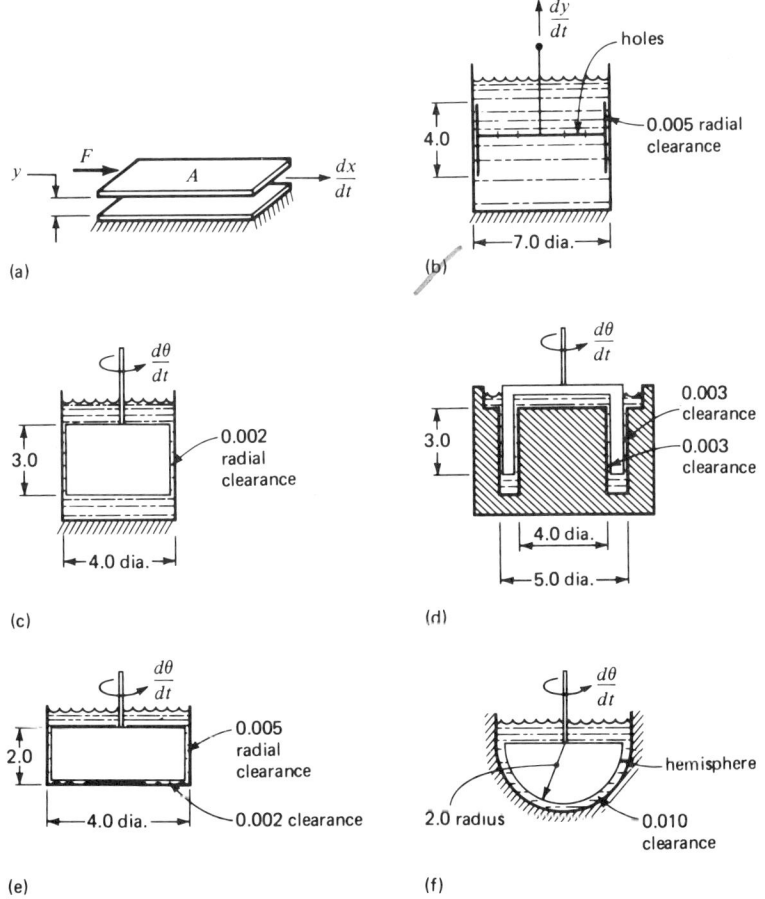

2-16. Obtain the differential equations of motion for the following systems in which x is the input and y is the response.

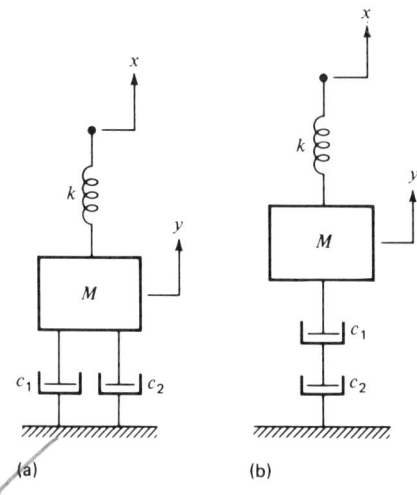

2-17. Using a loop analysis based on Kirchhoff's first law, obtain the differential equation relating current and voltage for each of the following systems. Assume the capacitors are initially discharged.

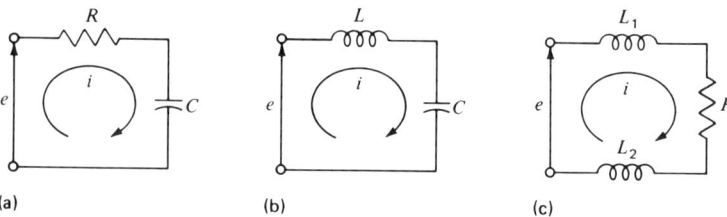

2-18. Using a node analysis based on Kirchhoff's second law, obtain the differential equation relating the node voltage e_1 and the applied voltage e for each of the following systems.

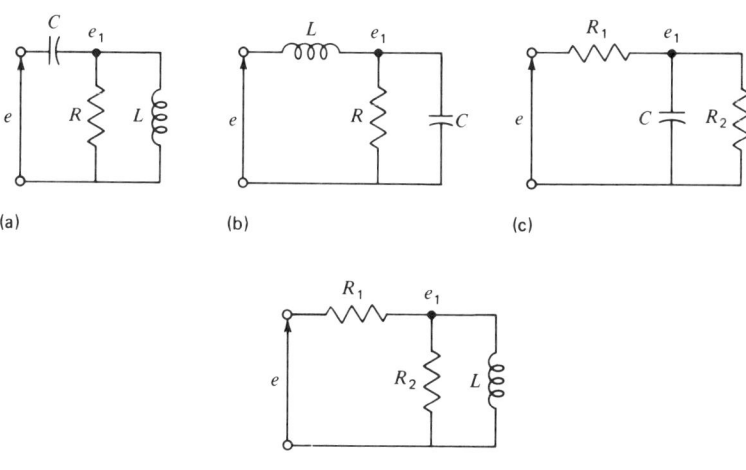

(a)   (b)   (c)

(d)

2-19. A swimming pool is heated by a gas-fired heater that supplies 200,000 Btu/h to the water. The pool is rectangular, 15 ft by 30 ft, with an average depth of 5 ft. The water circulating system keeps the water temperature quite uniform, even while the heater is operating. The heat loss from the pool has been found to be approximately equal to 5000 Btu/h for each degree Fahrenheit difference between the temperature of the water and that of the ambient air. Write the differential equation that describes how the water temperature varies with time when the heater is turned on and the ambient air temperature is 50°F.

2-20. Several aluminum parts, having a total mass of 10 kg, are placed in a steel oven for heat treatment. The steel of the oven has a mass of 150 kg. Its walls are insulated on the outside, and it can be assumed that there is never any appreciable difference in the temperatures of the steel, aluminum, and air inside the oven as they change during the process. It has been found that heat is lost from the oven at a rate of 200 W for each degree difference (in kelvins) between the oven temperature and that of the ambient air. The mass of air in the oven is small, so its heat capacitance can be neglected. The process begins with the oven and its contents at the ambient temperature of 310 K, and heat is supplied at a constant rate of 40 kW. The specific heats of steel and aluminum are 486 and 896 J/(kg · K), respectively. Develop the differential equation describing the temperature of the system as a function of time.

3
Solving Differential Equations

3-1 Introduction

Now that we have learned how to obtain differential equations, let us turn our attention to solving them. This chapter deals with the ordinary linear differential equation with constant coefficients—the basic type of equation found in this book. Before we discuss solutions, some preliminary remarks about differential equations will be made to provide a review and to put the material that follows into its proper perspective.

Suppose that $t, u, v, \ldots$ are independent variables and that $x, y, z, \ldots$ are dependent variables and are functions of $t, u, v, \ldots$. Then the functional dependence is shown as

$$x = x(t, u, v, \ldots)$$
$$y = y(t, u, v, \ldots)$$
$$z = z(t, u, v, \ldots)$$

3-1 INTRODUCTION

A differential equation shows the relationship between the independent variables, the dependent variables, and the derivatives of the dependent variables with respect to the independent variables. For a simple case, this relationship can be expressed as

$$f\left(t, x, \frac{dx}{dt}\right) = 0$$

Consider the following examples:

$$\frac{d^2x(t)}{dt^2} + \frac{dx(t)}{dt} + 3x(t) = 5t$$

$$\frac{dy(u)}{du} + y(u) = \cos u$$

$$\frac{\partial^2 x}{\partial t^2} + \frac{\partial^2 x}{\partial u^2} + \frac{\partial^2 x}{\partial v^2} = \frac{\partial x}{\partial \lambda}$$

where $x = x(t, u, v, \lambda)$.

If there is *one* independent variable, as in the first two examples, the differential equation is *ordinary*. If there are several independent variables so that the derivatives are partial derivatives, as in the third, the equation is a *partial* differential equation. Our concern is with ordinary differential equations.

The *order* of a differential equation is the order of the highest derivative. For example,

$$\frac{d^2x}{dt^2} + x = 5$$

is a second-order equation, whereas

$$\frac{d^5y}{dt^5} + \frac{d^3y}{dt^3} + y = \sin t$$

is a fifth-order equation. The *degree* of a differential equation is the power to which the highest derivative is raised. The equation

$$\left(\frac{dx}{dt}\right)^2 + x = 0$$

is an illustration of a second-degree equation.

Ordinary differential equations may be classified by the nature of the coefficients of the terms in the equation. These coefficients may be variable or constant. An example of an equation with *variable coefficients* is

$$t^3 \frac{d^2x}{dt^2} + t \frac{dx}{dt} + t^2 x = 15 \sin 3t$$

and an equation with *constant coefficients* is

$$8\frac{d^2y}{dt^2} + 4\frac{dy}{dt} + 2y = 0$$

Differential equations may be further classified as *linear* or *nonlinear*. The two preceding examples illustrate linear equations. These would become nonlinear if the variable coefficients were functions of the dependent variable or if any of the derivatives were raised to a power other than 1. Examples of nonlinear equations are

$$x^3\frac{dx}{dt} + x = 10$$

and

$$8\left(\frac{d^2y}{dt^2}\right)^2 + 2y = 0$$

Linear equations are characterized by the fact that the principle of superposition holds. Consider the linear equation

$$\frac{d^2y}{dt^2} + \frac{dy}{dt} + y = f(t)$$

where $f(t)$ is called the *driving function*, or the *forcing function*. If the solution of the equation with a forcing function $f_1(t)$ is $y_1(t)$ and with $f_2(t)$ is $y_2(t)$, then the solution with a forcing function $f_1(t) + f_2(t)$ is $y_1(t) + y_2(t)$. In the case of a linear physical system (one whose behavior can be defined by a linear differential equation), this means that the response of the system to the sum of two inputs is equal to the sum of the responses to the inputs taken one at a time. The principle of superposition *does not* hold for a nonlinear system.

This chapter presents a method for solving ordinary linear differential equations with constant coefficients. An operational method, which will prove to be very useful in solving differential equations, will be employed. An operator p (frequently given elsewhere as D or s) is defined such that

$$py \equiv \frac{dy}{dt} \qquad p^2y \equiv \frac{d^2y}{dt^2} \qquad p^n y \equiv \frac{d^n y}{dt^n} \qquad (3\text{-}1)$$

and

$$\frac{1}{p}y \equiv \int_0^t y\,dt \qquad (3\text{-}2)$$

Thus, p denotes differentiation with respect to time, and $1/p$ denotes integration with respect to time. The utility of the p operator is that, for the most part, it can be treated as an algebraic quantity. This permits the replacement of differential equations with algebraic equations which are more easily handled.

3-2 Solution of Homogeneous Equations

The solution of ordinary linear differential equations with constant coefficients will be presented in three parts. The first involves the solution of homogeneous equations. A homogeneous equation is one in which the forcing function is zero.

FIRST-ORDER EQUATIONS

Consider the general equation

$$\frac{dy}{dt} + ky = 0$$

where k is a constant. Using Eq. 3-1,

$$py + ky = 0 = y(p + k)$$

Because y is the required nontrivial solution, and therefore is not zero, then

$$p + k = 0$$

This expression is called the *characteristic equation*.[1] From it,

$$p = -k$$

The solution of all first-order equations is of the form

$$y = C_1 e^{pt} \qquad (3\text{-}3)$$

where C_1 is an arbitrary constant determined from a knowledge of the value of y when time is assumed to be zero. This value of y is called an *initial condition*. From Eq. 3-3, our solution is

$$y = C_1 e^{-kt}$$

But is it? We say we have a solution when, upon substitution, it reduces the original equation to an identity or, said in another way, it satisfies the equation. From our solution,

$$\frac{dy}{dt} = -kC_1 e^{-kt}$$

Substituting the expressions for dy/dt and y into the original equation yields

$$-kC_1 e^{-kt} + kC_1 e^{-kt} = 0$$

which shows that the solution is correct.

---

[1] The *characteristic equation*, so named because it indicates certain characteristics of the system that are independent of the dependent variable and the forcing function, is obtained from any system equation by taking the corresponding homogeneous equation, writing it in operator form, and eliminating the dependent variable.

Example 3-1. Solve the equation
$$\frac{dy}{dt} + 2y = 0$$
Solution: $py + 2y = 0 = y(p + 2)$ and $p = -2$. From Eq. 3-3,
$$y = C_1 e^{-2t} \qquad (Ans.)$$

SECOND-ORDER EQUATIONS

Consider the general equation
$$\frac{d^2 y}{dt^2} + g\frac{dy}{dt} + ky = 0$$
where g and k are constants. Using Eq. 3-1,
$$p^2 y + gpy + ky = 0 = y(p^2 + gp + k)$$
and the characteristic equation is
$$p^2 + gp + k = 0$$
Solving this equation leads to two values of p:
$$p = \frac{-g \pm \sqrt{g^2 - 4k}}{2}$$
There are three different possibilities for the values of p, depending on whether the quantity under the radical is positive, zero, or negative. These three cases will be discussed separately.

CASE 1. If the quantity under the radical is positive, the roots of the characteristic equation, p_1 and p_2, will be real and unequal numbers. The solution is then
$$y = C_1 e^{p_1 t} + C_2 e^{p_2 t} \qquad (3\text{-}4)$$
where C_1 and C_2 are arbitrary constants that are determined from the initial conditions.

Example 3-2. Solve the equation
$$\frac{d^2 y}{dt^2} + 3\frac{dy}{dt} + 2y = 0$$
Solution: $p^2 y + 3py + 2y = 0$ and $p^2 + 3p + 2 = 0$; $p_1 = -1$ and $p_2 = -2$. From Eq. 3-4,
$$y = C_1 e^{-t} + C_2 e^{-2t} \qquad (Ans.)$$
The reader may verify this solution by substituting it into the original equation.

3-2 SOLUTION OF HOMOGENEOUS EQUATIONS

CASE 2. If the quantity under the radical is zero, the roots of the characteristic equation will be real and equal numbers. In this special case of repeated roots, the solution is

$$y = C_1 e^{p_1 t} + C_2 t e^{p_2 t} \qquad (3\text{-}5)$$

where $p_1 = p_2$.

Example 3-3. Solve the equation

$$\frac{d^2y}{dt^2} + 2\frac{dy}{dt} + y = 0$$

Solution: The characteristic equation is $p^2 + 2p + 1 = 0$ and $p_1 = p_2 = -1$. From Eq. 3-5,

$$y = C_1 e^{-t} + C_2 t e^{-t} \qquad (Ans.)$$

This solution can be checked by substitution into the original equation.

CASE 3. If the quantity under the radical is negative, the roots of the characteristic equation will be complex numbers, actually complex conjugates. In this case the roots of the generalized characteristic equation ($p^2 + gp + k = 0$) are

$$p = \frac{-g \pm j\sqrt{4k - g^2}}{2}$$

where $j = \sqrt{-1}$. Letting $a = -g/2$ and $b = (4k - g^2)^{1/2}/2$, the roots are $p_1 = a + jb$ and $p_2 = a - jb$, and the solution is

$$y = A_1 e^{p_1 t} + A_2 e^{p_2 t}$$

where A_1 and A_2 are also complex conjugates. This solution can be modified into

$$y = e^{at}(C_1 \sin bt + C_2 \cos bt) \qquad (3\text{-}6)$$

where C_1 and C_2 are real numbers. A further modification will give the form[2]

$$y = C_3 e^{at} \sin(bt + \phi) \qquad (3\text{-}7)$$

in which

$$C_3 = (C_1^2 + C_2^2)^{1/2} \quad \text{and} \quad \phi = \tan^{-1}\frac{C_2}{C_1}$$

This is probably the most useful form of solution for Case 3 when

---

[2] See Appendix B for the derivation of Eqs. 3-6 and 3-7.

working with dynamic systems. It gives the output as a single sinusoidal curve of a particular amplitude and frequency, with the angle ϕ showing the position of the curve with respect to time.

Example 3-4. Solve the equation

$$\frac{d^2y}{dt^2} + 4\frac{dy}{dt} + 13y = 0$$

Solution: The characteristic equation is $p^2 + 4p + 13 = 0$.

$$p = \frac{-4 \pm (-36)^{1/2}}{2} = \frac{-4 \pm j6}{2} = -2 \pm j3$$

From Eq. 3-6,

$$y = e^{-2t}(C_1 \sin 3t + C_2 \cos 3t) \qquad (Ans.)$$

HIGHER-ORDER EQUATIONS

A third-order differential equation will have a third-degree characteristic equation that can be solved for three roots. In general, an nth-degree characteristic equation will result from an nth-order differential equation, and there will be n roots. If special provisions are made for repeated roots and if the roots and arbitrary constants are permitted to be real or complex numbers, the general solution for all ordinary linear differential equations with constant coefficients is

$$y = C_1 e^{p_1 t} + C_2 e^{p_2 t} + C_3 e^{p_3 t} + \cdots + C_n e^{p_n t} \qquad (3-8)$$

The solution of an nth-order differential equation will involve n arbitrary constants which can be determined from n initial conditions.

To illustrate the use of Eq. 3-8, consider the following examples: If $p_1 = -1, p_2 = -2,$ and $p_3 = -3,$

$$y = C_1 e^{-t} + C_2 e^{-2t} + C_3 e^{-3t}$$

If $p_1 = -2, p_2 = -3 + j2,$ and $p_3 = -3 - j2,$

$$y = C_1 e^{-2t} + e^{-3t}(C_2 \sin 2t + C_3 \cos 2t)$$

This result is obtained by using Eq. 3-6 as the basis for handling the pair of complex roots. If $p_1 = 0, p_2 = -1, p_3 = -1,$ and $p_4 = -2,$

$$y = C_1 + C_2 e^{-t} + C_3 t e^{-t} + C_4 e^{-2t}$$

This result is obtained by remembering that e^0 is unity and by using Eq. 3-5 as the basis for handling the repeated roots. Finally, if $p_1 = p_2 = p_3 = -2,$

$$y = C_1 e^{-2t} + C_2 t e^{-2t} + C_3 t^2 e^{-2t}$$

3-3 CONSIDERATION OF INITIAL CONDITIONS

This last example illustrates that, for m repeated roots, the solution is

$$y = C_1 e^{p_1 t} + C_2 t e^{p_2 t} + C_3 t^2 e^{p_3 t} + \cdots + C_m t^{m-1} e^{p_m t} \quad (3\text{-}9)$$

3-3 Consideration of Initial Conditions

The solutions for the differential equations presented in the preceding section involved one or more arbitrary constants. These constants are determined from the initial conditions, that is, the values of the dependent variable and its derivatives existing when time is assumed to be zero. The method for obtaining the constants is straightforward and will be illustrated by three examples involving first-order and second-order differential equations. The method is, of course, quite general and is applicable to equations of higher order.

Example 3-5. Solve the equation

$$\frac{dy}{dt} + 2y = 0$$

with $y = 3$ when $t = 0$.
  Solution: By the method of Section 3-2 (see Eq. 3-3),

$$y = C_1 e^{-2t}$$

This expression for y is valid for all positive values of t including $t = 0$. From the knowledge that $y = 3$ when $t = 0$, we write

$$3 = C_1 e^0 = C_1$$

The arbitrary constant C_1 is now evaluated, and the solution is

$$y = 3e^{-2t} \quad (Ans.)$$

Example 3-6. Solve the equation

$$\frac{d^2 y}{dt^2} + 5 \frac{dy}{dt} + 6y = 0$$

with $y = 3$ and $dy/dt = -7$ when $t = 0$.
  Solution: By the method of Section 3-2 (see Eq. 3-4),

$$y = C_1 e^{-3t} + C_2 e^{-2t}$$

Differentiation yields

$$\frac{dy}{dt} = -3C_1 e^{-3t} - 2C_2 e^{-2t}$$

By substituting the initial conditions, the result is

$$3 = C_1 + C_2 \quad \text{and} \quad -7 = -3C_1 - 2C_2$$

from which $C_1 = 1$ and $C_2 = 2$. Therefore, the required solution is

$$y = e^{-3t} + 2e^{-2t} \qquad (Ans.)$$

Note that the solution of the second-order equation involved two arbitrary constants and that two initial conditions were necessary for evaluation.

Example 3-7. Solve the equation

$$\frac{d^2y}{dt^2} + 4y = 0$$

with $y = 0$ and $dy/dt = 8$ when $t = 0$.
Solution: By the method previously described (see Eq. 3-6),

$$y = e^0(C_1 \sin 2t + C_2 \cos 2t) = C_1 \sin 2t + C_2 \cos 2t$$

Differentiation yields

$$\frac{dy}{dt} = 2C_1 \cos 2t - 2C_2 \sin 2t$$

Substituting the initial conditions,

$$0 = C_1(0) + C_2(1) \quad \text{and} \quad C_2 = 0$$
$$8 = 2C_1 - 0 \qquad \text{and} \quad C_1 = 4$$

Therefore, the solution is

$$y = 4 \sin 2t \qquad (Ans.)$$

Check: A check is always desirable. The expressions for y and its derivatives are

$$y = 4 \sin 2t \qquad \frac{dy}{dt} = 8 \cos 2t \qquad \frac{d^2y}{dt^2} = -16 \sin 2t$$

Substituting these expressions into the original equation yields

$$-16 \sin 2t + 4(4 \sin 2t) = 0$$

showing that the solution does indeed satisfy the equation. We can also check in the following manner to see that the initial conditions are satisfied:

$$y|_{t=0} = 4 \sin 2(0) = 0$$
$$\frac{dy}{dt}\bigg|_{t=0} = 8 \cos 2(0) = 8$$

These checks prove that the solution is correct.

3-4 Solution of Nonhomogeneous Equations

The nonhomogeneous differential equation is very important in the study of dynamic systems, because determination of the response of a system to some input is usually required. A nonhomogeneous equation is one in which the forcing function is not zero. The equation

$$\frac{d^2y}{dt^2} + g\frac{dy}{dt} + ky = f(t)$$

illustrates a second-order nonhomogeneous equation.

The solution for a nonhomogeneous equation is composed of the sum of two parts. The first part is the solution of the associated homogeneous equation (that is, let $f(t) = 0$), which is called the *complementary function*. The second part is a solution that satisfies the entire equation; this solution is called the *particular integral*. Letting y_c denote the complementary function and y_p the particular integral, the complete solution can be expressed as

$$y = y_c + y_p \tag{3-10}$$

The method for determining y_c (the solution for the homogeneous equation) has been discussed. All that remains is to learn how to obtain the particular integral y_p. Note—and this is very important—that the arbitrary constants associated with y_c must not be determined until the complete solution y is obtained.

The procedure for obtaining the particular integral y_p is to substitute an assumed solution into the entire equation for the purpose of evaluating the unknown constants. The appropriate assumed solutions for the forcing functions commonly encountered in dynamic systems are given in Table 3-1. Notice that the assumed solution has a form in each case that is very similar to that of the forcing function. Operators are not used in solving for y_p.

TABLE 3-1
Assumed Solutions for Obtaining the Particular Integral y_p
($K, A, B, C,$ and ω are constants)

Forcing Function	Assumed Solution (y_p)
K	A
Kt	$At + B$
Kt^2	$At^2 + Bt + C$
$K \sin \omega t$	$A \sin \omega t + B \cos \omega t$

Three examples will now be given to illustrate the method for obtaining the particular integral for a nonhomogeneous differential equation.

Example 3-8. Obtain the particular integral for the equation

$$\frac{d^2y}{dt^2} + 3\frac{dy}{dt} + 5y = 15$$

Solution: The forcing function is a constant, and, from Table 3-1, we assume that $y_p = A$. This assumed solution must be substituted into the entire equation so that the constant A can be determined. The derivatives of y_p are

$$\frac{dy_p}{dt} = 0 \quad \text{and} \quad \frac{d^2y_p}{dt^2} = 0$$

Upon substitution into the original equation, the result is $0 + 0 + 5A = 15$; $A = 3$. The particular integral is, therefore,

$$y_p = A = 3 \qquad (Ans.)$$

Note that $y = 3$ satisfies the equation.

Example 3-9. Obtain the particular integral for the equation

$$\frac{d^2y}{dt^2} + \frac{dy}{dt} + 2y = 4t$$

Solution: From Table 3-1, the assumed solution is $y_p = At + B$. The derivatives are $dy_p/dt = A$ and $d^2y_p/dt^2 = 0$. Substituting these values into the original equation yields

$$0 + A + 2At + 2B = 4t$$

Upon equating the coefficients of like terms (that is, the t terms and the constant terms), the results are $2A = 4$ and $A = 2$; $A + 2B = 0$ and $B = -1$. Therefore, the particular integral is

$$y_p = 2t - 1 \qquad (Ans.)$$

Example 3-10. Obtain the particular integral for the equation

$$\frac{dy}{dt} + 5y = 10 \sin 5t$$

Solution: From Table 3-1, the assumed solution is $y_p = A \sin 5t + B \cos 5t$. The first derivative is $dy_p/dt = 5A \cos 5t - 5B \sin 5t$. Substituting these expressions into the original equation yields

$$5A \cos 5t - 5B \sin 5t + 5A \sin 5t + 5B \cos 5t = 10 \sin 5t$$

By equating coefficients of like terms, $5A + 5B = 0$ and $5A - 5B =$

3-4 SOLUTION OF NONHOMOGENEOUS EQUATIONS

10, from which $A = 1$ and $B = -1$. The particular integral is

$$y_p = \sin 5t - \cos 5t \qquad (Ans.)$$

A fourth example will be given to summarize the material presented thus far and to illustrate a minor difficulty which can occur in assuming the form for the particular integral.

Example 3-11. Solve the equation

$$\frac{d^2y}{dt^2} + 2\frac{dy}{dt} = 4$$

with $y = 0$ and $dy/dt = 6$ when $t = 0$.

Solution: The complete solution is composed of the sum of the complementary function and the particular integral. To obtain the complementary function, the associated homogeneous equation is written as

$$\frac{d^2y}{dt^2} + 2\frac{dy}{dt} = 0$$

From Eq. 3-1,

$$p^2y + 2py = 0 = y(p^2 + 2p)$$

Since the solution y is not zero,

$$p^2 + 2p = p(p + 2) = 0$$

and this is the characteristic equation. Its roots are $p_1 = 0$ and $p_2 = -2$. These roots are real and unequal. Therefore, from Eq. 3-4, the complementary function y_c is

$$y_c = C_1 + C_2 e^{-2t}$$

The arbitrary constants C_1 and C_2 must not be evaluated until the complete solution has been obtained.

To obtain the particular integral, we note that the forcing function is a constant and, from Table 3-1, assume that $y_p = A$. The derivatives are $dy_p/dt = d^2y_p/dt^2 = 0$. Upon substitution into the original equation, the result is

$$0 + 0 = 4$$

This is not an acceptable equation, and the conclusion is that the assumed solution is at fault. However, an assumed solution whose first derivative is a constant would appear to work. Therefore, let us

assume[3] that
$$y_p = At$$
The derivatives are $dy_p/dt = A$ and $d^2y_p/dt^2 = 0$. Substituting these expressions into the original equation yields $0 + 2A = 4$; $A = 2$. Therefore,
$$y_p = 2t$$
From Eq. 3-10,
$$y = y_c + y_p = C_1 + C_2 e^{-2t} + 2t$$
This is the complete solution, and now the arbitrary constants can be determined. The derivative of the solution is
$$\frac{dy}{dt} = -2C_2 e^{-2t} + 2$$
Upon substitution of the initial conditions, the result is $0 = C_1 + C_2$ and $6 = -2C_2 + 2$, from which $C_1 = 2$ and $C_2 = -2$. Therefore,
$$y = 2 - 2e^{-2t} + 2t \quad (Ans.)$$

3-5 Transient and Steady-State Responses

Alternative terminology should be introduced at this point to conform with that used by most engineers. Although the mathematician speaks of the *complementary function*, the engineer usually speaks of the *transient solution*. Reference to Eq. 3-3 (which is typical of complementary functions) will indicate why this is so. The expression exhibits a decaying exponential term whose contribution to the total solution ultimately vanishes and therefore is transient in nature. In the future, the terms *transient solution* and *complementary function* will be used interchangeably for all stable systems (systems in which the complementary function *does* decay).

The particular integral is that part of the total solution that remains after the transient terms have ceased to contribute. Thus the particular integral is called the *steady-state solution*.

In terms of system response, the *transient response* is defined by the total solution of the system equation in which transients are considered. On the other hand, the *steady-state response* persists after the transients have disappeared (as will be the case with stable systems).

---

[3]This new assumption can also be based on the mathematical statement that, if the assumed particular integral duplicates any term in the complementary function, it must be multiplied by the lowest power of the independent variable sufficient to eliminate the duplication.

3-6 Combining Simultaneous Equations

There are times when it is desirable to combine a number of simultaneous differential equations into a single equation. Here the p operator is very useful, because the set of differential equations can be replaced by a set of algebraic equations. Once the algebraic equations have been solved, the desired differential equation can be recovered. The approach will be illustrated with an example.

Example 3-12. Combine the two given equations into one that can be solved for z. The forcing function is to be x.

$$\frac{dy}{dt} + 2y = 5x$$

$$3\frac{dz}{dt} + z = 2y$$

Solution: Introduction of the p operator (Eq. 3-1) yields

$$y(p + 2) = 5x \quad \text{or} \quad y = \frac{5x}{p + 2}$$

$$z(3p + 1) = 2y = \frac{10x}{p + 2}$$

allowing us to eliminate y, the variable of secondary importance in this problem. Multiplying both sides of the latter equation by $p + 2$,

$$z(3p + 1)(p + 2) = 10x$$

$$z(3p^2 + 7p + 2) = 10x$$

The required differential equation, then, is

$$3\frac{d^2z}{dt^2} + 7\frac{dz}{dt} + 2z = 10x \qquad (Ans.)$$

3-7 Conclusion

The concepts required for solving ordinary linear differential equations with constant coefficients have been developed. These are sufficient to permit us to continue our study of the dynamic behavior of physical systems.

Although primary attention in this chapter has been given to the solution of first- and second-order equations, the concepts can easily be extended to higher-order equations. The work involved, however, is greater. One troublesome aspect is the determination of the roots of the characteristic equation, but this can readily be done with the aid of a digital computer. If such facilities are not available, the techniques introduced in Appendix C may be of interest to the reader.

Problems

3-1. For each of the following equations, give (1) the order of the equation, (2) the degree of the equation, (3) whether or not it is linear.

(a) $\dfrac{d^2y}{dt^2} + y^2 = 10 \cos \omega t$

(b) $\dfrac{d^4y}{dt^4} + y = 8t^2$

(c) $\left(\dfrac{dy}{dt}\right)^2 + 4y = 5t$

(d) $3\dfrac{d^3x}{dy^3} - 4y\dfrac{d^2x}{dy^2} + 5\dfrac{dx}{dy} + 10x = y^2 + \cos y$

(e) $\dfrac{d^3x}{dt^3} + 2\dfrac{d^2x}{dt^2} + 5 \sin x = t$

(f) $2\dfrac{d^2y}{dt^2} + 3y\dfrac{dy}{dt} + 4y = \cos \omega t$

3-2. Solve the following differential equations:

(a) $\dfrac{dy}{dt} + 3y = 0$

(b) $\dfrac{dy}{dt} + \dfrac{y}{2} = 0$

(c) $\dfrac{d^2y}{dt^2} + 4\dfrac{dy}{dt} + 3y = 0$

(d) $\dfrac{d^2y}{dt^2} + 7\dfrac{dy}{dt} + 12y = 0$

(e) $\dfrac{d^2y}{dt^2} + 4\dfrac{dy}{dt} + 4y = 0$

(f) $\dfrac{d^2y}{dt^2} + 3\dfrac{dy}{dt} = 0$

(g) $\dfrac{d^2y}{dt^2} + 9y = 0$

PROBLEMS

(h) $\dfrac{d^2y}{dt^2} + 2\dfrac{dy}{dt} + 2y = 0$

(i) $\dfrac{d^2y}{dt^2} + 6\dfrac{dy}{dt} + 13y = 0$

(j) $\dfrac{d^3y}{dt^3} + 3\dfrac{d^2y}{dt^2} + 3\dfrac{dy}{dt} + y = 0$

(k) $\dfrac{d^3y}{dt^3} + 5\dfrac{d^2y}{dt^2} + 11\dfrac{dy}{dt} + 15y = 0$

3-3. Solve the following differential equations and determine the values of the constants from the initial conditions specified:

(a) $\dfrac{dy}{dt} + 5y = 0$; at $t = 0$, $y = 2$

(b) $\dfrac{d^2y}{dt^2} + 3\dfrac{dy}{dt} + 2y = 0$; at $t = 0$, $y = 1$ and $\dfrac{dy}{dt} = 2$

(c) $\dfrac{d^2y}{dt^2} + 6\dfrac{dy}{dt} + 9y = 0$; at $t = 0$, $y = 0$ and $\dfrac{dy}{dt} = 2$

(d) $\dfrac{d^2y}{dt^2} + 8\dfrac{dy}{dt} + 25y = 0$; at $t = 0$, $y = 1$ and $\dfrac{dy}{dt} = 8$

3-4. Obtain the particular integrals for the following differential equations:

(a) $\dfrac{d^2y}{dt^2} + 6\dfrac{dy}{dt} + 3y = 6$

(b) $\dfrac{d^2y}{dt^2} + 4\dfrac{dy}{dt} + 2y = 6t$

(c) $\dfrac{dy}{dt} + 4y = 10 \sin 2t$

(d) $2\dfrac{dy}{dt} + 5y = 6 + 8t^2$

(e) $3\dfrac{d^2y}{dt^2} + \dfrac{dy}{dt} + 2y = 5t + 2$

(f) $\dfrac{d^2y}{dt^2} + 3\dfrac{dy}{dt} + y = 3 \cos 5t$

3-5. Solve the following differential equations:

(a) $\dfrac{d^2y}{dt^2} + 8\dfrac{dy}{dt} + 12y = 12t$; at $t = 0$, $y = \dfrac{dy}{dt} = 0$

(b) $\dfrac{d^2y}{dt^2} + 4\dfrac{dy}{dt} + 8y = 4$; at $t = 0$, $y = \dfrac{dy}{dt} = 0$

(c) $2\dfrac{d^2x}{dt^2} - 5\dfrac{dx}{dt} + 2x = 16t$; at $t = 0$, $x = \dfrac{dx}{dt} = 0$

(d) $3\dfrac{dy}{dt} + 2y = t + \sin 2t$; at $t = 0$, $y = 2$

(e) $\dfrac{d^2y}{dt^2} + 7\dfrac{dy}{dt} + 10y = 10t$; at $t = 0$, $y = -2$, $\dfrac{dy}{dt} = 0$

(f) $0.5\dfrac{d^2y}{dt^2} + \dfrac{dy}{dt} + 0.5y = 1 + t^2$; at $t = 0$, $y = 2$, $\dfrac{dy}{dt} = 2$

3-6. (a) Combine the following three equations into one in which i is the response and g is the input (forcing function).

$$A\dfrac{df}{dt} + Bf = Cg$$

$$h = Df$$

$$E\dfrac{di}{dt} + i = Kh$$

(b) Combine the following three equations into one in which v is the response and r is the input.

$$B\dfrac{ds}{dt} + s = Ar$$

$$\dfrac{du}{dt} = Cs$$

$$D\dfrac{du}{dt} + Eu = v$$

4
Linear Equations for Modeling Nonlinear Systems

4-1 Introduction

The analytical techniques presented in this book (which is only an introduction to the broad field of dynamic systems) are all based on *linear equations*. The purpose of this chapter is to explain how linear equations may be used satisfactorily as models for many systems that are not truly linear.

There is nothing in this chapter that is actually a prerequisite for the remainder of the book, so it could be saved for later if desired, or even omitted, without making it more difficult to understand subsequent material. (Note, however, that some of the problems at the ends of later chapters do require the use of linearization techniques.) The material on linearization is included, however, to give the student a better appreciation of how to apply the theory of the book to realistic engineering problems. It is felt by the authors that an early introduction of linearization techniques will help to develop a better sense of engineering practicality.

In real life, few (if any) dynamic systems are completely linear. One approach to analyzing a nonlinear system is to write the nonlinear equation that describes it (if it is feasible to describe it by mathematical functions), and then to find an exact solution for the equation. In some cases this can be done, with a precise solution as the result. If a closed-form solution is not obtainable, the nonlinear equation may be solved by graphical techniques, analog techniques, or numerical methods to obtain an approximate (but normally sufficiently accurate) solution. Modern analog and digital computers make this approach practical for a great many nonlinear systems. A computer solution is for one specific case only, however, and does not allow one to generalize with any certainty about the nature of solutions for other cases.

The other approach is to model the nonlinear system by an approximate linear equation, applying linearization techniques so as to develop a sufficiently accurate representation. In many cases the nonlinearities are slight, and no great amount of thought is required for writing a linearized equation. In other cases the system will be found quite nonlinear, however, and a good deal of engineering judgment is necessary to develop the linearized equation that will give the most acccurate solution over the operating range of the system.

With the proper use of linearization techniques, useful information about the behavior of many real dynamic systems can be obtained by the application of linear theory. It should also be pointed out that a knowledge of linear system theory is a prerequisite to the study of nonlinear theory.

4-2 Examples of Nonlinearities

The mathematical distinction between linear and nonlinear differential equations was presented in Chapter 3. A nonlinear system equation is caused by one or more nonlinear elements within the system. At this point we should distinguish between linear and nonlinear *elements*. Consider a system element (e.g., spring, dashpot, capacitor) whose significance within the system is that it produces a certain output (x_2) that is a known function of some input (x_1),

$$x_2 = f(x_1) \qquad (4\text{-}1)$$

If the element is linear, a plot of x_2 versus x_1 will be a straight line; hence the term *linear*. If the plot is not a straight line, the element is nonlinear. We now consider the types of physical characteristics that cause various elements to be nonlinear.

NONLINEAR TRANSLATIONAL SPRING

A translational spring is a good example of a physical device that may be linear or nonlinear. A spring produces a force that is a function of its length; that is, referring to Fig. 4-1,

$$F_s = f(y) \tag{4-2}$$

The helical spring of Fig. 4-1a is designed to be linear over its working range, with the straight-line force-deflection characteristic illustrated in part (c) of the figure. The spring of Fig. 4-1b, however, has been purposely made nonlinear by providing different clearances between adjacent coils, with some of them almost touching when the spring is in the unloaded condition. With this design, a small amount of compression will cause some coils to touch and thereby become inactive. The number of active coils will continue to decrease as the spring compression is increased, causing the spring rate to progressively increase, as illustrated by the force-deflection curve of Fig. 4-1c.

NONLINEAR LIQUID-LEVEL SYSTEM

Analysis of the liquid-level system of Fig. 4-2a requires knowledge of the relationship between the outlet flow (q_o) and the head (h). With the nonlinear sharp-edged orifice, the flow-head relationship is given by the equation

$$q_o = G_v' \sqrt{h} \tag{4-3}$$

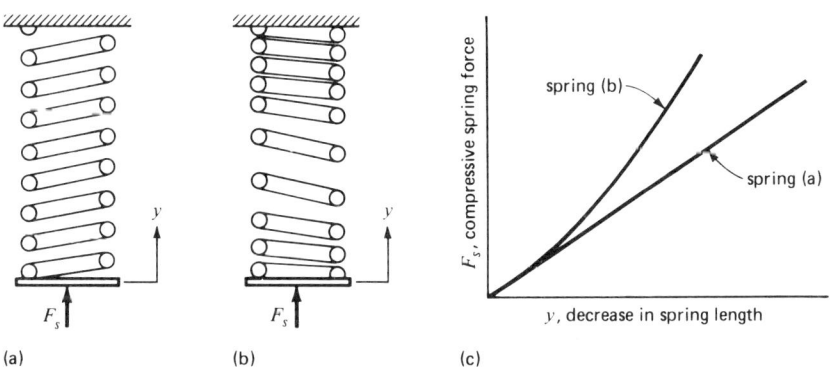

FIG 4-1. (a) Linear translational spring.
(b) Nonlinear translational spring.
(c) Force-deflection characteristics.

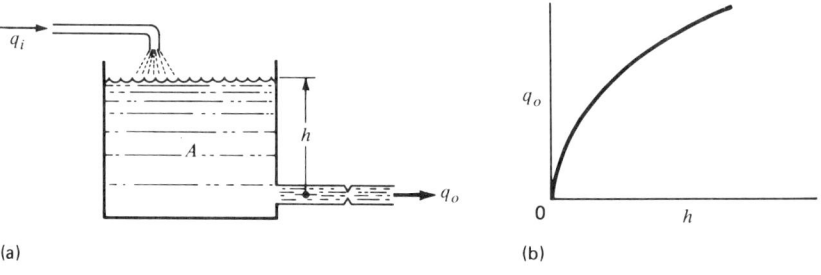

FIG 4-2. (a) A liquid-level system with a nonlinear restriction (sharp-edged orifice) in the outlet line.
(b) Flow-head characteristic of the orifice.

over the normal range of operation where flow is turbulent. Constant G'_v is determined primarily by the size of the orifice. (The prime is used to distinguish it from the coefficient G_v used for the corresponding linear relationship, Eq. 2-28.) The flow-head relationship, as given by Eq. 4-3 and illustrated in Fig. 4-2b, is not a straight line; the system therefore contains a nonlinearity. The nonlinear equation describing the system (note the derivation in Section 2-3) is

$$A \frac{dh}{dt} + G'_v \sqrt{h} = q_i \tag{4-4}$$

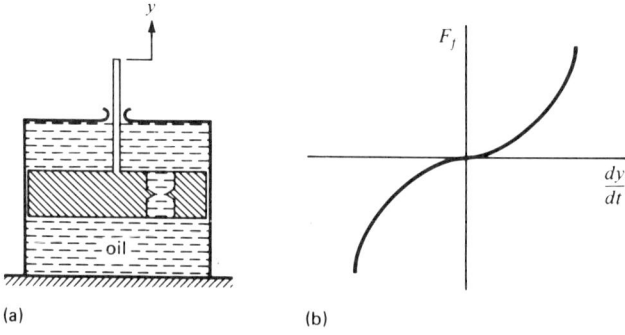

FIG 4-3. (a) A device that gives nonlinear damping by forcing oil through a sharp-edged orifice.
(b) Relationship between friction force and velocity.

NONLINEAR DAMPER

Figure 4-3a illustrates a device that produces damping by pushing oil back and forth through a sharp-edged orifice. The pressure drop across the orifice is proportional (when flow is turbulent) to the square of the flow rate through it; it follows directly that the magnitude of the damping force is proportional to the *square* of the velocity of the mass. The friction force in this case may be represented mathematically by the equation

$$F_f = c' \left| \frac{dy}{dt} \right| \frac{dy}{dt} \tag{4-5}$$

with c' a constant determined by the physical parameters of the damper (not to be confused with the coefficient c used for viscous damping). The form in Eq. 4-5, rather than one using the simple square of the velocity, is necessary to give the proper sign to the friction force so that it will always be shown to oppose relative motion between the ends of the damper. The nonlinear force-velocity characteristic of the damper is illustrated in Fig. 4-3b.

SIMPLE PENDULUM

The simple pendulum is illustrated in Fig. 4-4. The force of gravity acting on the mass produces a torque on the pendulum rod that is a function of the angle θ. A simple force analysis (as given in Section 2-2) shows the relationship to be

$$T = Mgl \sin \theta \tag{4-6}$$

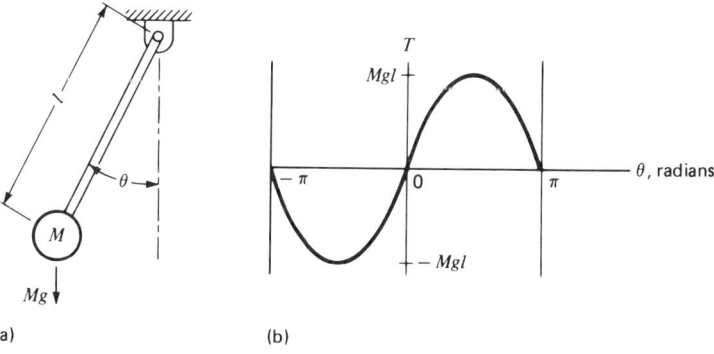

(a) (b)

FIG 4-4. (a) The simple pendulum.
(b) Relationship between torque on the rod and the pendulum angle.

The simple pendulum is therefore a nonlinear device, since the curve of Fig. 4-4b is harmonic. (Note, however, how closely the curve corresponds to a straight line if θ deviates only a small amount from 0.)

4-3 Principles of Linearization

The fundamental goal of linearization is quite simple. The linear differential equation that is chosen should represent the system more accurately, over the range of operation under consideration, than any other linear differential equation. *In practice, this means writing system equations as if all the system elements were linear, and then choosing "average" values for coefficients so that the solution will be as accurate as possible for the type of operation being analyzed.*

In most cases it is not possible to write a linearized equation that will adequately represent the system over all possible conditions of operation. One should therefore determine the *normal operating condition* for the situation under study and write the linearized equation to be accurate for small deviations about that condition. It is usually advantageous to write the linearized equation in terms of *deviations of the dependent variable* rather than in terms of the dependent variable itself. This involves an additional step in the development of the system equation (as illustrated in Example 4-2).

In order to write accurate linearized equations for systems containing nonlinear components, details of the nonlinearities must be known. This information may be in the form of experimental data or theoretical relationships. Except for those cases where theory is known to give precisely accurate relationships, good experimental data for the actual system components are preferred. Unfortunately, however, these data are not always readily available.

The techniques to be used for proper selection of the "average" coefficient values will vary for different types of nonlinearities, and a great deal of engineering judgment is sometimes required. One technique useful for cases in which the nonlinearity can be precisely represented by a mathematical function is to replace that function by a series expansion, retaining only those terms in the series that are linear, provided that the rest of the series terms can be shown to be sufficiently small. For example, the nonlinear relationship

$$x = \sin \theta$$

may be rewritten, by means of a series expansion [3],

$$x = \theta - \frac{\theta^3}{3!} + \frac{\theta^5}{5!} - \cdots$$

4-3 PRINCIPLES OF LINEARIZATION

If the system operation produces only small values of θ, all terms of the series beyond the first are negligible, and $\sin \theta$ may be replaced by θ. This is the approach used for linearizing the equation of the simple pendulum in Section 2-2. If, however, the value of θ should become large during system operation, another method of linearization would have to be used.

Another linearization technique is to study the plot of the nonlinear relationship and determine the straight line that most nearly coincides with the true curve over the expected range of operation. In many cases where a system operates within a limited, well-defined range, a tangent to the curve at the midpoint of the operating range is the best choice. The equation of the straight-line approximation is used in writing the linearized system equation. If the true nonlinear curve corresponds to an exact mathematical function, the equation for the tangent approximation can be found by taking the derivative. If the curve is based on experimental data, a graphical determination of the straight-line approximation must be made.

Several examples will now be given to illustrate the application of linearization techniques to various types of systems. The problems at the end of the chapter will then give the student practice in applying these techniques to a wide range of dynamic systems.

Example 4-1. A coil spring has been wound with variable spacing between its coils (and with some adjacent coils in preloaded contact with each other) so that it provides the force-deflection curve of Fig. 4-5a. Determine the proper linearized differential equation of motion to represent each of the three mass-spring-damper systems illustrated (Fig. 4-5b, c, and d) that use this spring. The mass is 0.1 lb · s²/in. (i.e., it weighs 38.6 lb) and the viscous damping coefficient c is 5.0 lb · s/in. for all three cases.

Solution: The three systems can be considered identical except for one important difference; that is, the position of *static equilibrium* occurs with three different values of spring tension: 0 lb, +38.6 lb, and −38.6 lb. The linearized equation of motion will have the same form for all three cases,

$$M\frac{d^2y}{dt^2} + c\frac{dy}{dt} + ky = 0 \qquad (4\text{-}7)$$

if y is defined for each case as the displacement relative to the position of static equilibrium. The problem is to find the proper "average" value of k to use in the above equation for each case.

The spring rate of a linear spring is defined by the equation $k = F_s/y$. When dealing with a nonlinear spring, however, it is necessary to use the more precise definition,

$$k = \frac{\delta F_s}{\delta y} \qquad (4\text{-}8)$$

in which δF_s is a deviation away from any given reference spring force F_s, and δy is the corresponding deviation in spring length caused by the change in force. For the limiting case of infinitesimal movements away from the position of static equilibrium, Eq. 4-8 becomes

$$k = \frac{dF_s}{dy} \tag{4-9}$$

and this definition is a good approximation to use for a real system in which the movements are finite but small. With the curve of experimental force-deflection data available (Fig. 4-5a), the derivative can be readily determined as the slope of the curve evaluated at the proper point. By drawing tangents to the curve at the points corresponding to static equilibria of the three systems under consideration, and measuring the slopes of these tangents (Fig. 4-5a), we obtain

$k = 20$ lb/in. for Fig. 4-5b
$k = 5$ lb/in. for Fig. 4-5c
$k = 40$ lb/in. for Fig. 4-5d

Use of the proper value of k in the linearized equation of motion will give accurate representation of each of the three systems. The smaller the amplitude of vibration in each case, the greater the accuracy one would expect from the use of the linearized equation. (A qualitative idea of the accuracy of this type of linearization is obtained by noting how nearly the tangent line coincides with the true spring rate curve over the range of motion that occurs).

Example 4-2. Obtain a linearized differential equation to accurately represent the liquid-level system of Fig. 4-2a. The outlet tube contains a sharp-edged orifice that has the following theoretical relationship between flow rate and head (for turbulent operation, which is assumed to exist here):

$$q_o = 12\sqrt{h}$$

The average inlet flow rate is 36 in.^{3}/s, with small deviations occurring on either side of this value. The chosen orifice size causes the average liquid level h to be 9 in. (i.e., equilibrium occurs with $h = 9.0$ in., $q_i = 36$ in.^{3}/s and $q_o = 36$ in.^{3}/s).

Solution: Two alternative solutions are presented to illustrate the advantage of using a linearized differential equation for *deviations* of the head from the normal operating point rather than using an equation written in terms of the total head.

1. *Solution based on total head*: If we assume a linear relationship between q_o and h, this system can be represented by the equation (see

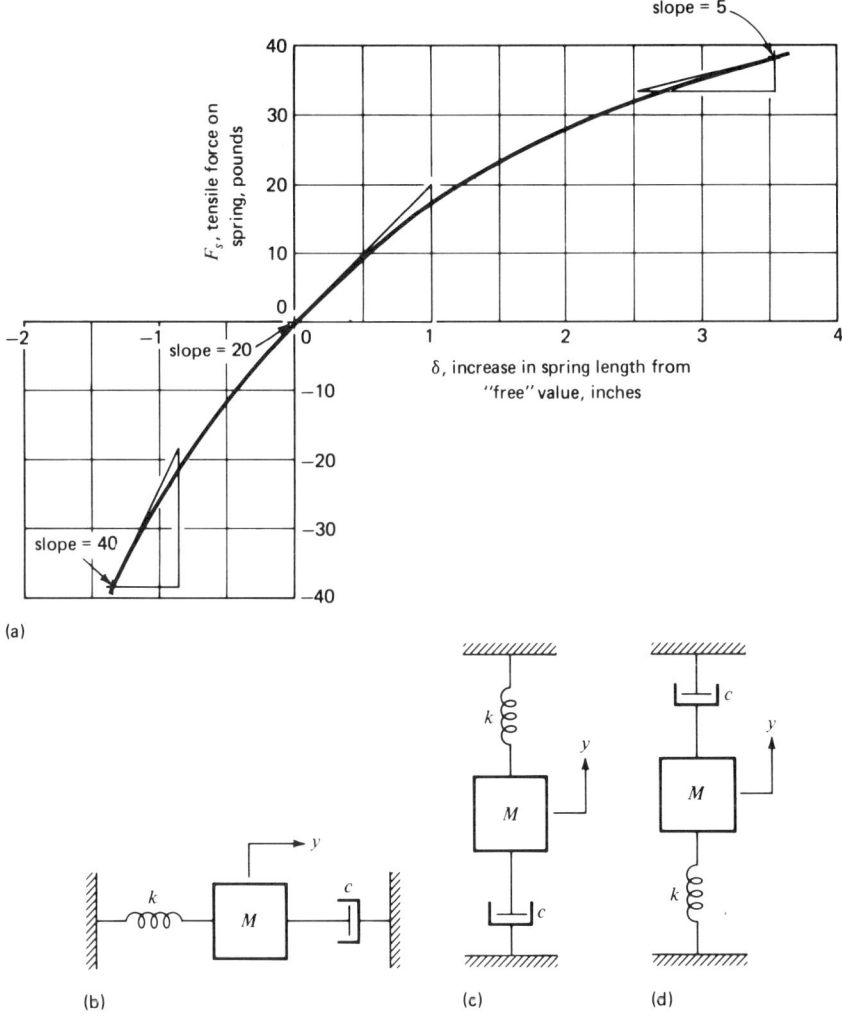

FIG 4-5. Vibrating systems with a nonlinear spring.
  (a) Spring characteristics.
  (b) System with zero spring force at position of static equilibrium.
  (c) System with tensile spring force at position of static equilibrium.
  (d) System with compressive spring force at position of static equilibrium.

Section 2-3)

$$A \frac{dh}{dt} + G_v h = q_i \quad (4\text{-}10)$$

For our nonlinear system we can use the above equation and determine the value of G_v which gives the most accurate answers. At equilibrium, $A\, dh/dt = 0$, so we may write

$$G_v(9) = 36$$

It is therefore necessary to use $G_v = 4$ to satisfy the system equation for normal operating conditions. The resulting equation, however, will be found rather inaccurate for excursions of any significant magnitude away from average conditions (i.e., the linearized equation will show a *change* in the outlet flow rate that is very inaccurate for any given *change* in head away from the 9 inch level). Figure 4-6a illustrates why this is true. The linearized equation is based on an assumed straight-line relationship between flow rate and head that intersects the true curve at two points only—at $h = 0$ and $h = 9$. At all other values of head, a significant error can be seen to exist.

2. *Solution based on deviation of head from the normal operating point*: To obtain better accuracy, it is necessary to consider *deviations* of head and flow rate away from normal or average conditions, and define the linearized valve coefficient in terms of these deviations; that is,

$$G_v = \frac{\delta q_o}{\delta h} \quad (4\text{-}11)$$

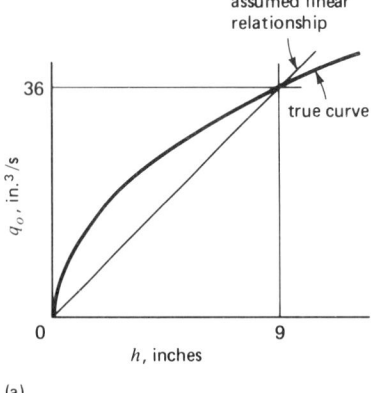
(a)

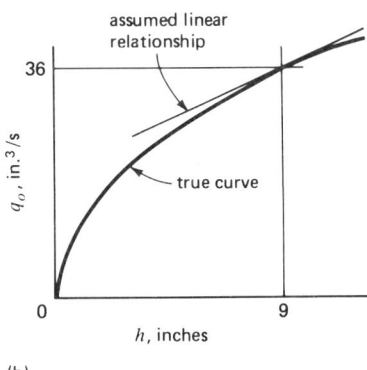
(b)

FIG 4-6. Linearization of a nonlinear flow-head relationship.
    (a) Using linearized equation $A\, dh/dt + G_v h = q_i$.
    (b) Using linearized equation $A\, d\delta h/dt + G_v\, \delta h = \delta q_i$.

4-3 PRINCIPLES OF LINEARIZATION

To develop the system equation based on deviations, we start with the more basic relationship for this system (from Eq. 2-27),

$$A \frac{dh}{dt} + q_o = q_i \quad (4\text{-}12)$$

Defining H, Q_o, and Q_i as the average or reference values of head, outlet flow rate, and inlet flow rate, respectively, we may write

$$h = H + \delta h$$
$$q_o = Q_o + \delta q_o$$
$$q_i = Q_i + \delta q_i$$

with δh, δq_o, and δq_i being deviations from the average values. Since H is, by definition, a constant,

$$\frac{dh}{dt} = \frac{d\delta h}{dt}$$

and the linearized system equation may now be written as

$$A \frac{d\delta h}{dt} + Q_o + \delta q_o = Q_i + \delta q_i \quad (4\text{-}13)$$

Since the reference values are based on a condition of equilibrium, $Q_o = Q_i$ and the equation reduces to

$$A \frac{d\delta h}{dt} + \delta q_o = \delta q_i \quad (4\text{-}14)$$

With G_v defined by Eq. 4-11, this becomes

$$A \frac{d\delta h}{dt} + G_v \, \delta h = \delta q_i \quad (4\text{-}15)$$

For this problem, G_v must be obtained mathematically since there are no experimental data. With infinitesimal excursions from the point of normal operation, Eq. 4-11 becomes

$$G_v = \frac{dq_o}{dh} \quad (4\text{-}16)$$

Thus G_v can be obtained by taking the derivative of the theoretical equation for the sharp-edged orifice in the following manner:

$$G_v = \frac{d(12\sqrt{h})}{dh} = \frac{12}{2\sqrt{h}}$$

Evaluation at the point of normal operation ($h = 9$) yields

$$G_v = 2 \frac{\text{in.}^3/\text{s}}{\text{in.}}$$

We could, of course, have plotted the theoretical curve of q_o versus h and found G_v as the slope of the tangent at $h = 9$, as illustrated in

Fig. 4-6b. For this example, however, taking the derivative produces the answer more quickly and accurately.

The linearized differential equation for the system may now be written in the final form

$$A \frac{d\delta h}{dt} + 2\delta h = \delta q_i \qquad (4\text{-}17)$$

This is a differential equation of the *deviation* of the head rather than the head itself, but this should present no particular problem so long as the meaning of each of the terms is clearly understood.

The linearized equation will give accurate answers when h stays close to 9 in. (Note, in Fig. 4-6b, that the tangent representing the linearized valve coefficient is very nearly coincident with the true theoretical curve for small deviations.)

The reader may wonder why it was necessary to write the linearized liquid-level equation in terms of deviation of head, while the mass-spring-damper systems of Example 4-1 were adequately represented by an equation in y rather than δy. The answer is that we judiciously defined y to be zero in each case for the position of static equilibrium. For that example, therefore, y and δy are the same. A similar approach could have been used for the liquid-level system, but it would probably be confusing to call the head h zero when in fact it was not.

Example 4-3. The pendulum system of Fig. 4-7 is shown in its position of static equilibrium, at which $\theta = 30°$. The pendulum mass weighs 5 lb, and the inertia of all other parts is negligible. The coil spring is linear, with $k = 2$ lb/in. Develop a linearized equation of motion for the system.

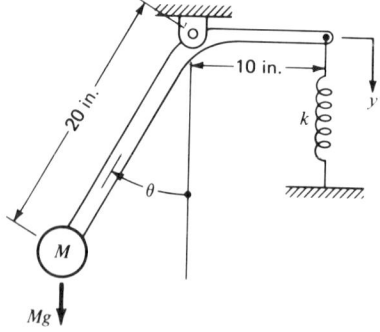

FIG 4-7. A pendulum system.

4-3 PRINCIPLES OF LINEARIZATION

Solution: For this problem, we will define the angle θ as shown in Fig. 4-7 and write the linearized system equation in terms of *deviations* from the position of static equilibrium ($\theta = 30°$).

The total effective torsional spring constant of the system has two components. One of them, which we will call G_1, is due to the translational spring, and the other, G_2, is caused by gravity acting upon the pendulum mass.

The system equation is derived by application of Newton's second law,

$$\Sigma T = I \frac{d^2\theta}{dt^2} = -(T_1 + T_2) \qquad (4\text{-}18)$$

where T_1 and T_2 are the torques of the two "torsional springs" discussed above. If we evaluate the torsional spring rates at the position of static equilibrium,

$$T_1 = T_{1_{\text{ref}}} + G_1 \, \delta\theta$$

$$T_2 = T_{2_{\text{ref}}} + G_2 \, \delta\theta$$

where $T_{1_{\text{ref}}}$ and $T_{2_{\text{ref}}}$ are the spring torques at static equilibrium. Substituting these equations into Eq. 4-18 and using the relationship

$$\frac{d^2\theta}{dt^2} = \frac{d^2 \, \delta\theta}{dt^2}$$

yields

$$I \frac{d^2 \, \delta\theta}{dt^2} + T_{1_{\text{ref}}} + G_1 \, \delta\theta + T_{2_{\text{ref}}} + G_2 \, \delta\theta = 0$$

We know that $T_{1_{\text{ref}}} = -T_{2_{\text{ref}}}$ because of the definition of "static equilibrium," so the system equation reduces to

$$I \frac{d^2 \, \delta\theta}{dt^2} + (G_1 + G_2) \, \delta\theta = 0 \qquad (4\text{-}19)$$

The basic equation for spring rate that is applicable to both linear and nonlinear torsional springs is

$$G = \frac{\delta T}{\delta \theta} \qquad (4\text{-}20)$$

Determination of the two components G_1 and G_2 will be based on Eq. 4-20.

To determine G_1 we note that for a small increase in displacement $\delta\theta$ of the pendulum arm away from the position of static equilibrium, the decrease in spring length is

$$\delta y \approx 10 \, \delta\theta \qquad (4\text{-}21)$$

Using the definition of translational spring rate given by Eq. 4-8,

$$k = \frac{\delta F_s}{\delta y}$$

and combining it with Eq. 4-21 yields the expression for the resultant change in spring force δF_s:

$$\delta F_s = 10k\,\delta\theta$$
$$= 20\,\delta\theta$$

The change in torque due to the change in spring force is

$$\delta T = 10\,\delta F_s$$
$$= 10(20\,\delta\theta) = 200\,\delta\theta$$

The torsional spring constant from the translational spring can now be determined by direct substitution into Eq. 4-20,

$$G_1 = \frac{\delta T}{\delta \theta} = \frac{10 \cdot k \cdot 10\,\delta\theta}{\delta\theta} = k\cdot 10^2 = k\,l^2$$
$$= \frac{200\,\delta\theta}{\delta\theta} = 200 \text{ in.}\cdot\text{lb/rad}$$

(It should be noted that no linearization was required in determining G_1.)

We now turn our attention to the second torsional spring constant G_2. The torque caused by the force of gravity acting on the pendulum mass is

$$T = 5(20 \sin\theta) = 100 \sin\theta$$

For small motions, the definition of torsional spring constant (Eq. 4-20) can be approximated by the differential expression

$$G \approx \frac{dT}{d\theta} \qquad (4\text{-}22)$$

For this problem, therefore,

$$G_2 = \frac{d}{d\theta}(100 \sin\theta) = 100 \cos\theta$$

Since G_2 must be evaluated at the condition of static equilibrium,[1] where $\theta = 30°$,

$$G_2 = 100 \cos 30° = 86.6 \text{ in.}\cdot\text{lb/rad}$$

---

[1] Note that Eq. 2-23 for the effective torsional spring rate of a simple pendulum

$$G = \frac{Mgl \sin\theta}{\theta} \approx Mgl$$

is valid only when the pendulum arm is vertical at the position of static equilibrium. It is therefore not applicable to this example problem.

4-3 PRINCIPLES OF LINEARIZATION

The moment of inertia of the pendulum is

$$I = M(20^2) = \frac{5}{386}(400)$$
$$= 5.18 \text{ lb} \cdot \text{in.} \cdot \text{s}^2$$

Substitution of the spring rates and the moment of inertia into Eq. 4-19 yields the linearized differential equation of the system,

$$\boxed{5.18 \frac{d^2 \delta\theta}{dt^2} + 286.6 \, \delta\theta = 0} \qquad (Ans.)$$

It should be noted that we could have defined $\theta = 0$ at the position of static equilibrium rather than $\theta = 30°$. The linearized system equation would then have been written in terms of θ, rather than $\delta\theta$. The proper approach to use is mostly a matter of personal preference. It would be instructive for the reader to rework the problem with the definition $\theta = 0$ at the position of static equilibrium.

Example 4-4. The vibrating system of Fig. 4-8a contains a linear spring, but damping is obtained by forcing oil back and forth through a sharp-edged orifice. The parameters of the damper are such as to cause the damping force (for all operations except where dy/dt is very small and flow through the orifice is laminar) to be given by the

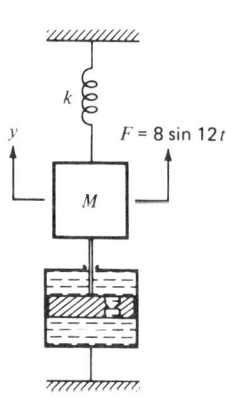

(a)

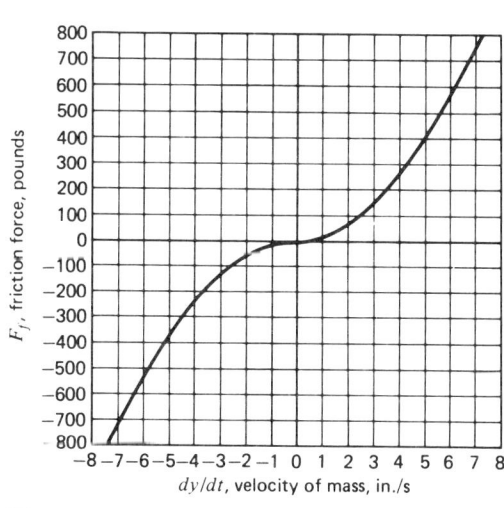

(b)

FIG 4-8. (a) Vibrating system with nonlinear damper.
(b) Characteristics of the damper.

equation

$$F_f = 15 \left| \frac{dy}{dt} \right| \frac{dy}{dt} \qquad (4\text{-}23)$$

Develop a linearized differential equation of motion for the system with the given sinusoidal forcing function.

Solution: The linearized equation of motion for this mass-spring-damper system is

$$M \frac{d^2y}{dt^2} + c \frac{dy}{dt} + ky = 8 \sin 12t \qquad (4\text{-}24)$$

The task at hand is to find an "average" value for the damping coefficient c that will make this equation an accurate representation of the system with the given forcing function.

Figure 4-8b shows the damping force as a function of the velocity of the mass, plotted from Eq. 4-23. With the sinusoidal force applied to the mass as shown, we would expect its velocity to alternate between some positive value and a negative value of equal magnitude, with an average value of zero. We might be inclined to choose our linearized damping coefficient to correspond to this average velocity of zero, but this would result (using the given equation for F_f) in a coefficient that was itself zero, indicating no damping at all, whereas we know that the damper does dissipate energy for any finite motion. If, on the other hand, we were to choose the position of static equilibrium ($y = 0$) for evaluating c, the damping would undoubtedly be too large, since the velocity of the mass can be shown to be a maximum when $y = 0$. It should be apparent that we need to use a value of c that is somewhere between zero and the value corresponding to maximum velocity. The value corresponding to one-half the maximum velocity would be a reasonably good choice.[2]

To determine this "average" or "equivalent" viscous damping coefficient, it is necessary to know the details of the motion of the mass, including the *amplitude* of the vibration. But we need the differential equation of motion (with numerical values for the coefficients) to solve for the amplitude. This presents a dilemma; without additional information (which is unlikely to be available) we cannot find the proper value of c to use except by trial-and-error methods. We must guess what the motion of the mass is, select a value of c to agree with this, and then solve for the motion to see if it is what we had assumed. If it is not, then a new value of c must be chosen and the process repeated. After a few iterations we would have the proper

---

[2] A detailed analysis based on the amount of energy dissipated per cycle will lead to a more precise "average" value of c. It is not felt worthwhile to pursue the point here, however, because the intention is merely to provide an *introduction* to the principles of linearization.

value of c to give us a linearized equation that accurately represents the system with the given externally applied force. If the magnitude of this force or its frequency were changed, however, it would be necessary to use a different value of c even though the system itself remained the same.

This example points out some of the difficulties that can occur in writing linear equations to represent nonlinear systems, and we will not carry out the details of the solution of this particular problem.

4-4 Conclusion

The accuracy with which a linearized differential equation describes the action of a nonlinear dynamic system depends upon

1. the degree of nonlinearity of the system
2. the accuracy of available data, experimental or analytical, on system parameter characteristics
3. the range over which the system operates

If component data are very accurately known, and the system operates with only small excursions away from equilibrium values, a linearized equation can give an accurate representation of the dynamic action of the system even if it is quite nonlinear.

Linearization is a very important part of dynamic system analysis, particularly if we restrict our analytical techniques to those using linear differential equations.

Problems

4-1. A 10-lb weight is suspended from the ceiling by a coil spring as shown. The spring is nonlinear, the relationship between the applied tensile force F_s and its elongation Δ (from its free length) being $F_s = 8\Delta^2$. Derive a linearized equation of motion for the system, assuming small vertical movements about the position of static equilibrium.

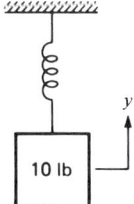

4-2. Variables R and P are related by the equation $R = 0.60P^3$. If operation normally occurs with R at or near 4.8 and we want to use linearized equations assuming $\delta R = K\delta P$, what is the proper value of K?

4-3. Write linearized differential equations of motion for the two different mass-spring systems illustrated. All springs are identical, having the force-deflection characteristics given in part (c) of the figure.

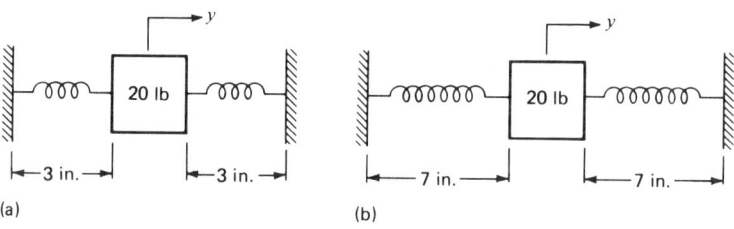

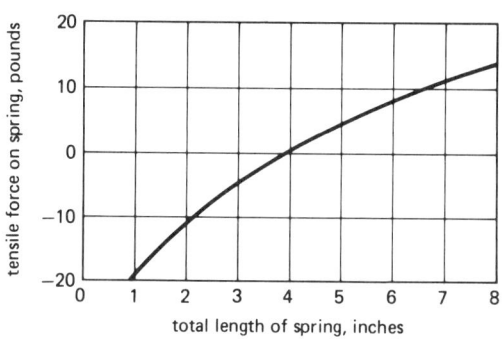

4-4. A 20-kg mass is suspended by a spring having the given nonlinear characteristics. Write the linearized differential equation of motion for small vertical movements about the position of static equilibrium.

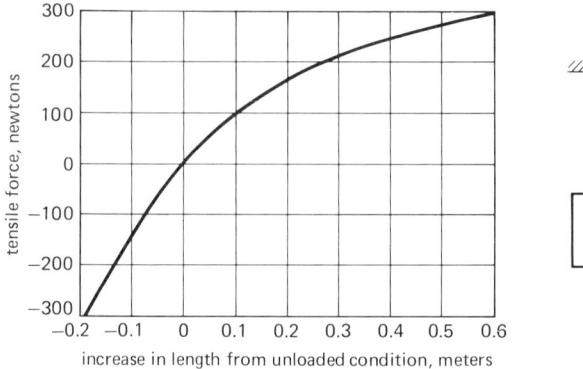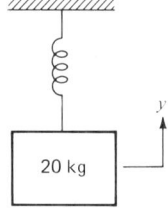

PROBLEMS

4-5. Springs 1 and 3 are identical, whereas spring 2 has quite different characteristics. (See the given force-deflection curves.) The free lengths of all three springs are 0.8 m. Obtain the linearized differential equation that describes the system.

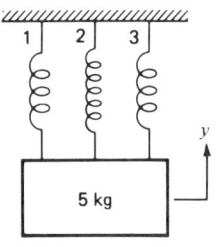
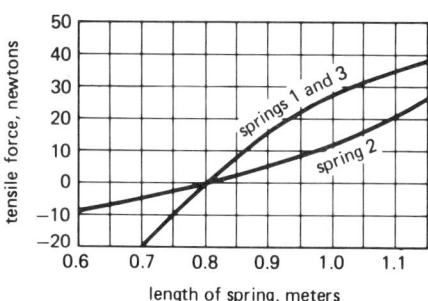

4-6. A nonlinear torsional spring with the given characteristics is used to support the cantilevered mass as shown. Write the linearized differential equation that describes the system if an initial twist is given to the spring so that static equilibrium occurs with the rod horizontal (as shown). Assume the rod has negligible mass and that the motion is limited to small distances away from the position of static equilibrium.

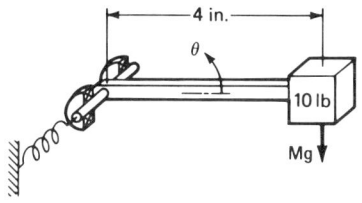
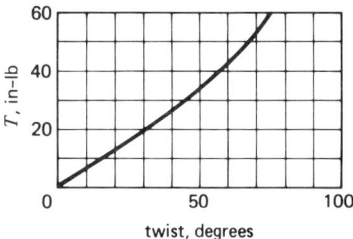

4-7. The double pendulum is suspended from the ceiling as shown. Determine the linearized differential equation of motion for small movements about the position of static equilibrium. Assume that all mass is negligible except for M_1 and M_2. Values are as follows: $l_1 = 0.5$ m, $l_2 = 0.3$ m, $M_1 = 2$ kg, $M_2 = 1$ kg.

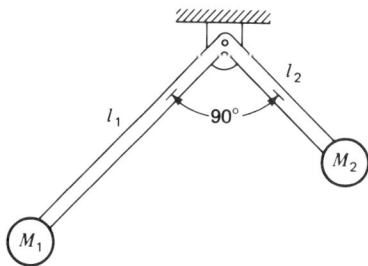

4-8. Four air springs are used as shown to provide isolation between a vibrating machine and the floor. The machine weighs 10,000 lb, the piston area of each spring is 10 in.^{2}, and sufficient air is pumped into each spring to give it a volume of 100 in.^{3} at static equilibrium. Determine the proper linearized spring constant (for each individual air spring) to be used for a system analysis based on small movements away from the position of static equilibrium. Assume that isentropic conditions prevail in the cylinder. *Hint*: Use the thermodynamic relationship for isentropic compression and expansion of air, $PV^{1.4} = $ constant, and find k as

$$k = \frac{dF}{dy} = A\frac{dP}{dy}$$

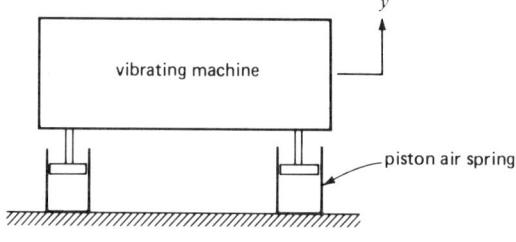

4-9. Rubber matting 1/2 in. thick has the nonlinear pressure-deflection characteristics shown. It is to be used for vibration isolation of a machine weighing 1000 lb, the bottom of which is a flat rectangle measuring 24 × 48 in. Vibration calculations have shown that a rubber pad placed between the machine and the floor should have an effective spring constant of 2000 lb/in. Determine the pad area and the number of layers needed to give the desired spring characteristics. It is desirable to have a pad that is nearly as large as the bottom of the machine. (Note that a certain amount of trial and error is necessary.)

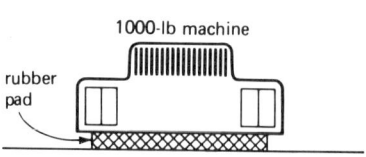 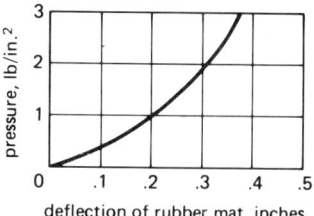

4-10. Obtain the differential equation that relates $\delta\theta_o$ to small variations in the input angular velocity ω_i (i.e., $\delta\omega_i$) about a mean input velocity of 25 rad/s. (Note that $\omega_o = d\theta_o/dt$.)

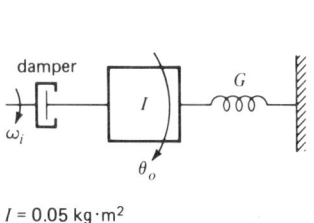

$I = 0.05$ kg·m²
$G = 8.0$ N·m/rad

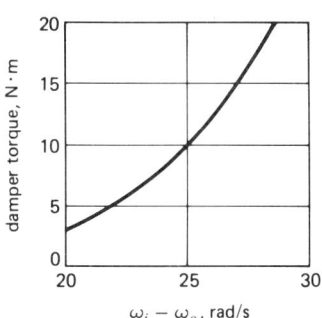

4-11. The linearized equation $\delta q = C\delta(\Delta P)$ is to be used to describe the sharp-edged orifice illustrated in the figure, with $\Delta P = P_1 - P_2$. The exact pressure-flow relationship is known to be $P_1 - P_2 = 4q^2$, with pressure in pounds per square inch and flow rate in cubic inches per second.
(a) Determine the proper value of C for the linearized equation if operation is known to occur near the condition of $\Delta P = 30$ lb/in.².
(b) Determine the linearized equation that gives the *change in pressure drop* caused by a small *change in flow rate* away from a normal value of 10 in.³/s.

4-12. Two sides of a tank are tapered as shown so that the area of the tank varies linearly from 1000 cm² at the bottom to 1500 cm² at the top. The outlet valve gives a linear relationship between q_o and h, with

$$G_v = 150 \frac{\text{cm}^3/\text{s}}{\text{cm}}$$

(a) Write a linearized differential equation relating h to q_i for operation about an average liquid height of 80 cm.
(b) Repeat for operation about an average liquid height of 30 cm.

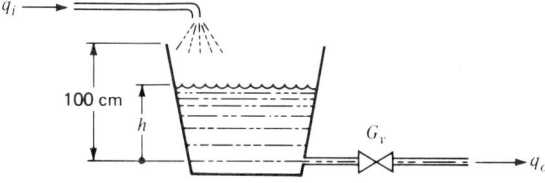

4-13. The flow rates through the inlet line and outlet line are both controlled by sharp-edged orifices having the same flow-pressure relationship, $q = 10\sqrt{\Delta P}$ in.³/s, with ΔP the pressure drop across the orifice in pounds per square inch. Determine a linearized differential equation for the change in fluid level, δh, as a function of the change in inlet line pressure, δP_1. The system uses water, which weighs 62.4 lb/ft³. The area of the tank is 500 in.². The normal value of P_1 is 1.0 lb/in.², gage.

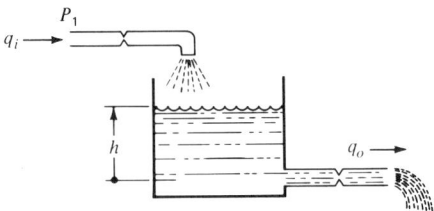

4-14. The characteristics of the two valves are given in graphical form. Determine the linearized differential equation for δh as a function of δP_1, if P_1 is normally 2.0 lb/in.², gage.

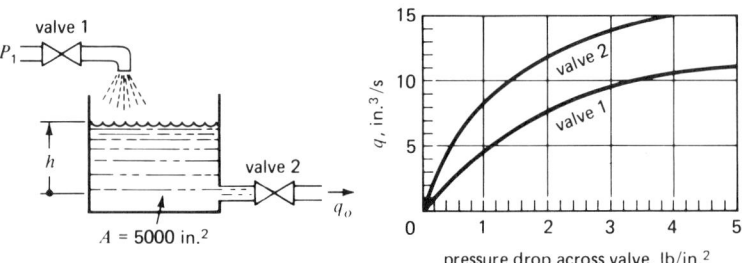

4-15. A simple water-level system is being built, and a choice must be made as to whether a sharp-edged orifice or a special linear valve should be used as the outlet line valve. The normal value of q_i is 10 in.³/s, and the cross-sectional area of the tank is 1000 in.². Either valve would be sized to give a normal water level of 20 in.

Which of the two valves would make the water level less sensitive to small changes in q_i? Give a quantitative measure of the difference in sensitivity of the two cases.

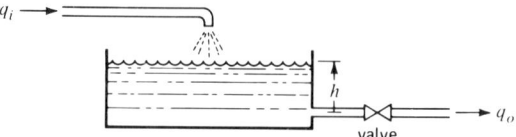

4-16. A resistor is nonlinear, its resistance R (ohms) varying with current i (amperes) as given by the equation

$$R = 10 + 1.5i$$

It is desired to represent the change in voltage as a linear function of a change in the applied current,

$$\delta e = K \, \delta i$$

Determine the proper value of K if operation normally occurs with a current of about 2 amperes. Use Ohm's law: $e = iR$.

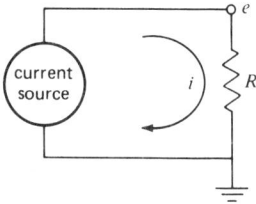

5
Introduction to Vibrations

5-1 Introduction

Almost everyone has been annoyed at some time or another by a vibration in some vehicle, appliance, or machine. Even those who lack sufficient engineering or scientific background to understand what a vibration really is, or what might cause it, can still appreciate many of the problems caused by excessive vibration. Vibration, however, is not always undesirable. It is often purposely created to perform a useful function. A radio or phonograph speaker works through vibration, for example, and conveyor systems for small machine parts often employ vibration to align the parts and to move them in the proper direction. Vibrating machinery is used for such tasks as stirring paint and cutting hair.

A vibration can be described in simple terms as a periodic oscillation, or movement back and forth, of some object. As normally defined, a vibration can occur only in a device having both mass and a flexible element or spring, even though a device such as a swinging pendulum with

5-1 INTRODUCTION

no spring is usually considered a vibrating system. A very simple type of vibrating system is illustrated in Fig. 5-1. Shown is a single mass supported by a mechanical spring; it is assumed that the mass is constrained to move in a straight line. If the spring is linear, the vibrating mass will move up and down with harmonic motion about the point of static equilibrium.

The question now arises as to what action is necessary to cause the simple system of Fig. 5-1 to vibrate. There are actually a number of different ways in which vibration can be induced in such a system. The mass might be pulled down to a certain position and then abruptly released. The subsequent action of the mass would then be referred to as *free vibration*. The same type of vibration could also be induced by striking the mass underneath with a sharp blow or by otherwise imparting a velocity to the mass. A free vibration produced in any one of these ways will eventually die out because of energy dissipation resulting from hysteresis effects within the spring and friction forces produced by the movement of the mass through the surrounding air.

In order to have a continuing, or *steady-state*, vibration it is necessary to have some type of continuous excitation. The excitation might be a *periodic force* on the mass, applied through an external device in contact with the mass, or by some other means such as magnetism or sound waves. An externally applied excitation could alternatively be a *periodic movement* such as a vertical oscillation of the top of the spring of Fig. 5-1. Some excitations occur within the vibrating system itself. For example, the mass of Fig. 5-1 could represent a motor with an unbalanced rotor or a machine with a piston that reciprocates during operation.

Vibration is defined as a periodic motion of some object or system, but there are many variations of the form such motion can take. In addition to the simple translational vibration that occurs with the system of Fig. 5-1, some systems will vibrate rotationally, with rotation back and forth about the position of static equilibrium. In the case of large complex devices, in which the mass and elasticity are distributed in a complex manner throughout the total structure, a state of vibration may find some parts moving up while others are moving down, and, in many cases, rotational vibration will be superimposed upon translational vibration.

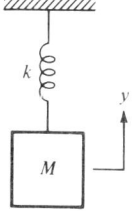

FIG 5-1. Simple vibrating system.

In many cases, vibration is undesirable and can cause one or more serious problems. The loading of metal components during vibration causes stresses that can produce failure. Even if stress levels are below the yield point, they will normally alternate between tension and compression. The stress cycles accumulate rapidly under a condition of steady-state vibration, so fatigue failures can occur in a matter of days or even hours.

Vibration can make it impossible to maintain accuracy with even the most precise instruments and tools. It is impossible, for example, to maintain a smooth finish on a workpiece in a lathe if the tool is vibrating. Vibration commonly causes bolts and other fasteners to loosen or fail. A vibrating device will often produce unpleasant noise.

5-2 Basic Principles and Definitions

Our study of mechanical vibrations will begin with a discussion of some basic principles. In addition, it is important that the definitions of a number of vibration terms be clearly understood. Definitions of the more basic terms are presented here; others will be presented as they occur in the development of the theory.

Vibrations are characterized by *periodic* motion; that is, the motion repeats itself after a certain increment of time, called the *period* (given the symbol P in this book). The movement of the mass over a single period is called a *cycle* of vibration.

Frequency is defined as the number of cycles that occur within a given increment of time. It is most commonly measured in cycles per second and normally referred to by the equivalent term *hertz*, in honor of the famous German physicist. The symbol f will be used in this book for frequency in hertz (Hz). It should be noted that frequency is the reciprocal of the period,

$$f = \frac{1}{P} \tag{5-1}$$

Harmonic motion is the most important type of periodic motion in the study of vibrations. In its simplest form, harmonic motion may be represented by the equation

$$y = Y \sin \omega t \tag{5-2}$$

where ω is the frequency in radians per second. (The significance of ω and its units will be considered in detail a little later.) If the displacement y is plotted against time t, the well-known sine-wave curve of Fig. 5-2 is obtained. The *amplitude* of the vibration is Y, and the *peak-to-peak amplitude* is equal to $2Y$.

If y in Eq. 5-2 is the position of a vibrating object, then we may determine expressions for its velocity and acceleration by taking the

5-2 BASIC PRINCIPLES AND DEFINITIONS

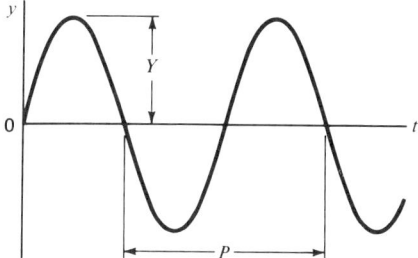

FIG 5-2. Harmonic motion as given by the equation $y = Y \sin \omega t$.

derivatives

$$\frac{dy}{dt} = \omega Y \cos \omega t \qquad (5\text{-}3)$$

$$\frac{d^2y}{dt^2} = -\omega^2 Y \sin \omega t \qquad (5\text{-}4)$$

These are plotted, together with the position curve, in Fig. 5-3. Note that the relative amplitudes of the three curves depend upon the value of ω.

In Eq. 5-2, ωt is the argument of the sine function. Therefore, ωt must be an angle. With t in seconds, ω must be measured in radians per second in order that the product ωt be in radians.

The question now arises as to the significance of ω, which has the units of an angular velocity, and ωt, which has the units of an angle. It is not immediately obvious that an angle and an angular velocity have any

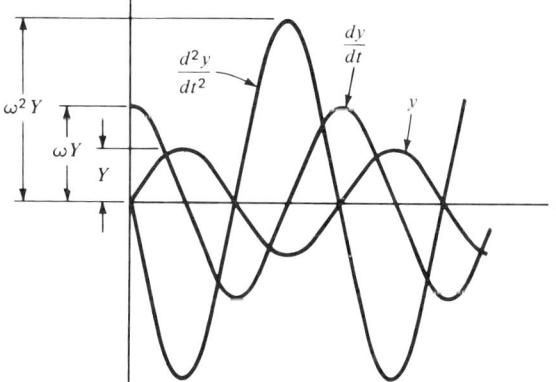

FIG 5-3. Position, velocity, and acceleration of a vibrating object with motion given by the equation $y = Y \sin \omega t$.

significance when we are discussing a translating object that moves with harmonic motion in a straight line (as the mass in Fig. 5-1). Such harmonic motion can, however, be conveniently represented by a rotating vector, a representation that allows the units of ω to take on a greater significance. If the vector Y in Fig. 5-4 is rotated counterclockwise at the angular velocity ω rad/s (starting out at $t = 0$ pointing to the right on the horizontal axis), the vertical projection is given by Eq. 5-2, $y = Y \sin \omega t$, and the horizontal projection by the equation $y = Y \cos \omega t$. (Either projection can be used, depending upon whether the harmonic motion is represented by a sine or a cosine expression).

One cycle of harmonic motion is represented by a rotation of the vector through 360°—that is, when the angle ωt (as shown on Fig. 5-4) is equal to 2π radians. This gives the required relationship between ω and f,

$$\omega = 2\pi f \qquad (5\text{-}5)$$

In view of the required relationship of Eq. 5-5, ω may be considered not only as the angular velocity of a rotating vector, but also as a frequency of vibration, comparable to f, except that it has the units of radians per second. The frequency ω is sometimes given the name *circular frequency* because of the representation of Fig. 5-4, in which the rotating vector traces out a circle. In this book, however, the simple term *frequency* will be used for both f and ω. With consistency in the use of the symbols and with the different units for each, there should be no reason for confusion.

Two harmonic functions of the same frequency may not always be in phase; that is, they may pass through values of zero amplitude at different instants of time. If they are not in phase, then their phase relationship may be conveniently expressed by means of a *phase angle* (given the symbol ϕ

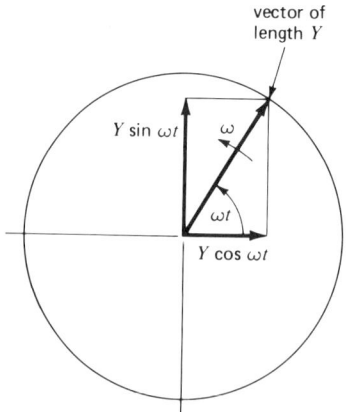

FIG 5-4. Rotating-vector representation of harmonic motion.

5-2 BASIC PRINCIPLES AND DEFINITIONS

in this book) measured in radians or degrees. If one vibrating mass has motion given by the equation

$$y_1 = Y_1 \sin \omega t$$

and another by the equation

$$y_2 = Y_2 \sin(\omega t + \phi)$$

they will appear as in Fig. 5-5a when plotted together with respect to time. With one cycle of vibration equivalent to an angle ωt of 2π, the curve of y_2 is found to *lead* that of y_1 by the angle ϕ in radians. The rotating vectors whose vertical projections represent y_1 and y_2 must be displaced by the phase angle ϕ as illustrated in Fig. 5-5b. Note that if ϕ were negative, y_2 would *lag* behind y_1. With reference to Fig. 5-3, we see that with pure harmonic motion the velocity leads the displacement by $\pi/2$ radians (90°), and the acceleration leads the displacement by π radians (180°).

Free vibration is the term applied to vibration that occurs in a device or system in the absence of any external force or other excitation. (An external excitation may be necessary to start the vibration, but it is free vibration once it is removed.) Free vibration will always take place at the system's natural frequency. For systems with more than one natural frequency, free vibration may occur at one of those alone, or be the superposition of motion at two or more natural frequencies. In line with the above, *natural frequency* is defined as a frequency at which free vibration can occur.

Forced vibration applies to a vibration that occurs as a result of a continuously applied periodic external force or other excitation. *Forced vibration takes place at the frequency of the excitation. Resonance* is the condition of large vibratory amplitude that occurs at the *resonance frequency*, which is defined as the input frequency that will produce the

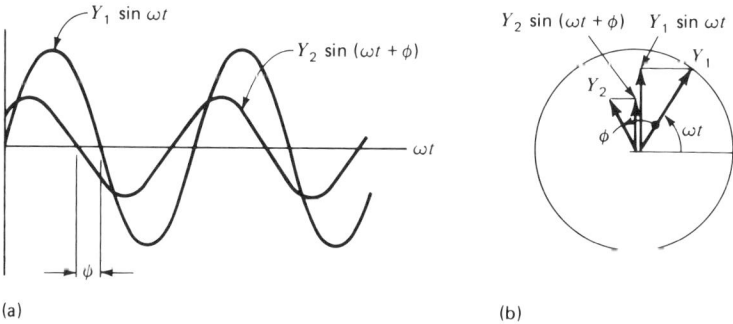

FIG 5-5. (a) Two harmonic functions of the same frequency separated in time by the phase angle ϕ.
(b) Rotating-vector representation.

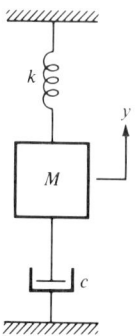

FIG 5-6. A single-degree-of-freedom system.

maximum response amplitude under steady-state conditions. (This will be a local maximum for systems with more than one resonance frequency.)

Degrees of freedom is another important term, giving some indication of the mathematical complexity of any particular vibrating system. *The number of degrees of freedom is defined for mechanical systems as the number of coordinates that must be specified in order to completely describe the position and/or motion of a given system.* The mass-spring-damper system of Fig. 5-6 has a single degree of freedom, since knowing how y and its derivatives vary with time tells us everything we need to know about its vibration characteristics.

For the two-mass system of Fig. 5-7, knowledge of both y_1 and y_2 and their derivatives is necessary to completely understand the vibratory motion of the system; it is therefore a two-degree-of-freedom system.

Figure 5-8 illustrates another type of system. A single mass is attached to the end of a rod, which is then suspended by a universal joint from the ceiling, forming a type of pendulum. The possible range of movements of this device will cause components of displacement in the x, y, and z directions, all of which must be known as a function of time in order to define the motion of the system. The system does not have three degrees of freedom, however, but merely two, since knowledge of x and z will allow us, by geometry, to determine y.[1] (Note that for this particular system knowledge of y and x is not sufficient for determining z, and likewise knowledge of y and z does not allow x to be determined, so x and z must be the two coordinates used).

---

[1] If, however, the universal joint were replaced by a ball joint that allowed the mass to rotate about the axis of the rod, then this rotation would be part of the dynamic behavior of the system. A third coordinate, the angle of rotation of the rod, would have to be added to completely specify the action of the system, which would in this case have three degrees of freedom.

5-2 BASIC PRINCIPLES AND DEFINITIONS

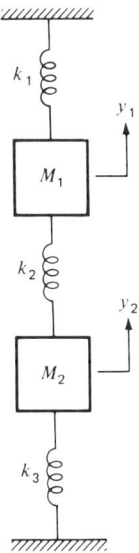

FIG 5-7. A two-degree-of-freedom system.

A rigid body with no constraints at all on its motion (none, at least, for small movements) will have *six* degrees of freedom. As illustrated in Fig. 5-9, the three Cartesian coordinates x, y, and z can be used to define the position of the center of gravity of a rigid body in three-dimensional space (with the derivatives dx/dt, dy/dt, and dz/dt defining the velocity). The three angular coordinates θ_x, θ_y, and θ_z are used to define the body's attitude (i.e., its three-dimensional angular orientation). A satellite in space

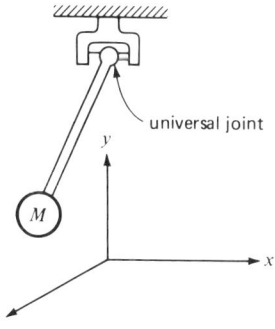

FIG 5-8. A pendulum having two degrees of freedom (three if rotation about rod is allowed).

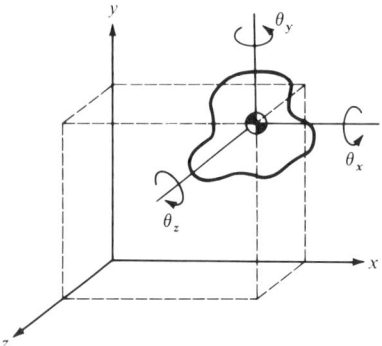

FIG 5-9. Coordinates that completely define the position and/or motion of a rigid body having six degrees of freedom.

is an obvious example of a body with six degrees of freedom, but many objects mounted on springs will also have six degrees of freedom since such systems often have no truly rigid constraints on their motion.

5-3 Units

The differential equations that occur in the theoretical analyses of mechanical vibrating systems contain terms having units of mass, force, displacement, and time. In order to obtain correct answers when using these equations, a compatible set of units must be used. The units of force, displacement, and time are very straightforward and should require no discussion. The units of mass are not quite so obvious, however, and often cause a great deal of difficulty. One source of confusion is that the same word *pound* is often used as a unit of mass as well as a unit of force.

The current emphasis on a worldwide system of units[2] may cause all our engineering work to be carried out with a single consistent set of units based on the metric system before too many years have passed. Until that time, and even after it, however, a good engineer should be capable of working with the other systems of units. Problems in this book have been set up using both the U.S. and the metric systems. In general, different systems of units are not mixed within a single problem, since it is felt that manipulation of conversion factors is of limited educational value for the type of student who will use this book.

---

[2] Referred to as SI units, from the French name "Le Système International d'Unités."

TRANSLATIONAL SYSTEMS

Newton's second law of motion, used to derive the differential equations for all vibrating systems, states that force is proportional to the product of mass and acceleration. With the proper choice of units, however, the law may be written as the *equality*

$$F = M \frac{d^2y}{dt^2} \qquad (5\text{-}6)$$

The consistent set of SI units for Eq. 5-6 is force in newtons (N), mass in kilograms (kg), displacement in meters (m), and time in seconds (s).

To determine the proper combination of U.S. units, the effect of gravity on a given mass should be considered. The equation

$$W = Mg \qquad (5\text{-}7)$$

comes from the application of Eq. 5-6 to a free-falling mass with no force other than gravity acting upon it. Equation 5-7 shows that the weight of the free-falling mass (the *force* exerted on it by gravity) will cause it to have the *acceleration g*. The value of the *gravity* constant g at the surface of the earth has been found experimentally to be 32.2 ft/s², or 386 in./s². Equation 5-7 may be rewritten

$$M = \frac{W}{g} \qquad (5\text{-}8)$$

Since each side of Eq. 5-8 must have the same units, it is evident that the proper units of mass for use in Eq. 5-6 are lb · s²/ft for F in pounds and y in feet, and lb · s²/in. for F in pounds and y in inches.[3] Note that mass in these units may be readily obtained by simply dividing the weight by the appropriate gravity constant g, as shown by Eq. 5-8.

ROTATIONAL SYSTEMS

For rotational vibrating systems, Newton's second law of motion states that torque is proportional to the product of the moment of inertia and angular acceleration, or, with compatible units,

$$T = I \frac{d^2\theta}{dt^2} \qquad (5\text{-}9)$$

Any of the following combinations of units may be used in Eq. 5-9 (with θ in radians in each case):

---

[3]The alternative term "slug" is often used for lb · s²/ft, but its use tends to obscure the required compatibility of units in Eq. 5-6. There is no term (like slug) that can be used in place of lb · s²/in.

1. Torque in N · m, moment of inertia in kg · m², time in seconds
2. Torque in ft · lb, moment of inertia in lb · ft · s² (i.e., slug · ft²), time in seconds
3. Torque in in.·lb, moment of inertia in lb · in.·s², time in seconds

The moment of inertia I, as used in Eq. 5-9, is defined by the fundamental equation

$$I = \int_{\substack{\text{total}\\\text{mass}}} r^2 \, dM \tag{5-10}$$

with r the radius from the axis of rotation to each differential element of mass dM. It is important not to confuse this *mass moment of inertia* with the similar but different *area moment of inertia* (given the symbol I_a in this book but more commonly denoted by simply I) used in beam calculations (see Appendix A). Note that the *area moment of inertia* has different units (e.g., in.⁴) than the mass moment of inertia (lb · in.·s²).

5-4 Conclusion

This chapter is meant to serve as an introduction to the study of mechanical vibrations. The basic principles of vibrations have been presented, and a number of vibration terms have been introduced and defined.

The vibration equations developed in this book are all based on the use of the equalities $F = M \, d^2y/dt^2$ and $T = I \, d^2\theta/dt^2$ for Newton's second law of motion. A set of compatible units must therefore be used if correct answers are to be obtained. The proper units to use in both the U.S. and SI systems of measurement have been presented and discussed. When checking the units of an equation, it should be remembered that *radian* is dimensionless.

A clear understanding of the basic material presented in this chapter is a prerequisite for the study of the vibration topics covered in subsequent chapters.

Problems

5-1. Determine the frequency (in hertz and also in radians per second) and the period of the given harmonic function.

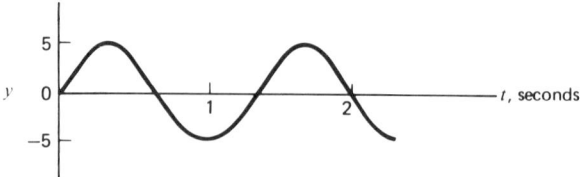

PROBLEMS

5-2. If the given harmonic motion is represented by the equation $y = A \sin(\omega t + \phi)$, what is ϕ?

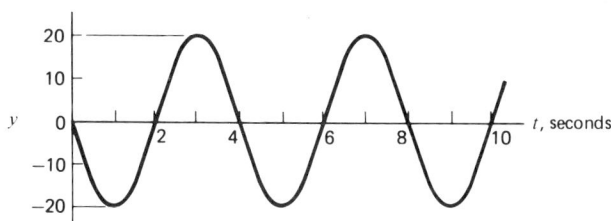

5-3. Represent the given harmonic curve (a) by a sine function and (b) by a cosine function.

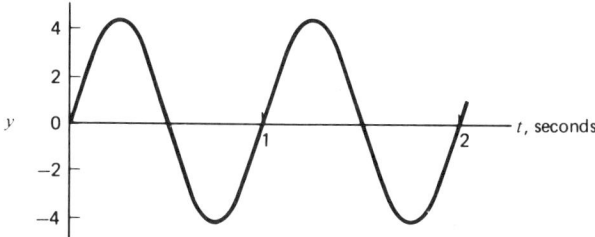

5-4. List several values of ϕ (positive as well as negative) that will allow the given harmonic curve to be represented by the equation $y = 2\cos(0.785t + \phi)$.

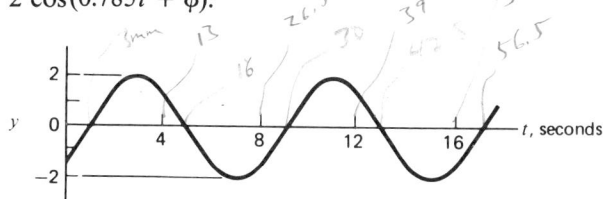

5-5. Determine (a) the maximum velocity and (b) the maximum acceleration of an object having the harmonic motion of Prob. 5-1.
5-6. Repeat Prob. 5-5 for the harmonic motion of Prob. 5-2.
5-7. Repeat Prob. 5-5 for the harmonic motion of Prob. 5-3.
5-8. Repeat Prob. 5-5 for the harmonic motion of Prob. 5-4.

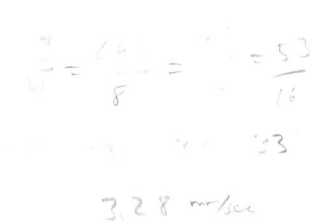

5-9. A block of steel is found to weigh 30 lb on the earth. Determine the linearized equation of motion if it is taken to the moon and suspended from a spring having the given nonlinear force-deflection characteristics. Gravity on the moon is 5.3 ft/s².

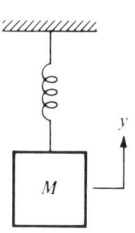
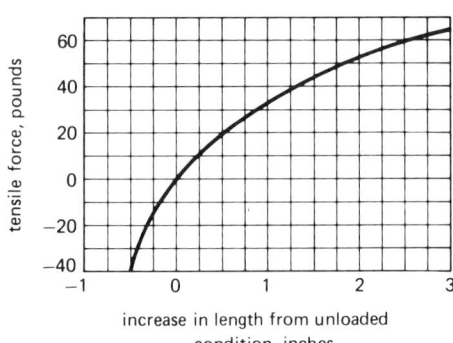

increase in length from unloaded condition, inches

5-10. A block of steel is found to weigh 10 lb on the surface of the moon, where $g = 5.3$ ft/s². If it is suspended from a spring having the nonlinear characteristics given in Prob. 5-9, determine the equation of motion (a) on the moon, (b) on the earth, and (c) in a spacecraft coasting in space.

5-11. The two masses shown are vibrating with harmonic motion at the same frequency. If the velocity of M_2 is downward when the masses are in the positions shown in the figure, y_2 can be said to ———— (lead or lag) y_1 by ———— degrees.

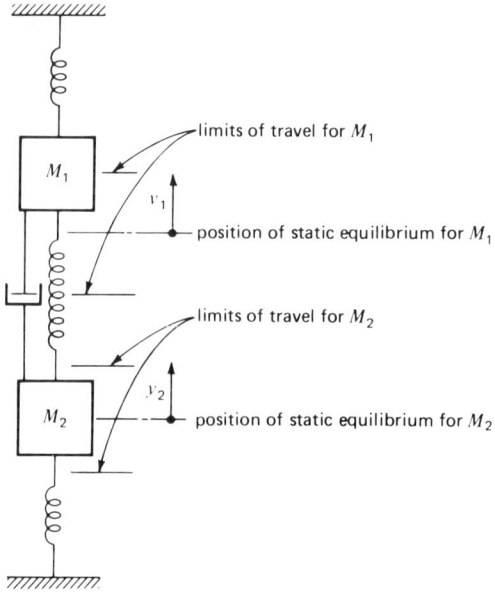

PROBLEMS

5-12. A crank-connecting rod drive causes the input arm to oscillate with a peak-to-peak amplitude of 40°. A viewer looking down on the system saw the two given views occurring at the same instant of time. If the input motion is to be represented by the equation $\theta_i = \Theta_i \sin \omega t$ and the output by $\theta_o = \Theta_o \sin(\omega t + \phi)$, determine the magnitude and sign of ϕ. (There is a fixed kinematic relationship between the rotating crank and the input arm; the latter can therefore be considered as the input to the dynamic system.)

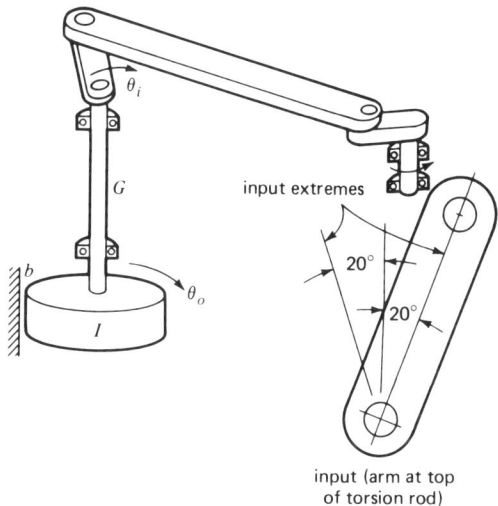

input (arm at top of torsion rod)

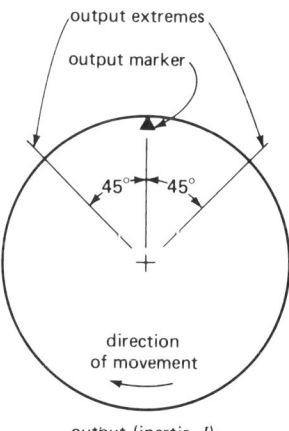

output (inertia, I)

5-13. Repeat Prob. 5-12 for the different set of input and output views given.

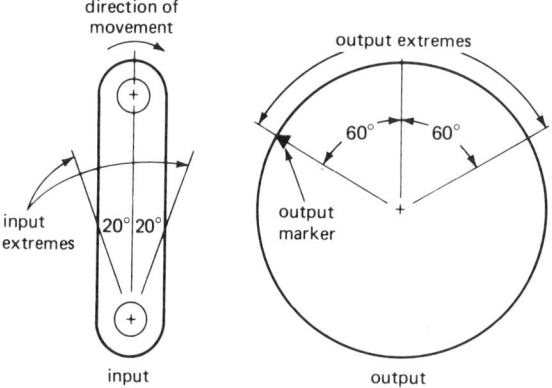

5-14. If the velocity of M_1 is to the left and the velocity of M_2 is to the right when the masses are in the positions shown, y_2 can be said to ———— (lead or lag) y_1 by ———— degrees.

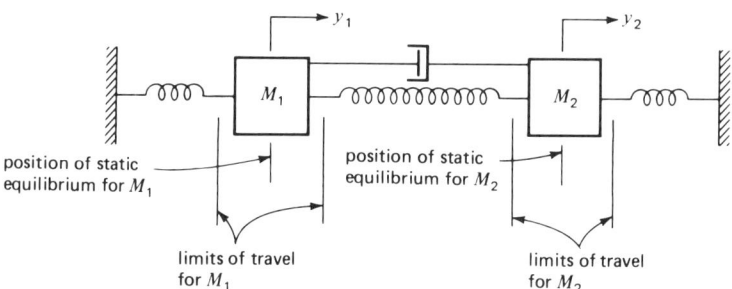

5-15. Repeat Prob. 5-12 for the different set of input and output views given.

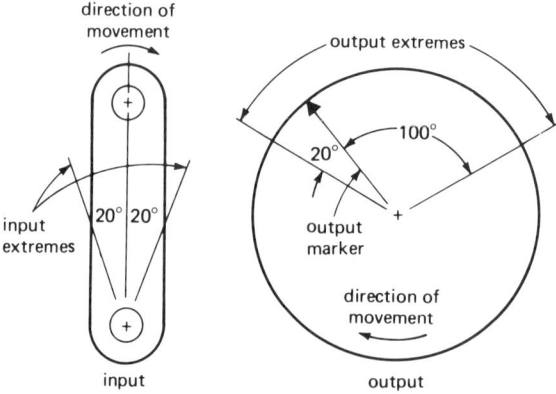

PROBLEMS

5-16. If the velocity of M_2 is to the right when the masses are in the positions shown, y_2 can be said to _____ (lead or lag) y_1 by _____ degrees.

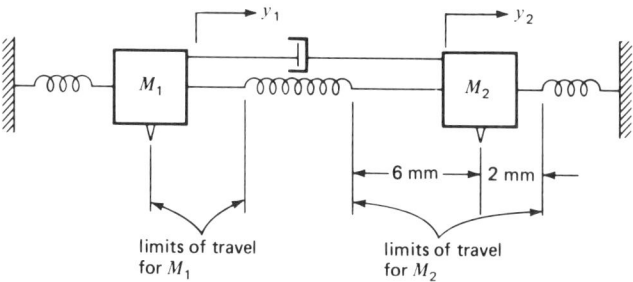

5-17. Determine the degrees of freedom of each of the following systems.

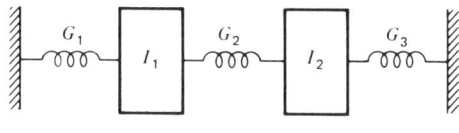

(a)

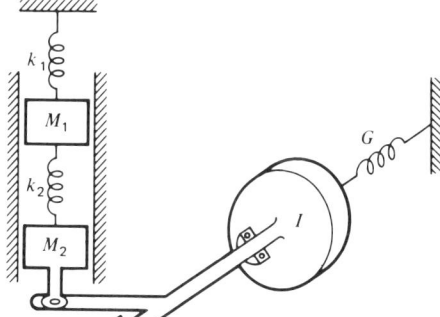

(b)

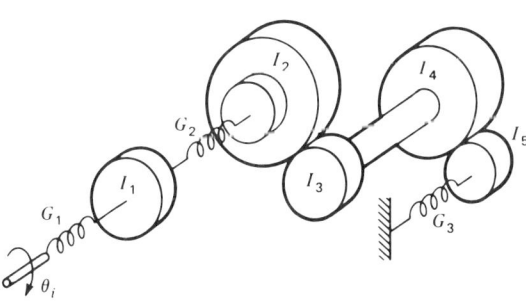

(c)

Free Vibration: Systems with a Single Degree of Freedom

6-1 Introduction

The purpose of this chapter is to illustrate the free-vibration characteristics of simple single-degree-of-freedom mechanical systems. Free vibration can be characterized by two parameters: *undamped natural frequency* and *damping ratio*.

Let us consider a simple single-degree-of-freedom translational system composed of a mass, a spring, and a viscous damper (Fig. 6-1). The equation describing this system is readily derived to be the second-order homogeneous differential equation

$$M \frac{d^2y}{dt^2} + c \frac{dy}{dt} + ky = 0 \qquad (6\text{-}1)$$

Since there is no forcing function, the vibratory action of this system is determined by initial conditions alone, and it is therefore *free vibration*.

6-2 UNDAMPED FREE VIBRATION

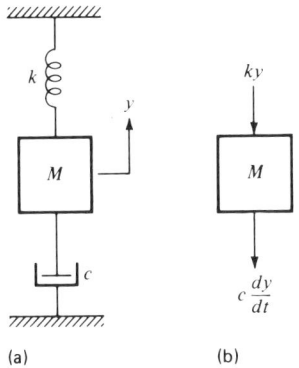

FIG 6-1. (a) A simple translational vibrating system.
(b) Free-body diagram of the mass.

6-2 Undamped Free Vibration

Let us first examine in detail the case in which $c = 0$, that is, a system with zero damping. (A real mechanical vibrating system cannot have zero damping; if it did, it would be a perpetual-motion machine. It may closely approach this condition, however, and much useful work with actual systems can be done with the assumption of no damping.) With this simplification, the equation of motion reduces to

$$M \frac{d^2 y}{dt^2} + ky = 0 \qquad (6\text{-}2)$$

Following the method of solution presented in Chapter 3, the characteristic equation is found to be

$$p^2 + \frac{k}{M} = 0$$

giving the two roots

$$p_1 = +j\sqrt{\frac{k}{M}} \quad \text{and} \quad p_2 = -j\sqrt{\frac{k}{M}}$$

The solution to Eq. 6-2 is therefore (see Section 3-2)

$$y = C_1 \sin \sqrt{\frac{k}{M}}\, t + C_2 \cos \sqrt{\frac{k}{M}}\, t \qquad (6\text{-}3)$$

which may also take the alternative form (see Appendix B)

$$y = C_3 \sin\left(\sqrt{\frac{k}{M}}\, t + \phi\right) \qquad (6\text{-}4)$$

The undamped system will therefore vibrate with sinusoidal motion of constant amplitude. The natural frequency at which this free vibration occurs is given the symbol ω_n. From Eqs. 6-3 and 6-4 it can be seen that

$$\omega_n = \sqrt{\frac{k}{M}} \qquad (6\text{-}5)$$

A more specific term for ω_n is *undamped natural frequency*. Although the addition of damping causes free vibration to occur at a slightly lower frequency, the undamped natural frequency as given by Eq. 6-5 is still an important parameter for characterizing the system.

The amplitude of vibration—that is, the values of C_1 and C_2 in Eq. 6-3 (or C_3 in Eq. 6-4)—depends upon the initial conditions. If y and dy/dt are both zero at $t = 0$, then $C_1 = C_2 = 0$, and no motion at all occurs. If $y = y_0$ and $dy/dt = 0$ at $t = 0$, then the free vibration is described by the equation

$$y = y_0 \cos \omega_n t$$

If, on the other hand, we have the initial conditions $y = 0$ and $dy/dt = v_0$ at $t = 0$, the motion is

$$y = \frac{v_0}{\omega_n} \sin \omega_n t$$

which is a sine function instead of the cosine function produced by the other set of initial conditions. If there are both position and velocity nonzero initial conditions, the resultant motion is the superposition of sine and cosine functions; that is, harmonic motion occurs that can be described as a sine wave displaced on the time axis by the angle ϕ (as represented by Eq. 6-4).

Actually, when considering free vibration of a *real* mechanical system, we seldom care whether the amplitude is a maximum, zero, or something in between at some instant of time we define as $t = 0$. (This distinction may be very important for other types of dynamic systems, however.) The more significant factors are the natural frequency, unaffected by the initial conditions, and the amplitude of the motion. Note also that we are normally free to define the instant at which $t = 0$ in any manner we please. If we give the mass the displacement y_0, and release it at $t = 0$, the displacement, as a function of time, will be $y_0 \cos \omega_n t$. If we do the same thing, but define $t = 0$ to be the instant occurring $1/4$ cycle later (i.e., when $y = 0$), the initial conditions will be $y = 0$ and $dy/dt = -\omega_n y_0$, and the equation $y = -y_0 \sin \omega_n t$ must be used to describe the motion. The

6-2 UNDAMPED FREE VIBRATION

situation is identical in both cases, but the form of the equation is determined by our arbitrary choice of zero time.

Although the basic definition of free vibration restricts the term to situations in which there is no input (or forcing function), the case of the *step input* should be considered at this point since it produces system oscillation that is indistinguishable from free vibration. It will be shown, in fact, that by defining a few things differently, the response of a system to a step input can be made to match the definition of free vibration.

The system of Fig. 6-2a is caused to oscillate by giving the top end of the spring an abrupt vertical displacement, called a *step input*, which can be described mathematically

$$x = 0 \quad t < 0$$
$$x = X \quad t \geq 0$$

The differential equation describing the response of the system to the step input is

$$M \frac{d^2y}{dt^2} + ky = kX$$

Since this is a nonhomogeneous equation, the solution consists of two parts, the complementary solution and the particular integral, the total solution being

$$y = C_1 \sin \omega_n t + C_2 \cos \omega_n t + X$$

If the mass is motionless at $t = 0$, the solution becomes

$$y = X - X \cos \omega_n t$$
$$= X(1 - \cos \omega_n t)$$

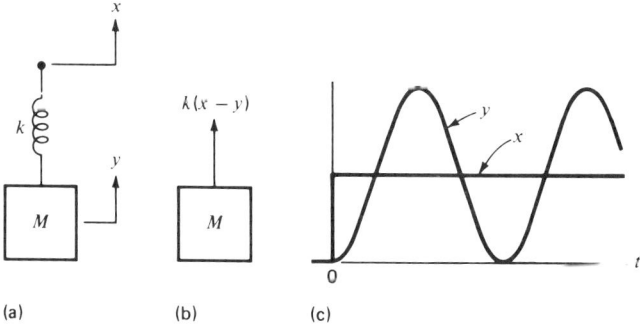

FIG 6-2. (a) A mass-spring system excited by a position step input.
(b) Free-body diagram of the mass.
(c) The step input x and response y.

The action of a system to a step input of this type is most precisely called *system response to a step input*. To make the response match the definition of free vibration, however, we could arbitrarily define y as the displacement away from the position of static equilibrium based on the condition of the system for $t > 0$. By further defining $x = 0$ as the position of the top of the spring for $t \geqslant 0$ (i.e., saying that $x = -X$ for $t < 0$), we could say that the system is excited by the initial conditions $y = -X$ and $dy/dt = 0$ at $t = 0$ rather than by a step input. (Response to a step input is considered in more detail in Chapter 14.)

Example 6-1. A 100-lb weight is attached to a simply supported steel beam as illustrated in Fig. 6-3. The beam has a uniform cross section with area moment of inertia $I_a = 0.02$ in.^{4}. The other beam dimensions are given in the figure.
(a) Determine the natural frequency of lateral vibration in hertz.
(b) If the system is put into a state of free vibration by giving the mass a lateral displacement of 1/4 inch and then releasing it abruptly, what is the maximum acceleration of the weight?
Solution: The mass of the beam is assumed to be insignificant in comparison with that of the 100-lb weight. Using the applicable beam equation in Appendix A, we calculate the spring constant of the beam (i.e., the spring constant corresponding to a force applied at the weight) to be

$$k = \frac{F_s}{y} = \frac{3EI_a l}{a^2 b^2}$$

$$= \frac{3(30 \times 10^6)(0.02)(30)}{(13^2)(17^2)}$$

$$= 1105 \text{ lb/in.}$$

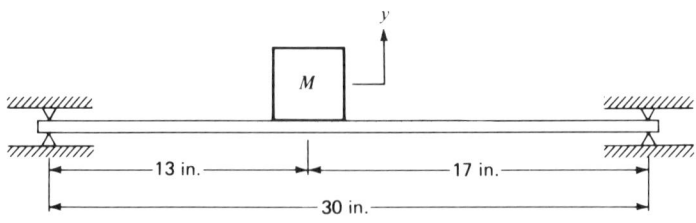

FIG 6-3. A translational vibrating system consisting of a mass supported by a beam spring.

6-2 UNDAMPED FREE VIBRATION

From Eq. 6-5,

$$\omega_n = \sqrt{\frac{k}{M}} = \sqrt{\frac{1105(386)}{100}} = \sqrt{4270}$$
$$= 65.3 \text{ rad/s}$$

Using Eq. 5-5,

$$f_n = \frac{\omega_n}{2\pi} = \frac{65.3}{2\pi}$$
$$= 10.4 \text{ Hz} \qquad (Ans.)$$

From Eq. 5-4,

$$\left|\frac{d^2y}{dt^2}\right|_{max} = \omega_n^2 y_0 = (65.3)^2(0.25)$$
$$= 1067 \text{ in./s}^2 \qquad (Ans.)$$

ROTATIONAL SYSTEMS

If the single-degree-of-freedom rotational vibrating system without damping illustrated in Fig. 6-4 is analyzed, the free-vibration characteristics are found to be identical to those of the analogous translational system. Free vibration is described by the equation

$$\theta = C_1 \sin \sqrt{\frac{G}{I}}\, t + C_2 \cos \sqrt{\frac{G}{I}}\, t \qquad (6\text{-}6)$$

or by the alternative equivalent equation

$$\theta = C_3 \sin\left(\sqrt{\frac{G}{I}}\, t + \phi\right) \qquad (6\text{-}7)$$

Since the free vibration will occur at the undamped natural frequency, we can write

$$\omega_n = \sqrt{\frac{G}{I}} \qquad (6\text{-}8)$$

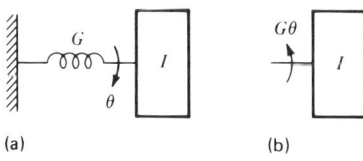

(a) (b)

FIG 6-4. (a) An undamped single-degree-of-freedom rotational vibrating system. (b) Free-body diagram of the inertia.

The similarity between Eq. 6-8 and Eq. 6-5 should be noted; the natural frequency is in both cases equal to the square root of the system's spring rate divided by its inertia parameter. All the previous discussion on initial conditions and response to a step input can be directly applied to this rotational system.

When analyzing rotational vibrating systems, it is important to distinguish between the frequency of vibration ω and the angular velocity $d\theta/dt$ of the vibrating component, both of which have the same units, radians per second. It is fairly easy to overlook the distinction between these two parameters, but Example 6-2 should help clarify this point. There is no likelihood of having this type of confusion with translational systems, since the velocity term, dy/dt, and the frequency, ω, are expressed in different units.

Example 6-2. A flywheel having a mass moment of inertia $I = 4.0$ kg · m² is attached to an axle 3 meters long. The other end of the axle is rigidly secured against rotation. (This system is represented by the schematic diagram of Fig. 6-4). A static torque of 100 N · m, applied at the flywheel, is found to result in a flywheel rotational displacement of 12°.

(a) Calculate the natural frequency of rotational vibration for this system.

(b) If the flywheel is twisted 12° and then abruptly released, what is its maximum angular velocity as it oscillates back and forth?

Solution: The torsional spring constant G is calculated by means of the basic spring definition

$$G = \frac{T}{\theta} = \frac{100}{12/57.3}$$
$$= 478 \text{ N} \cdot \text{m/rad}$$

Assuming the mass moment of inertia of the axle to be negligible, the natural frequency is found from Eq. 6-8,

$$\omega_n = \sqrt{\frac{G}{I}} = \sqrt{\frac{478}{4.0}} = 10.9 \text{ rad/s} \quad (Ans.)$$

Since the flywheel will vibrate with harmonic motion,

$$\theta = \theta_0 \cos \omega_n t$$

with $\theta_0 = 12/57.3$ rad. We find $d\theta/dt$ by differentiation,

$$\frac{d\theta}{dt} = -\omega_n \theta_0 \sin \omega_n t$$

6-2 UNDAMPED FREE VIBRATION

Therefore,

$$\left|\frac{d\theta}{dt}\right|_{max} = \omega_n \theta_0 = 10.9\left(\frac{12}{57.3}\right)$$

$$= 2.28 \text{ rad/s} \qquad (Ans.)$$

The distinction between the angular velocity $d\theta/dt$ and the frequency of vibration ω_n should be carefully noted.

Example 6-3. A pendulum consists of a light rod 20 inches long with a 10-lb weight attached to the end (Fig. 6-5).

(a) Determine the natural frequency (in hertz) and the period of oscillation, assuming a massless rod.

(b) If we take into account the inertia of the rod, which weighs 1.0 lb, has $I = 0.30$ lb · in.·s² about the pivot point, and has its center of gravity at the midpoint, what will the natural frequency be?

Solution: The equation of motion for the simple pendulum was developed in Section 2-2. Since the pendulum can be considered a simple rotational vibrating system, however, its natural frequency may be determined by direct application of Eq. 6-8. If the inertia of the rod is neglected and the weight is assumed to be concentrated at a point, the moment of inertia about the pivot point is (from Eq. 2-22),

$$I = Ml^2$$

The torsional spring constant (Eq. 2-23).

$$G = \frac{Mgl \sin\theta}{\theta}$$

can be linearized (by noting that $\sin\theta \approx \theta$) to

$$G \approx Mgl$$

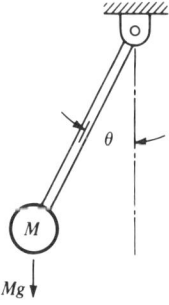

FIG 6-5. A pendulum.

Substituting these expressions into Eq. 6-8 produces

$$\omega_n = \sqrt{\frac{Mgl}{Ml^2}} = \sqrt{\frac{g}{l}}$$

Since the pendulum rod is 20 inches long,

$$\omega_n = \sqrt{\frac{386}{20}} = 4.4 \text{ rad/s}$$

From Eq. 5-5

$$f_n = \frac{\omega_n}{2\pi} = 0.70 \text{ Hz} \qquad (Ans.)$$

From Eq. 5-1,

$$P = \frac{1}{f_n} = \frac{1}{0.70} = 1.43 \text{ seconds} \qquad (Ans.)$$

If the inertia of the rod is included in the analysis, the moment of inertia is

$$I = Ml^2 + 0.30$$
$$= \frac{10}{386}(20)^2 + 0.30$$
$$= 10.66 \text{ lb} \cdot \text{in.} \cdot \text{s}^2$$

The effective torsional spring constant will also consist of two components; that is,

$$G = M_1 g l_1 + M_2 g l_2$$
$$= (10)(20) + (1)(10)$$
$$= 210 \text{ in.} \cdot \text{lb/rad}$$

The natural frequency is

$$f_n = \frac{1}{2\pi}\sqrt{\frac{G}{I}} = \frac{1}{2\pi}\sqrt{\frac{210}{10.66}}$$
$$= 0.706 \text{ Hz} \qquad (Ans.)$$

6-3 Damped Free Vibration

If the viscous damping coefficient c is not zero for the system of Fig. 6-1, the characteristics of free vibration may be quite different from those of the undamped system. The exact characteristics of damped free vibration depend upon the system parameters, in particular upon the effectiveness of the damping.

6-3 DAMPED FREE VIBRATION

The differential equation describing the system of Fig. 6-1 with damping is found to be

$$M\frac{d^2y}{dt^2} + c\frac{dy}{dt} + ky = 0 \tag{6-9}$$

For discussion purposes, a more suitable form of Eq. 6-12 is obtained by dividing through by k, so that the coefficient of the displacement term is 1; that is,

$$\frac{M}{k}\frac{d^2y}{dt^2} + \frac{c}{k}\frac{dy}{dt} + y = 0 \tag{6-10}$$

The characteristic equation is therefore

$$\frac{M}{k}p^2 + \frac{c}{k}p + 1 = 0 \tag{6-11}$$

The value of the damping coefficient c cannot be used to predict system damping characteristics without also knowing M and k. Another parameter, the *damping ratio* ζ, is used as a measure of the *effectiveness* of the damping present, and therefore is a unifying factor for all single-degree-of-freedom linear vibrating systems (and for a great number of other types of second-order dynamic systems).

For the vibrating system under discussion, the damping ratio is the ratio of the damping actually present to that which would produce critical damping. In equation form,

$$\zeta \equiv \frac{c}{c_c} \tag{6-12}$$

where c_c is the value of damping required to produce critical damping. But what is critical damping? Critical damping is defined mathematically as the value of system damping required to produce real and equal roots from the system characteristic equation. (The physical manifestation of critical damping will be explored later.) The roots of the characteristic equation (Eq. 6-11) for the system being considered are

$$p = -\frac{c}{2M} \pm \frac{(c^2 - 4Mk)^{1/2}}{2M} \tag{6-13}$$

For real and equal roots,

$$c^2 = 4Mk$$

Therefore,

$$c_c = 2(Mk)^{1/2} \quad \text{critical damping coefficient} \tag{6-14}$$

Substituting Eq. 6-14 into Eq. 6-12 yields the damping ratio for the vibrating system. It is

$$\zeta = \frac{c}{2(Mk)^{1/2}} \tag{6-15}$$

The differential equation for the mechanical vibrating system (Eq. 6-10 repeated for convenience) is

$$\frac{M}{k}\frac{d^2y}{dt^2} + \frac{c}{k}\frac{dy}{dt} + y = 0$$

This equation can now be put into more general form. From Eq. 6-5,

$$\frac{M}{k} = \frac{1}{\omega_n^2}$$

The coefficient of the velocity term can be manipulated in the following manner:

$$\frac{c}{k} = \frac{2c\sqrt{M}}{2\sqrt{k}\sqrt{k}\sqrt{M}} = \frac{2}{\omega_n} \cdot \frac{c}{2\sqrt{Mk}}$$

Combining with Eq. 6-15 yields the desired coefficient form,

$$\frac{c}{k} = \frac{2\zeta}{\omega_n} \tag{6-16}$$

Equation 6-10 can now be written as

$$\frac{1}{\omega_n^2}\frac{d^2y}{dt^2} + \frac{2\zeta}{\omega_n}\frac{dy}{dt} + y = 0 \tag{6-17}$$

At this time it should be pointed out that Eq. 6-17 is a general form that is very useful for the study not only of mass-spring-damper systems, but of all types of second-order dynamic systems. If a dynamic system differential equation is put into the form of Eq. 6-17 (by rewriting so that the coefficient of the zeroth derivative is made unity), then it becomes equivalent to Eq. 6-17, and the undamped natural frequency ω_n and the damping ratio ζ can be obtained directly from the values of the coefficients of the first and second derivatives.

Another significant aspect of the general second-order equation (Eq. 6-17) is the signs of the three terms. For real mechanical vibrating systems composed of inertia elements, springs, and dampers, all three terms will have positive signs. This fact can often serve as a useful check on the validity of a derivation.

To determine the solution of Eq. 6-17, the characteristic equation is obtained as

$$p^2 + 2\zeta\omega_n p + \omega_n^2 = 0$$

6-3 DAMPED FREE VIBRATION

from which

$$p = -\zeta\omega_n \pm \omega_n(\zeta^2 - 1)^{1/2}$$

The roots of the characteristic equation (and therefore the free vibration characteristics of the system) depend on the value of the damping ratio ζ.

For $\zeta < 1$: Roots are complex, and free vibration has the form (see Section 3-2)

$$y = e^{-\zeta\omega_n t}\left(C_1 \sin \omega_n\sqrt{1-\zeta^2}\, t + C_2 \cos \omega_n\sqrt{1-\zeta^2}\, t\right) \quad (6\text{-}18)$$

For $\zeta = 1$: Roots are real and equal, and we find that

$$y = C_1 e^{-\zeta\omega_n t} + C_2 t e^{-\zeta\omega_n t} \quad (6\text{-}19)$$

For $\zeta > 1$: Roots are real and unequal, and

$$y = C_1 \exp\left(-\zeta\omega_n + \omega_n\sqrt{\zeta^2-1}\right)t + C_2 \exp\left(-\zeta\omega_n - \omega_n\sqrt{\zeta^2-1}\right)t \quad (6\text{-}20)$$

The constants C_1 and C_2 depend, in each case, upon initial conditions.

The free-vibration characteristics of the damped system, produced by the initial conditions $y = y_0$ and $dy/dt = 0$ at $t = 0$, are shown in dimensionless form in Fig. 6-6. With $\zeta < 1$, true oscillation occurs. With $\zeta = 1$ (a critically damped system), the system reaches static equilibrium in the least amount of time and with no overshoot.[1] With $\zeta > 1$, a longer period of time is required for the system to return to static equilibrium. Note that if $\zeta \geq 1.0$, "free vibration" is not a good descriptive term for the phenomenon.

Although the undamped natural frequency ω_n appears in the solution in all three cases, in none of them is there oscillation at that frequency (except at the limiting case where $\zeta = 0$). For damped systems of this type in which true vibration does occur (i.e., systems with $\zeta < 1$), it takes place at the *damped natural frequency* ω_d, with

$$\omega_d = \omega_n(1 - \zeta^2)^{1/2} \quad (6\text{-}21)$$

This result can be seen by noting the form of Eq. 6-18. The damped natural frequency ω_d is always less then ω_n. If ζ is small, however, the difference between the two is insignificant. The term is meaningless, of course, if ζ is one or greater.

The curves of Fig. 6-6 can be applied to a single-degree-of-freedom rotational system, such as that of Fig. 6-7, by merely changing the ordinate from y/y_0 to θ/θ_0. Solutions for this rotational system have exactly the

---

[1]If it is desired to have a system reach and remain within a few percent of the steady-state value in the shortest possible time, the damping ratio should be set at about 0.7.

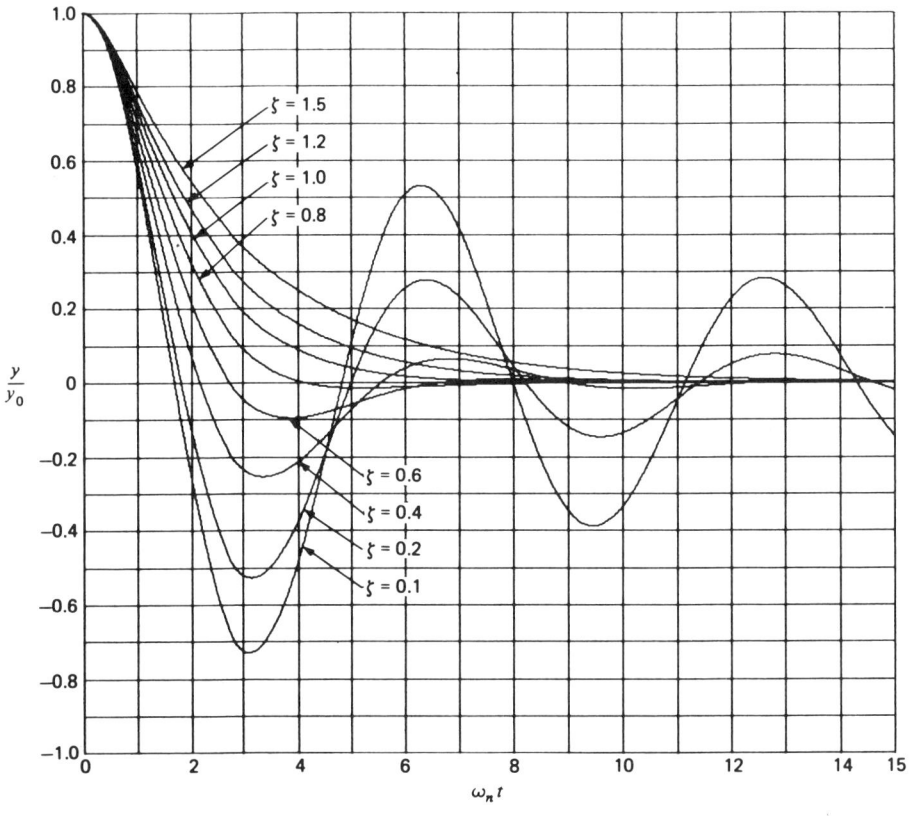

FIG 6-6. Free-vibration characteristics of a single-degree-of-freedom mass-spring-damper system and of other second-order dynamic systems.

same form as those of the analogous translational system, with the damping ratio defined as follows:

$$\zeta = \frac{b}{b_c}$$
$$= \frac{b}{2(IG)^{1/2}} \quad (6\text{-}22)$$

and the undamped natural frequency given by Eq. 6-8, $\omega_n = \sqrt{G/I}$.[2]

The curves of Fig. 6-6 are also applicable to all other dynamic systems represented by homogeneous second-order equations of the general form of Eq. 6-17.

---

[2] As an exercise, the reader should write the equation for Fig. 6-7 and then put it into the form of Eq. 6-17 to develop the expressions for ω_n and ζ.

6-3 DAMPED FREE VIBRATION

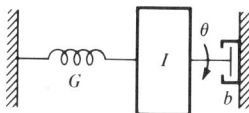

FIG 6-7. A single-degree-of-freedom rotational vibrating system.

Example 6-4. In the electric circuit of Fig. 6-8, $e = 12$ volts, $R = 500$ ohms, $C = 10$ μF, and $L = 0.10$ henry. Determine the manner in which the voltage e_1 varies with time after the switch is closed.

Solution: The equation describing the circuit is obtained from a node analysis by applying Kirchhoff's second law,

$$\Sigma i = 0$$

to the single node. The result is

$$-C\frac{de_1}{dt} - \frac{1}{R}e_i - \frac{1}{L}\int_0^t e_1\, dt = -\frac{1}{R}e$$

The integral term is removed by performing a single differentiation to obtain

$$C\frac{d^2 e_1}{dt^2} + \frac{1}{R}\frac{de_1}{dt} + \frac{1}{L}e_1 = \frac{1}{R}\frac{de}{dt}$$

The equation is now put into the general form of Eq. 6-17 (noting that $de/dt = 0$),

$$LC\frac{d^2 e_1}{dt^2} + \frac{L}{R}\frac{de_1}{dt} + e_1 = 0 \tag{6-23}$$

We now compare the coefficients of Eq. 6-17 and 6-23. The undamped natural frequency is obtained by setting

$$\frac{1}{\omega_n^2} = LC$$

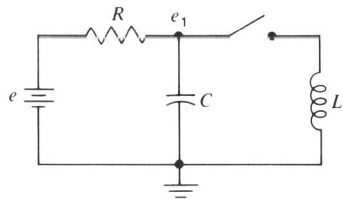

FIG 6-8. The electric circuit of Example 6-4.

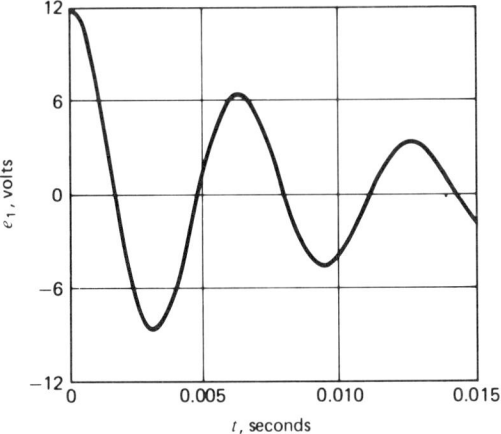

FIG 6-9. The free-vibration characteristics of the electric circuit of Fig. 6-8.

from which

$$\omega_n = \sqrt{\frac{1}{LC}} = \sqrt{\frac{1}{(0.10)(10 \times 10^{-6})}} = 1{,}000 \text{ rad/s}$$

Similarly,

$$\frac{2\zeta}{\omega_n} = \frac{L}{R}$$

$$\zeta = \frac{L\omega_n}{2R} = \frac{(0.10)(1000)}{(2)(500)} = 0.1$$

We can now apply the curve in Fig. 6-6 corresponding to $\zeta = 0.1$. The voltage e_1, starting at 12 volts at $t = 0$, is found to oscillate in a decaying manner as shown in Fig. 6-9, a curve taken directly from Fig. 6-6, but plotted on coordinates applying to this problem.

6-4 Determination of Damping Ratio from Experimental Data

If the damping ratio is considerably less than 1, and thus the system oscillates several times or more when set into free vibration, the *logarithmic decrement method* may be used to determine the damping ratio ζ from experimental data. Considering the elementary mass-spring-damper system of Fig. 6-1, or any equivalent second-order system, the solution is given by Eq. 6-18, which is more conveniently used here in its alternative form

$$y = C_3 e^{-\zeta\omega_n t} \sin(\omega_d t + \phi) \tag{6-24}$$

6-4 DETERMINATION OF DAMPING RATIO FROM EXPERIMENTAL DATA

with C_3 and ϕ determined by initial conditions. This is the equation of a curve that is a harmonic function multiplied by a decaying term, as illustrated in Fig. 6-10. Note that the curve is tangent to the envelope of the exponential term at points where $\sin(\omega_d t + \phi)$ is equal to 1.

An expression will now be developed to relate the amplitude of any two successive peaks. If we apply Eq. 6-24 to any two such peaks (points of tangency to the envelope, to be more precise) where $\sin(\omega_d t + \phi) = 1$ and take the ratio of their amplitudes, we obtain the expression

$$\frac{y_n}{y_{n+1}} = \frac{e^{-\zeta\omega_n t_n}}{e^{-\zeta\omega_n(t_n + P)}} = e^{\zeta\omega_n P} \qquad (6\text{-}25)$$

Equation 6-25 shows that the ratio of the amplitudes is independent of which two adjacent peaks are chosen. Next, the *logarithmic decrement* δ is defined as the *natural logarithm* of that ratio,

$$\delta = \ln \frac{y_n}{y_{n+1}} \qquad (6\text{-}26)$$

A comparison of Eqs. 6-25 and 6-26 shows that

$$\delta = \zeta \omega_n P \qquad (6\text{-}27)$$

The period of the damped vibration is

$$P = \frac{2\pi}{\omega_d} = \frac{2\pi}{\omega_n \sqrt{1 - \zeta^2}} \qquad (6\text{-}28)$$

The equation for the logarithmic decrement can therefore be written

$$\delta = \frac{2\pi\zeta}{\sqrt{1 - \zeta^2}} \qquad (6\text{-}29)$$

If damping is small, the radical in this equation is approximately equal to

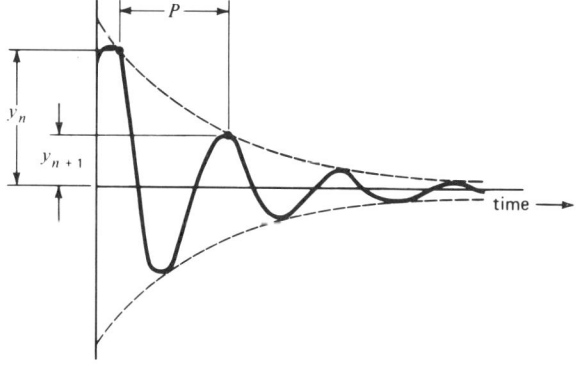

FIG 6-10. Damped free vibration, showing parameters used in logarithmic decrement method for determining ζ.

1.0, and we have

$$\zeta \approx \frac{\delta}{2\pi} \qquad (6\text{-}30)$$

If ζ is fairly large, then a more precise value can be found from the exact equation, which can be written in the form

$$\zeta = \frac{\delta}{\sqrt{4\pi^2 + \delta^2}} \qquad (6\text{-}31)$$

MODIFIED LOG DECREMENT METHOD FOR LOW DAMPING

If ζ is very small, it may be difficult to perceive any appreciable difference in the amplitudes of adjacent peaks of the curve of Fig. 6-10. With this situation, a *modified log decrement* δ' should be used, based on the following definition

$$\delta' = \ln\left(\frac{y_n}{y_{n+m}}\right) \qquad (6\text{-}32)$$

with m being the number of successive cycles over which the change in amplitude is to be measured. Following the same form of derivation, the equation for ζ is found to be

$$\zeta \approx \frac{\delta'}{2\pi m} \qquad (6\text{-}33)$$

Although this equation is not mathematically precise (being based on the relationship $\sqrt{1-\zeta^2} \approx 1.0$, in practice it should always be more than adequate, since this modified log decrement method would only be used for systems in which ζ is very small.

6-5 Conclusion

In this chapter the free-vibration characteristics of several mechanical systems have been investigated. Free vibration takes place at the damped natural frequency ω_d, but the characteristics of the motion can be obtained by knowing only the undamped natural frequency ω_n and the damping ratio ζ. The family of dimensionless curves in Fig. 6-6 shows how the vibratory amplitude decreases with time for different values of ζ and is very useful for determining the free vibration motion of any linear single-degree-of-freedom mass-spring-damper system (or any dynamic system described by an equation of the general form of Eq. 6-17). For systems in which $\zeta \geq 1.0$, vibration in the normal sense of the word will not occur because the system inertia does not overshoot its position of static equilibrium.

The amplitude of free vibration is determined by initial conditions. Although free vibration is defined as that occurring when there is no external forcing function acting on the system, the response of a mass-spring-damper system to a step input has been demonstrated to be, for all practical purposes, the same as free vibration.

Problems

6-1. Determine ω_n, ω_d, and ζ for each of the given systems.

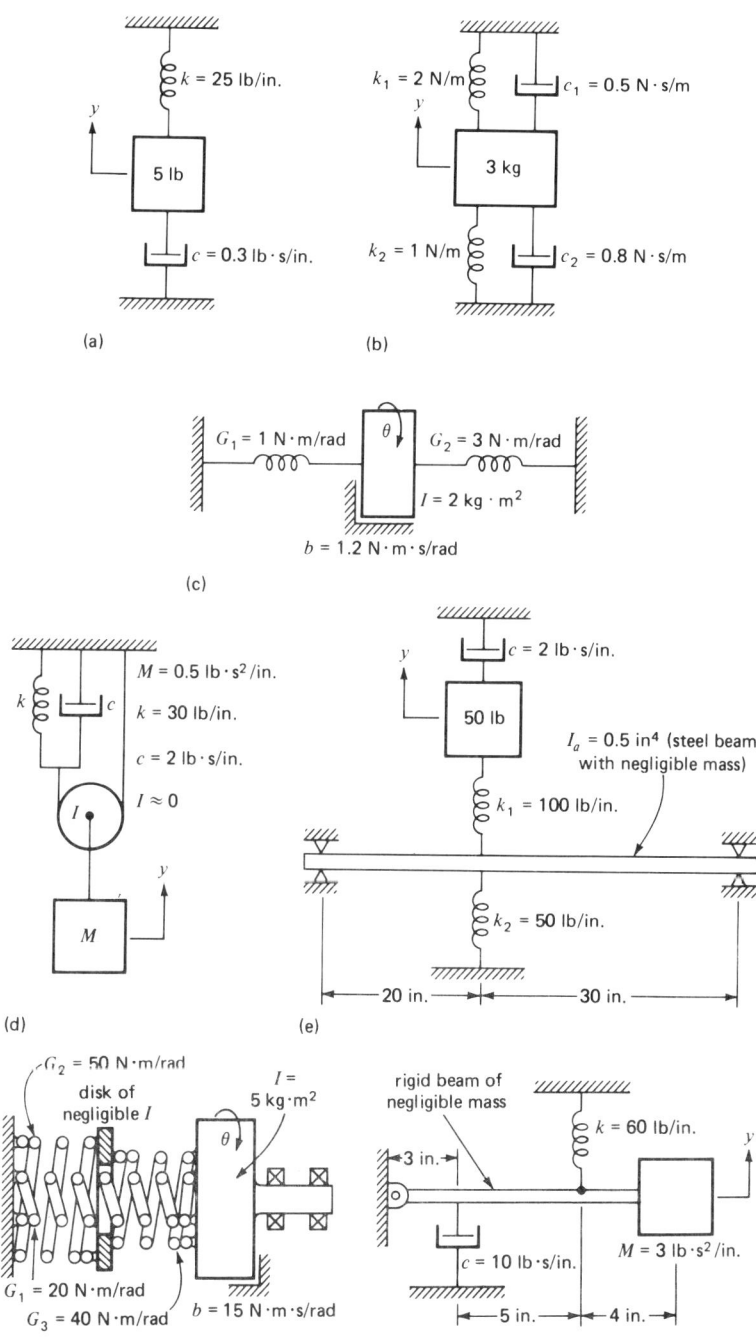

6-2. Using Fig. 6-6 as a guide, sketch the approximate free vibration characteristics of the system for each of the following sets of initial conditions:

$$(a)\ y_0 = 0.05\ \text{m},\ \left.\frac{dy}{dt}\right|_0 = 0$$

$$(b)\ y_0 = 0,\ \left.\frac{dy}{dt}\right|_0 = -0.50\ \text{m/s}$$

$$(c)\ y_0 = 0.05\ \text{m},\ \left.\frac{dy}{dt}\right|_0 = 0.50\ \text{m/s}$$

$$(d)\ y_0 = 0.05\ \text{m},\ \left.\frac{dy}{dt}\right|_0 = -0.50\ \text{m/s}$$

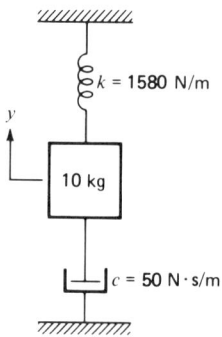

6-3. Determine, for small amplitude oscillation of each of the pendulum systems given, expressions for ω_n, ω_d, and ζ. Assume in each case that the mass M is the only significant inertia.

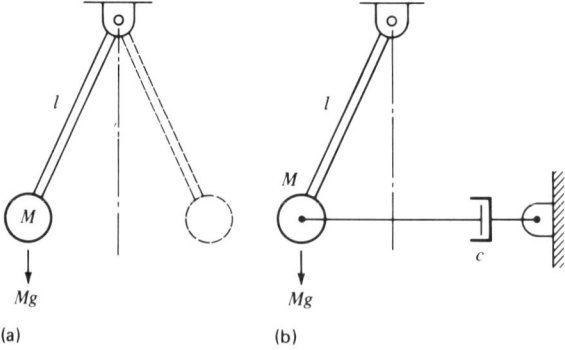

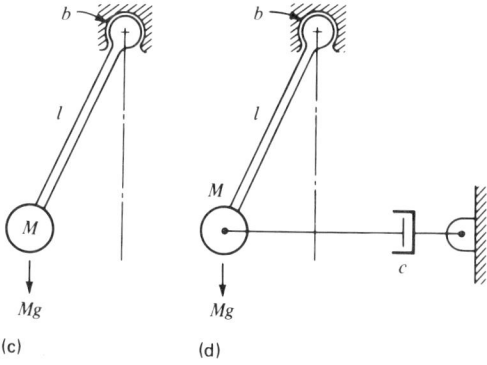

(c) (d)

6-4. The rotational system consists of a cylinder with a mass moment of inertia of 0.02 lb · in.·s² fixed to two plastic rods as shown. The rod material has a compliance of 1.5°/in.·lb per inch of length. The viscous damper is known to produce 2 in.·lb of torque when rotated at 3 rad/s. Determine ω_n, ω_d, and ζ.

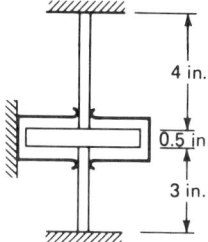

6-5. Using a log decrement method, determine the damping ratio of each of the systems whose free-vibration characteristics are given by the strip chart recordings shown.

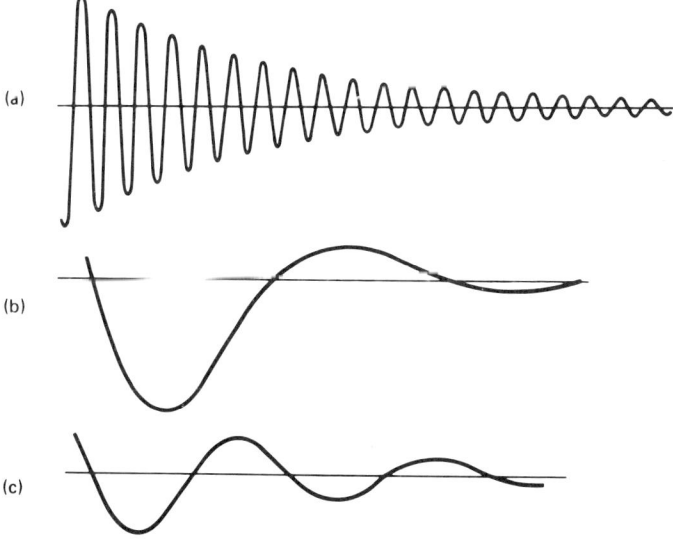

6-6. The damping of a tuning fork is nonlinear, produced by metal hysteresis and air friction. Its free-vibration characteristics are given by the strip chart recording in the figure. If it is to be modeled by a linear differential equation, determine the proper damping ratio to use when (a) the peak-to-peak amplitude is about 0.10 mm, (b) the peak-to-peak amplitude is about 0.05 mm.

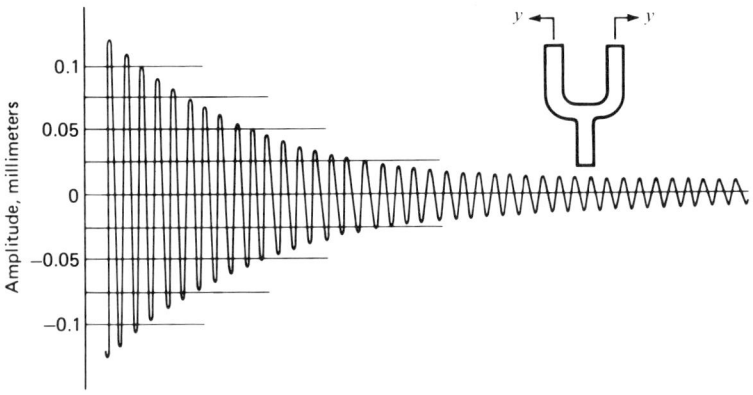

6-7. What value of the damping coefficient c would just cause the system, in free vibration, to change from an oscillatory response to a response that is exponential?

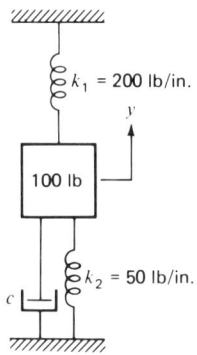

6-8. If the mass is given an initial displacement of -20 mm, how long will it take before it comes and remains within 1.0 mm of the position of static equilibrium? (Use Fig. 6-6.)

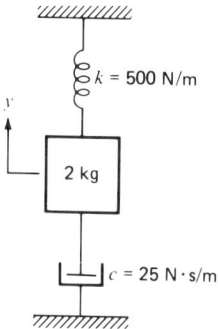

6-9. Determine, for the system of Prob. 6-8, the minimum value to which c could be reduced without an overshoot exceeding 2.0 mm.

6-10. If the mass is given an initial displacement of 2.0 inches and quickly released, what period of time is required before it remains within an error band (with respect to the position of static equilibrium) of ± 0.2 inch? (Use Fig. 6-6.)

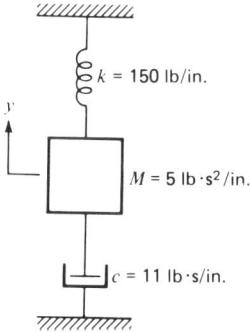

6-11. If the flywheel is given an initial twist of $40°$ and quickly released, determine the period of time required before it remains within an error band (with respect to the position of static equilibrium) of $\pm 5.0°$. (Use Fig. 6-6.)

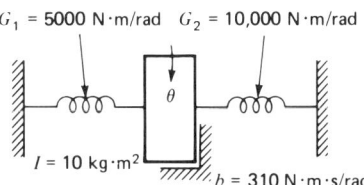

6-12. Two-inch-thick plastic foam, tested and found to have the nonlinear force-deflection characteristics given in the figure, is used to package fragile devices as shown. Determine for each case the natural frequency (undamped). Note that the foam is given an initial compression.

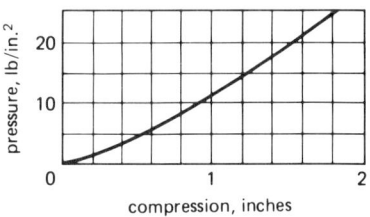

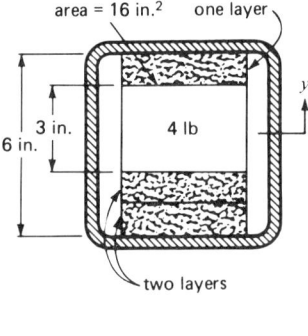

(a)

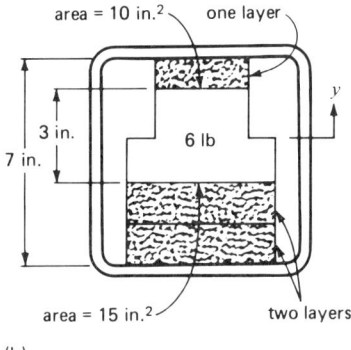

(b)

6-13. Three springs, each having a free length of 0.25 meter, are used to support a 50-kg mass. Using the given nonlinear force-deflection characteristics, determine the natural frequency for small-amplitude vibration.

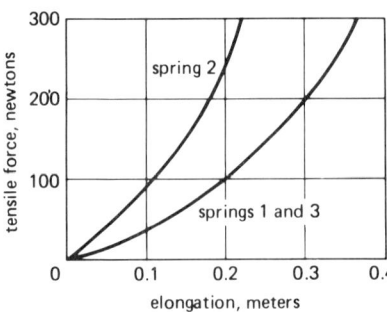
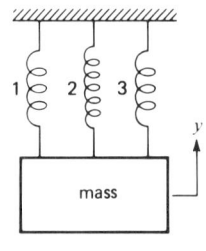

PROBLEMS

6-14. During use, a disk is driven at a constant speed of 50 r/min through a cylindrical rod. When the device is turned off, θ_i is stopped abruptly by a mechanical brake. Sketch the vibration of the disk (θ_0 versus t) that occurs after the braking, defining $t = 0$ as the instant at which the brake is applied.

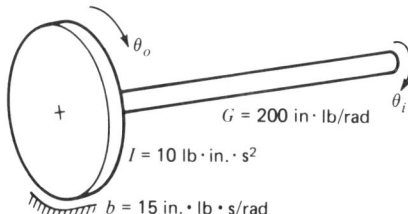

6-15. A 20-lb weight is hung by a spring having the given nonlinear characteristics. If $F = 0$ for $t < 0$ and $F = 10$ lb for $t \geq 0$, determine the frequency at which the weight will vibrate at $t > 0$. (Give units.)

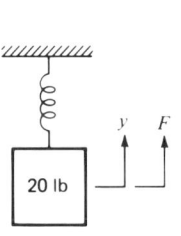 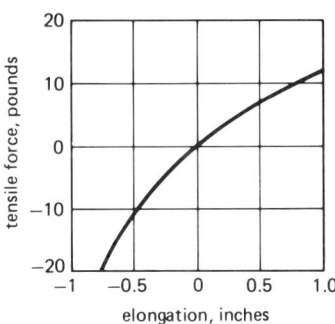

6-16. Using the loop analysis technique, write the differential equation describing each electric circuit. Determine for each system ω_n, ω_d, and ζ. Sketch the manner in which the current varies with time after the switch is closed. Assume the capacitors are initially discharged.

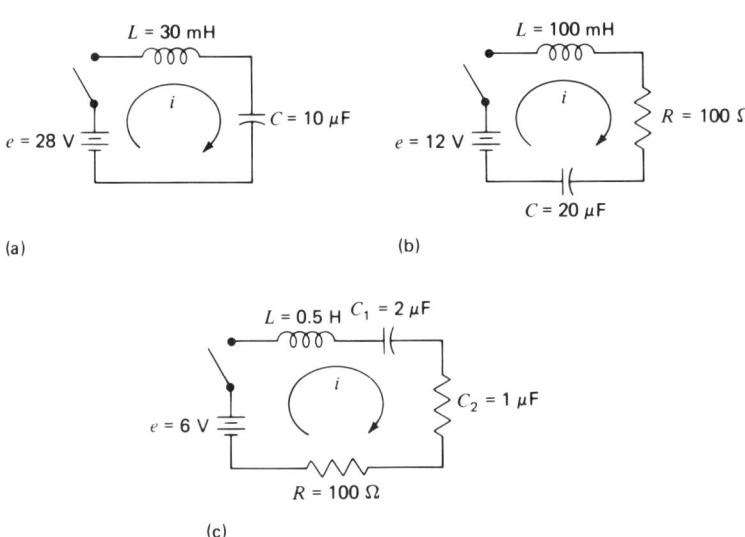

6-17. Using the node analysis technique, write the differential equation describing each electric circuit. Determine for each system ω_n, ω_d, and ζ. Sketch the manner in which the voltage e_1 varies with time after the switch is closed.

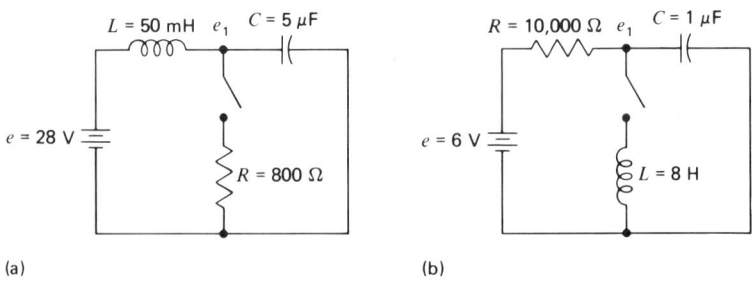

6-18. Derive an expression for the undamped natural frequency of a mass M suspended by a linear spring as a function of the static deflection δ_{static} caused by gravity.

7
Forced Vibration: Systems with a Single Degree of Freedom

7-1 Introduction

Knowledge of how a mechanical system responds to a continuous sinusoidal input can be very valuable for engineering design and development work. These data alone can often be used to tell whether or not the system will be satisfactory for a particular application. The expression *system response to a sinusoidal input* is a general term that can be used for any type of dynamic system. *Forced vibration* is the term more commonly used when the system is a mechanical vibrating system composed of mass, spring, and damper elements.

Data on the response of a vibrating system to a sinusoidal input at a single frequency are usually of limited value, being really significant only when the frequency is the only one that the input is ever likely to have. System response to sinusoidal inputs at all frequencies over some range of interest is, on the other hand, extremely useful information and tells a great deal about the system. Curves giving such response characteristics are standard tools for those who design, analyze, and use such dynamic systems. These curves are most commonly developed on the basis of having a *fixed amplitude* for the sinusoidal input as its frequency is varied over the range of interest. The term *frequency response* is used for data developed with this restriction. Frequency-response data can be developed both analytically and experimentally, depending upon whether or not the system or structure has already been built or is still in the design stage. Frequency-response data, although most useful in predicting the response of the system to harmonic inputs, can also be used to determine other system characteristics.

Only two output parameters are necessary to describe the response of a linear system to a given sinusoidal forcing function: (1) the ratio of the system's steady-state output amplitude to that of the forcing function, generally called the *amplitude ratio* or sometimes, for vibrating systems, the *magnification factor*, and (2) the phase difference between the sinusoidal signal of the input and that of the output, referred to as the *phase angle* or *phase lag*. The frequency of the steady-state output will always be the same as that of the input, as is shown mathematically in the solution of the corresponding differential equations.

When a continuing sinusoidal excitation is first applied to a system, or when its frequency is changed, there are transients (as given by the complementary solution to the applicable differential equation) that are superimposed upon the steady-state output (the particular integral). These transients die out quickly, however, in any real system with damping, so they are generally ignored when analyzing systems with this type of forcing function.

7-2 Examples of Mechanical Vibrating Systems with Sinusoidal Inputs

Even though this discussion is limited to sinusoidal inputs, there are still several different forms that this type of excitation can assume. A simple mass-spring-damper system might have a force applied to its mass that varies sinusoidally with time, as shown in Fig. 7-1. The equation of motion for this second-order system is readily derived to be

$$M \frac{d^2y}{dt^2} + c \frac{dy}{dt} + ky = F_0 \sin \omega t \qquad (7\text{-}1)$$

A sinusoidal input can also be applied as a displacement, with Fig. 7-2 showing one way in which this might occur. With this situation, the equation of motion is

$$M \frac{d^2y}{dt^2} + c \frac{dy}{dt} + ky = kX \sin \omega t \qquad (7\text{-}2)$$

7-2 EXAMPLES OF MECHANICAL VIBRATING SYSTEMS WITH SINUSOIDAL INPUTS

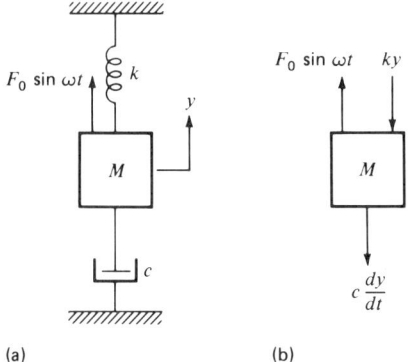

FIG 7-1. (a) Translational vibrating system excited by a sinusoidal force applied to the mass.
(b) Free-body diagram of the mass.

Figure 7-3 shows another way in which a displacement type of input can occur. In this case the periodic displacement transmits a force to the mass not through a spring alone, but through a spring and dashpot in parallel. The equation of motion is readily derived to be

$$M\frac{d^2y}{dt^2} + c\frac{dy}{dt} + ky = c\frac{dx}{dt} + kx$$

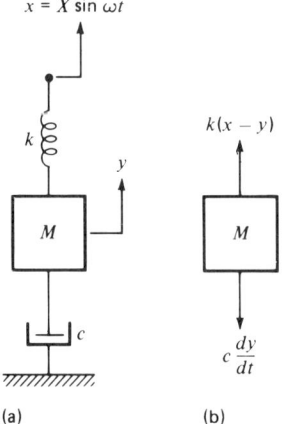

FIG 7-2. (a) Vibrating system with sinusoidal position excitation.
(b) Free-body diagram of the mass.

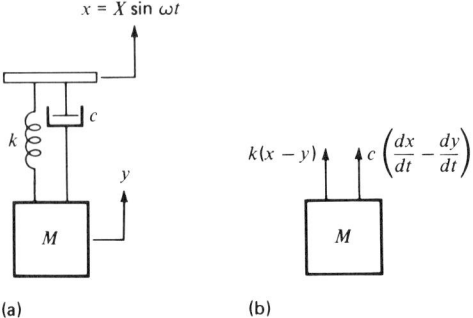

FIG 7-3. (a) Vibrating system with sinusoidal position excitation acting through a dashpot and spring in parallel.
(b) Free-body diagram of the mass.

With $x = X \sin \omega t$, and $dx/dt = X\omega \cos \omega t$, the equation can also be written

$$M \frac{d^2 y}{dt^2} + c \frac{dy}{dt} + ky = cX\omega \cos \omega t + kX \sin \omega t \quad (7\text{-}3)$$

Note that in this case the mathematical forcing function has two terms, even though the input is a single physical action.

Another very common type of excitation for a mass-spring-damper vibrating system is that caused by the centrifugal force of a rotating unbalance within the mass element itself, as illustrated in Fig. 7-4a. For this system, M is defined as the sum of the main mass and the rotating unbalance mass; that is, the total mass supported by the spring. The unbalance mass m located at a distance r from its center of rotation will have an absolute motion that is the superposition of its position with respect to the main mass and the motion of the main mass itself. With the unbalance mass rotating at the speed ω, the vertical component of its absolute position is

$$y_m = y + r \sin \omega t$$

The vertical force necessary to give it this motion is, from Newton's second law,

$$F_y = m \frac{d^2}{dt^2} (y + r \sin \omega t)$$

A free-body diagram of the unbalance mass is given in Fig. 7-4b. The horizontal component of the centrifugal force does not have to be considered for this particular example since the main mass is constrained to allow motion only in the vertical direction.

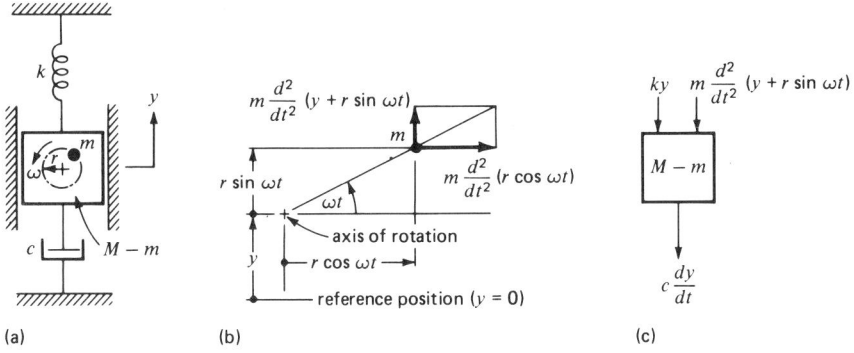

FIG 7-4. (a) Vibrating system excited by a rotating unbalance within the mass, constrained to have a single degree of freedom.
(b) Free-body diagram of the unbalance mass.
(c) Modified free-body diagram of the main mass (considering vertical forces only).

A modified free-body diagram of the main mass (considering only vertical forces) is given in Fig. 7-4c. Note that the vertical force on the unbalance mass is accompanied by an equal and opposite reaction on the main mass. Applying Newton's second law to the main mass yields

$$(M - m)\frac{d^2y}{dt^2} = -c\frac{dy}{dt} - ky - m\frac{d^2}{dt^2}(y + r\sin \omega t)$$

This reduces to

$$M\frac{d^2y}{dt^2} + c\frac{dy}{dt} + ky = mr\omega^2 \sin \omega t \qquad (7\text{-}4)$$

7-3 Frequency Response

From a comparison of Eqs. 7-1, 7-2, and 7-4, it is apparent that different types of harmonic inputs can be represented by a single sinusoidal term. The general second-order equation

$$\frac{1}{\omega_n^2}\frac{d^2y}{dt^2} + \frac{2\zeta}{\omega_n}\frac{dy}{dt} + y = K_1 \sin \omega t \qquad (7\text{-}5)$$

can be used to describe the forced-vibration characteristics of many single-degree-of-freedom systems. Note that this equation can be obtained from Eqs. 7-1 or 7-2 by (1) dividing through by the spring constant k, so that the coefficient of y is 1.0, (2) using the equations previously developed for ω_n and ζ (Eqs. 6-5 and 6-15), and (3) defining the constant K_1 in each case to fit the form of Eq. 7-5. With the general equation written in this form, *the units of K_1 will be the same as those of y*. Equation 7-5 cannot be obtained from Eq. 7-3. It can be obtained from Eq. 7-4, however, if we allow K_1 to be a function of the variable ω.

The general form of second-order equation (Eq. 7-5) is useful for the study of many types of second-order dynamic systems with sinusoidal inputs, inasmuch as it helps to characterize them by means of the undamped natural frequency (ω_n) and damping ratio (ζ). (The comments given for Eq. 6-17, the homogeneous version of the general second-order equation, also apply to Eq. 7-5.)

To obtain the steady-state solution to Eq. 7-5, the method presented in Chapter 3 is used.[1] We assume the solution

$$y_{ss} = A \sin \omega t + B \cos \omega t$$

which has the derivatives

$$\frac{dy_{ss}}{dt} = \omega A \cos \omega t - \omega B \sin \omega t$$

$$\frac{d^2 y_{ss}}{dt^2} = -\omega^2 A \sin \omega t - \omega^2 B \cos \omega t$$

Substitution of these expressions into Eq. 7-5 yields

$$-\left(\frac{\omega}{\omega_n}\right)^2 A \sin \omega t - \left(\frac{\omega}{\omega_n}\right)^2 B \cos \omega t$$

$$+ 2\zeta\left(\frac{\omega}{\omega_n}\right) A \cos \omega t - 2\zeta\left(\frac{\omega}{\omega_n}\right) B \sin \omega t$$

$$+ A \sin \omega t + B \cos \omega t = K_1 \sin \omega t$$

By equating coefficients of like terms, the single equation is converted into a pair of simultaneous linear equations:

$$-\left(\frac{\omega}{\omega_n}\right)^2 A - 2\zeta\left(\frac{\omega}{\omega_n}\right) B + A = K_1$$

$$-\left(\frac{\omega}{\omega_n}\right)^2 B + 2\zeta\left(\frac{\omega}{\omega_n}\right) A + B = 0$$

---

[1] Two alternative methods of solution for differential equations of this type, the *rotating-vector* and the *complex-number* techniques, are presented in Appendix D.

7-3 FREQUENCY RESPONSE

Simultaneous solution of the two equations yields

$$A = \frac{\left[1-(\omega/\omega_n)^2\right]K_1}{\left[1-(\omega/\omega_n)^2\right]^2 + 4\zeta^2(\omega/\omega_n)^2}$$

$$B = -\frac{2\zeta(\omega/\omega_n)K_1}{\left[1-(\omega/\omega_n)^2\right]^2 + 4\zeta^2(\omega/\omega_n)^2}$$

Substitution of these expressions into the assumed solution produces

$$y_{ss} = \frac{(\omega_n^2 - \omega^2)K_1\omega_n^2}{(\omega_n^2 - \omega^2)^2 + 4\zeta^2\omega_n^2\omega^2} \sin \omega t - \frac{2\zeta\omega K_1\omega_n^3}{(\omega_n^2 - \omega^2)^2 + 4\zeta^2\omega_n^2\omega^2} \cos \omega t$$

This can be manipulated (using Eq. 3-7) into the more useful form

$$y_{ss} = Y \sin(\omega t + \phi) \qquad (7\text{-}6)$$

with

$$\frac{Y}{K_1} = \frac{1}{\sqrt{\left[1-(\omega/\omega_n)^2\right]^2 + 4\zeta^2(\omega/\omega_n)^2}} \qquad (7\text{-}7)$$

and

$$\phi = -\tan^{-1}\frac{2\zeta\omega/\omega_n}{1-(\omega/\omega_n)^2} \qquad (7\text{-}8)$$

The angle ϕ is the phase angle between the excitation and response. A family of curves based on Eqs. 7-6 through 7-8 (Fig. 7-5) can be used to obtain the frequency-response characteristics of all vibrating systems (and, incidentally, of many other types of dynamic systems) that are represented by Eq. 7-5. Since a separate pair of amplitude ratio and phase angle curves is required for each value of the damping ratio ζ, some interpolation is normally necessary when using these curves. Application of the curves has been simplified by the use of normalized coordinates—that is, by using the dimensionless parameter Y/K_1 for the ordinate and ω/ω_n for the abscissa. With proper application of Fig. 7-5, one can avoid having to make separate calculations for each different single-degree-of-freedom vibrating system. For problems in which forced-vibration characteristics are required for a number of different frequencies and/or system parameter variations, the curves can save a great deal of labor. (Note, however, that Fig. 7-5 can be used only for systems whose equations of motion can be put in the form of Eq. 7-5.)

Although linear coordinate scales are often used for frequency-response curves, they are in most cases easier to use when plotted with logarithmic coordinates as has been done in Fig. 7-5 (with a linear scale being used, however, for the phase angle). This gives the advantage, first of

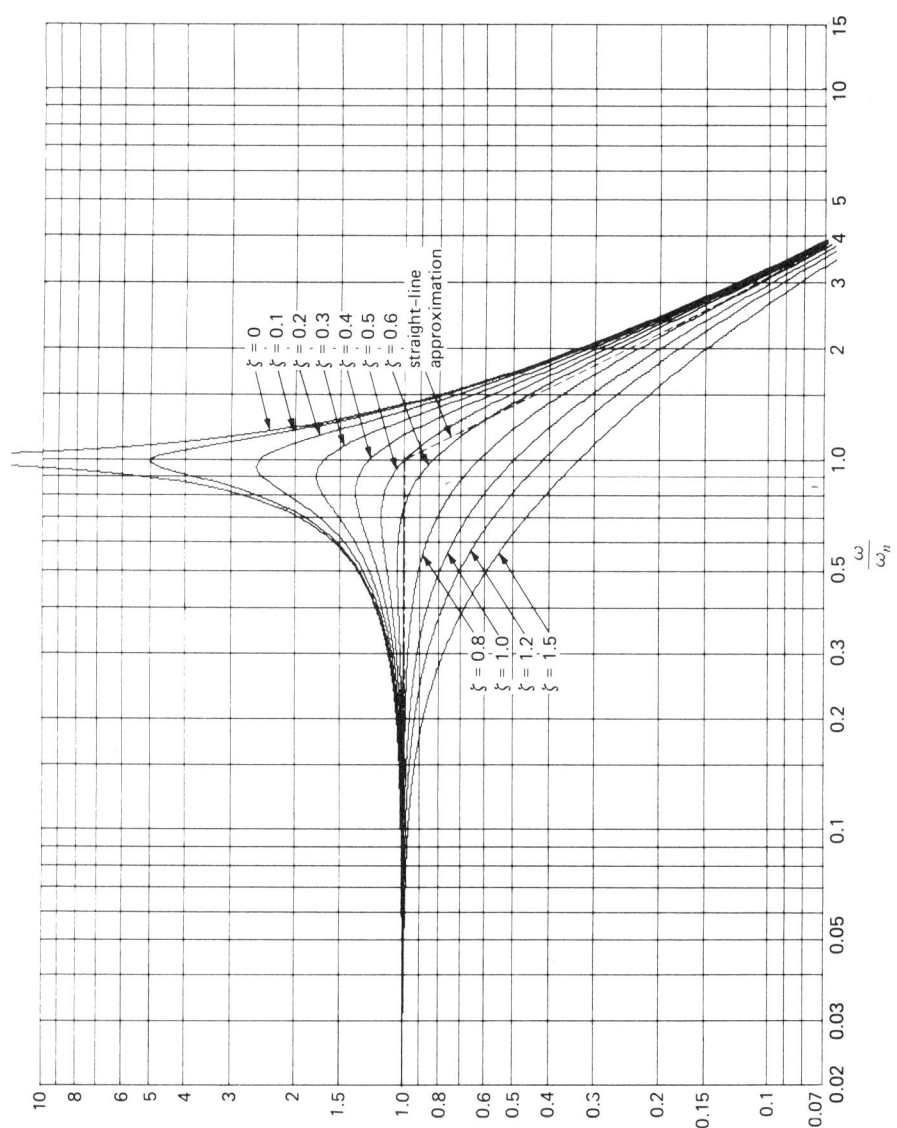

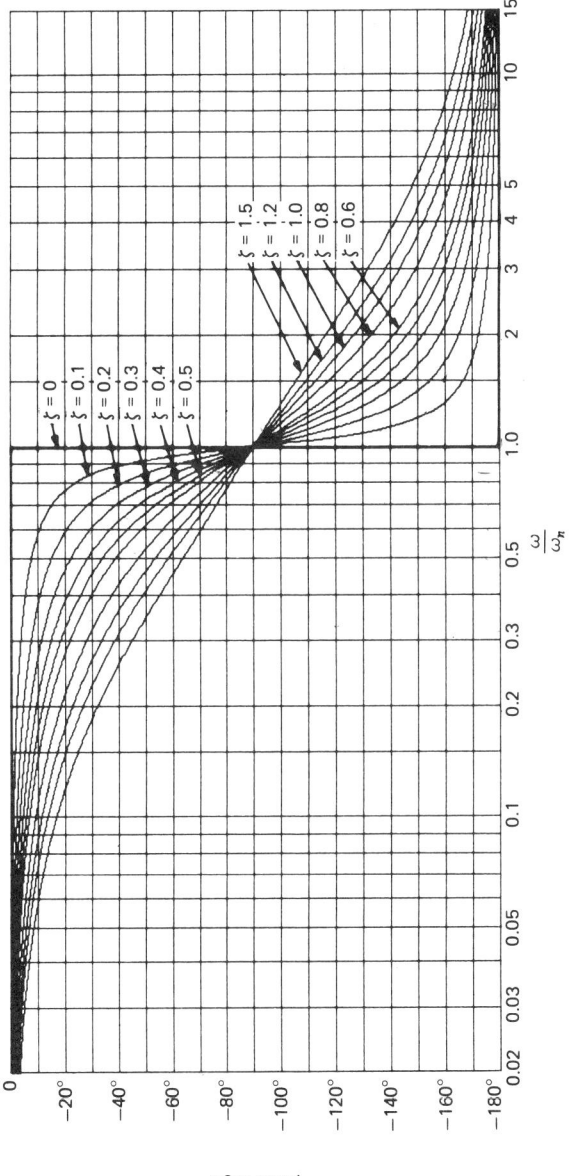

FIG 7-5. Frequency response characteristics of second-order dynamic systems described by Eq. 7-5.

all, of allowing a greater range of frequencies and amplitude ratios to be covered by a plot of reasonable size. In addition, it makes possible a straight-line approximation (to be described later) to the family of amplitude ratio curves, which is often useful.

A study of Fig. 7-5 will reveal some interesting characteristics of forced vibration of systems represented by the general form of Eq. 7-5. With zero damping ($\zeta = 0$), which can be closely approached but never quite achieved in a real mechanical vibrating system, the amplitude ratio is theoretically infinite with the forcing frequency at ω_n. The phase angle for $\zeta = 0$ is $0°$ at forcing frequencies below ω_n and $-180°$ at frequencies above ω_n, switching abruptly at the natural frequency. With an increase in damping ratio, the height of the resonance peak is reduced, with resonance occurring at a frequency somewhat less than ω_n, and the increase in phase lag takes place more gradually as the excitation frequency is raised. The phase angle curves are symmetrical about ω_n when plotted semilogarithmically as in Fig. 7-5, passing through $-90°$ at $\omega/\omega_n = 1.0$ for all values of ζ.

The dimensionless frequency ratio at which the amplitude ratio is a maximum is[2]

$$\frac{\omega}{\omega_n} = \sqrt{1 - 2\zeta^2} \qquad (7\text{-}9)$$

It should be noted that Eq. 7-9 is *not* the same as that of the damped natural frequency ω_d (Eq. 6-21), and the resonance peak of forced vibration *does not* occur at $\omega = \omega_d$, as one might be inclined to assume.

STRAIGHT-LINE APPROXIMATION

To develop a straight-line approximation to the amplitude ratio curve of Fig. 7-5, we note first of all that at frequencies appreciably below ω_n, Eq.

---

[2]Equation 7-9 can be derived by minimizing the denominator of Eq. 7-7, equating the derivative of the expression under the radical to zero:

$$\frac{d\left\{\left[1-(\omega/\omega_n)^2\right]^2 + 4\zeta^2(\omega/\omega_n)^2\right\}}{d(\omega/\omega_n)} = 0$$

$$= \frac{d\left[1 - 2(\omega/\omega_n)^2 + (\omega/\omega_n)^4 + 4\zeta^2(\omega/\omega_n)^2\right]}{d(\omega/\omega_n)}$$

$$= -4\left(\frac{\omega}{\omega_n}\right) + 4\left(\frac{\omega}{\omega_n}\right)^3 + 8\zeta^2\left(\frac{\omega}{\omega_n}\right)$$

Division by $4(\omega/\omega_n)$ yields

$$\left(\frac{\omega}{\omega_n}\right)^2 = 1 - 2\zeta^2$$

from which Eq. 7-9 follows directly.

7-3 FREQUENCY RESPONSE

7-7 reduces to

$$\frac{Y}{K_1} \approx 1.0 \qquad (7\text{-}10)$$

which is the equation of a straight line of zero slope whether plotted on proportional or logarithmic scales. At frequencies well above ω_n, Eq. 7-7 approaches the value

$$\frac{Y}{K_1} \approx \left(\frac{\omega_n}{\omega}\right)^2 \qquad (7\text{-}11)$$

Taking the logarithm of each side of Eq. 7-1, we obtain

$$\log \frac{Y}{K_1} \approx 2(\log \omega_n - \log \omega) \qquad (7\text{-}12)$$

Equation 7-12 indicates that the amplitude ratio curves become straight lines at high values of ω/ω_n when logarithmic scales are used for both coordinates. The straight line will always have a slope of -2 as a log-log plot; that is, if ω is increased by a factor of 10, Y/K_1 will decrease by a factor of 1/100. The intersection of the two straight lines occurs at ω_n. This straight-line approximation is most accurate in the area of $\omega/\omega_n = 1$ for values of ζ near 0.6, as can be seen from Fig. 7.5.

The straight-line approximation is particularly useful for showing how to extend the curve of Fig. 7-5 to lower or higher frequencies. The approximation becomes more accurate the farther away from the point of resonance it is applied.

ROTATIONAL SYSTEMS

The general second-order equation (Eq. 7-5) and Fig. 7-5 may be directly applied to analogous rotational systems, with the dependent variable y corresponding to an angle of rotation. Example 7-1 below demonstrates the application of Fig. 7-5 to a rotational problem.

> **Example 7-1.** In the paint mixing machine illustrated in Fig. 7-6, a drive unit (consisting of an electric motor, gear train, and other kinematic devices) is used to cause the top end of the shaft to oscillate rotationally with harmonic motion having a peak-to-peak amplitude of 180°. The speed of the electric motor can be varied to give any oscillation frequency between 0 and 10 Hz. The action of the impeller in paint of average consistency is found to give the approximate damping coefficient
>
> $$b = 2.4 \text{ in.} \cdot \text{lb}/(\text{rad}/\text{s})$$
>
> (a) Write the differential equation for this system, considering the motion of the shaft at the drive unit to be the input and the motion of the impeller to be the output.

FORCED VIBRATION: SYSTEMS WITH A SINGLE DEGREE OF FREEDOM

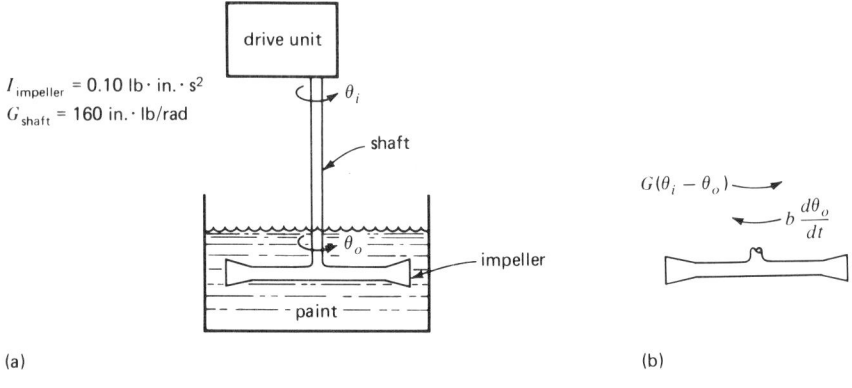

FIG 7-6. (a) A paint mixing machine.
(b) Free-body diagram of the impeller.

(b) What is the peak-to-peak amplitude of the impeller at its maximum frequency of 10 Hz, and what is the corresponding phase angle between the two ends of the shaft?

(c) At what frequency will the vibration amplitude of the impeller be a maximum? What is this maximum value, and what is the corresponding value of ϕ?

(d) Fatigue due to twisting of the shaft can be a serious problem in a system of this type. Determine the peak value of *angular twist* between the two ends of the shaft for (i) $\omega \ll \omega_n$, (ii) $\omega \gg \omega_n$, (iii) $\omega = \omega_n$.

Solution: (a) The equation of motion for this type of system is readily derived from Newton's second law to be

$$I\frac{d^2\theta_o}{dt^2} + b\frac{d\theta_o}{dt} + G\theta_o = G\theta_i$$

Putting in the values corresponding to the paint mixer, we obtain

$$0.10\frac{d^2\theta_o}{dt^2} + 2.4\frac{d\theta_o}{dt} + 160\theta_o = 160\left(\frac{90}{57.3}\right)\sin \omega t$$

Manipulating this into the general form of Eq. 7-5 produces

$$\frac{1}{\omega_n^2}\frac{d^2\theta_o}{dt^2} + \frac{2\zeta}{\omega_n}\frac{d\theta_o}{dt} + \theta_o = K_1 \sin \omega t$$

7-3 FREQUENCY RESPONSE

with

$$\omega_n = \sqrt{\frac{G}{I}} = \sqrt{\frac{160}{0.1}} = 40 \text{ rad/s}$$

$$\zeta = \frac{b}{2\sqrt{IG}} = \frac{2.4}{2\sqrt{16}} = 0.30$$

$$K_1 = \frac{90}{57.3} = \frac{\pi}{2} \text{ rad} \qquad (Ans.)$$

(b) From Fig. 7-5, the amplitude ratio at 10 Hz ($\omega = 62.8$ rad/s and $\omega/\omega_n = 1.57$) is found to be

$$\frac{\Theta_o}{K_1} = 0.5$$

so the output of the shaft is

$$\theta_o = 0.5\left(\frac{\pi}{2}\right)\sin(\omega t + \phi) = \frac{\pi}{4}\sin(62.8t + \phi)$$

The peak-to-peak output amplitude is therefore $\pi/2$ radians, or 90°. The phase angle is read directly from the curve as

$$\phi = -145° \qquad (Ans.)$$

(c) From Eq. 7-9 we find, for $\zeta = 0.3$, that the maximum output amplitude will occur at

$$\omega = \omega_n\sqrt{1 - 2(.09)}$$
$$= 0.905\omega_n = 0.905(40) = 36 \text{ rad/s} \qquad (Ans.)$$

(This answer could have also been obtained directly from Fig. 7-5.) The amplitude of the output at this frequency is read from the curve as

$$\Theta_o = 1.7\Theta_i = 1.7\left(\frac{\pi}{2}\right) = 2.67 \text{ rad} \qquad (Ans.)$$

(This is a peak-to-peak output amplitude of 5.34 rad, or 306°.) The phase angle is read from the curve as

$$\phi = -75° \qquad (Ans.)$$

(d) (i) At $\omega \ll \omega_n$, we see from Fig. 7-5 that $\Theta_o \approx \Theta_i$ and $\phi \approx 0°$. The twist in the shaft is therefore zero.

(ii) At $\omega \gg \omega_n$, $\Theta_o \approx 0$, and $\phi = -180°$. The angle of twist is therefore equal to the input motion, $\pm 90°$.

(iii) At $\omega = \omega_n$, $\Theta_o = 1.6\Theta_i = 144°$, and $\phi = -90° = -\pi/2$ rad.

Figure 7-7 shows how the input and output angles vary with time. By measurement on the figure, it is found that the difference between these two angles will vary approximately from $+170°$ to $-170°$

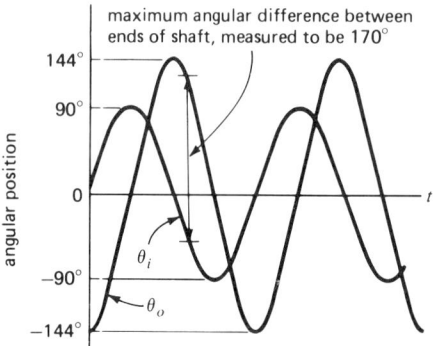

FIG 7-7. Input and response of paint mixing machine of Example 7-1.

during each cycle. This is actual torsional twist of the shaft, which could cause serious fatigue problems. (*Ans.*)

A more precise value for the angular twist at $\omega = \omega_n$ can be obtained mathematically by setting the derivative equal to zero, as follows:

$$\theta_i - \theta_o = 90° \sin 40t - 144° \sin\left(40t - \frac{\pi}{2}\right)$$

$$\frac{d(\theta_i - \theta_o)}{dt} = 0 = 3600 \cos 40t - 5760 \cos\left(40t - \frac{\pi}{2}\right)$$

$$\cos 40t = \frac{5760}{3600} \cos\left(40t - \frac{\pi}{2}\right)$$

$$= 1.6\left(\cos 40t \cos \frac{\pi}{2} + \sin 40t \sin \frac{\pi}{2}\right)$$

$$= 1.6 \sin 40t$$

$$\frac{1}{1.6} = \frac{\sin 40t}{\cos 40t} = \tan 40t$$

$$40t = \tan^{-1} 0.625$$

$$= 0.5586 \text{ rad}$$

$$t = 0.01396 \text{ second}$$

At $t = 0.01396$,

$$(\theta_i - \theta_o)_{max} = 90° \sin(0.5586) - 144° \sin\left(0.5586 - \frac{\pi}{2}\right)$$

$$= 169.8°$$

Torsional twist of the shaft is therefore $\pm 169.8°$. (*Ans.*)

7-3 FREQUENCY RESPONSE

Example 7-2. The valves of the two-tank liquid-level system shown in Fig. 7-8 are linear (i.e., the output flow of each tank is proportional to its head), with $G_{v_1} = G_{v_2} = 0.10$ m^2/s. The cross-sectional areas of the two tanks are also equal, with $A_1 = A_2 = 25$ m^2. The input flow rate varies sinusoidally with time about an average value, and can be represented by the equation

$$q_i = 0.7 + 0.2 \sin 0.002t \ (\text{m}^3/\text{s})$$

(a) Considering the change in head of the second tank (δh_2) to be the output, determine the undamped natural frequency and damping ratio of the system and the response of δh_2 to the given input.

(b) Demonstrate mathematically that the damping ratio of the system can never be less than 1, regardless of the values of G_{v_1}, G_{v_2}, A_1, and A_2.

Solution: (a) The equation for the first tank, considering its deviation of head from the normal operation point, is readily derived (as in Example 4-2) to be

$$A_1 \frac{d\delta h_1}{dt} + G_{v_1} \delta h_1 = \delta q_i$$

Similarly, for the second tank,

$$A_2 \frac{d\delta h_2}{dt} + G_{v_2} \delta h_2 = \delta q_1 = G_{v_1} \delta h_1$$

Rewriting in operator form,

$$\delta h_1 = \frac{\delta q_i}{A_1 p + G_{v_1}}$$

$$\delta h_2 (A_2 p + G_{v_2}) = G_{v_1} \delta h_1$$

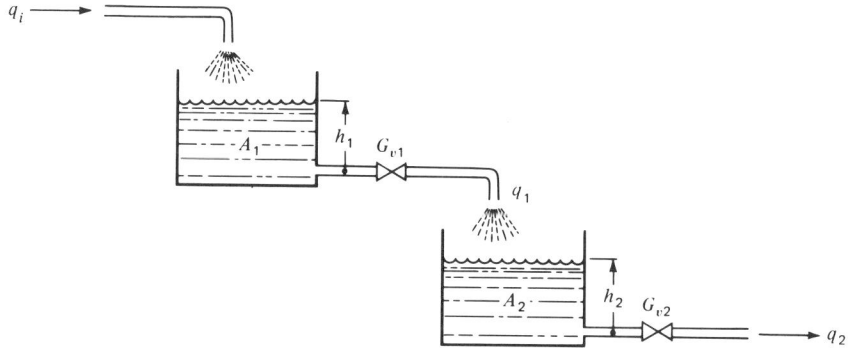

FIG 7-8. Two-tank liquid-level system of Example 7-2.

Combining,

$$\delta h_2 (A_2 p + G_{v_2}) = \frac{G_{v_1} \delta q_i}{A_1 p + G_{v_1}}$$

$$\left[A_1 A_2 p^2 + (A_1 G_{v_2} + A_2 G_{v_1}) p + G_{v_1} G_{v_2} \right] \delta h_2 = G_{v_1} \delta q_i$$

Converting back to differential form and manipulating into the general form of Eq. 7-5, we obtain

$$\frac{A_1 A_2}{G_{v_1} G_{v_2}} \frac{d^2 \delta h_2}{dt^2} + \frac{A_1 G_{v_2} + A_2 G_{v_1}}{G_{v_1} G_{v_2}} \frac{d \delta h_2}{dt} + \delta h_2 = \frac{1}{G_{v_2}} \delta q_i$$

$$= \frac{\hat{Q}_i}{G_{v_2}} \sin \omega t$$

with $\hat{Q}_i = 0.2$ m³/s and $\omega = 0.002$ rad/s for the system under consideration.

We can now solve for ω_n and ζ:

$$\frac{1}{\omega_n^2} = \frac{A_1 A_2}{G_{v_1} G_{v_2}} = \frac{(25)^2}{(0.10)^2} = 62{,}500$$

$$\omega_n = \sqrt{\frac{1}{62{,}500}} = 0.0040 \text{ rad/s} \qquad (Ans.)$$

$$\frac{2\zeta}{\omega_n} = \frac{A_1 G_{v_2} + A_2 G_{v_1}}{G_{v_1} G_{v_2}} = \frac{2(25)(0.10)}{(0.10)^2} = 500$$

$$\zeta = \frac{(500)(.0040)}{2} = 1.00 \qquad (Ans.)$$

The system response will have the form

$$\delta h_2 = \hat{H}_2 \sin(0.002 t + \phi)$$

where $\hat{H}_2$ is the amplitude of the head variation from the average value. In order to obtain $\hat{H}_2$ and ϕ from Fig. 7-5, we first determine the frequency ratio ω / ω_n:

$$\frac{\omega}{\omega_n} = \frac{0.002}{0.004} = 0.50$$

From Fig. 7-5,

$$\frac{\hat{H}_2}{K_1} = 0.8$$

$$\phi = -55° = -0.9599 \text{ rad}$$

For our system,

$$K_1 = \frac{\hat{Q}_i}{G_{v_2}} = \frac{0.2}{0.10} = 2.0 \text{ meters}$$

so that

$$\hat{H}_2 = 0.8(2) = 1.6 \text{ meters}$$

The system response is therefore

$$\delta h_2 = 1.6 \sin(0.002t - 0.9599) \qquad (Ans.)$$

The liquid level in the second tank will rise (and fall) from its average value with a peak-to-peak amplitude of 3.2 meters, lagging the input flow variations by 55°.

(b) The damping ratio of the system is

$$\zeta = \frac{\omega_n}{2}\left[\frac{A_1 G_{v_2} + A_2 G_{v_1}}{G_{v_1} G_{v_2}}\right]$$

$$= \frac{1}{2}\left[\frac{G_{v_1} G_{v_2}}{A_1 A_2}\right]^{1/2}\left[\frac{A_1 G_{v_2} + A_2 G_{v_1}}{G_{v_1} G_{v_2}}\right]$$

$$= \frac{1}{2}\sqrt{\frac{G_{v_1} G_{v_2}}{A_1 A_2} \frac{(A_1 G_{v_2} + A_2 G_{v_1})^2}{(G_{v_1} G_{v_2})^2}}$$

$$= \frac{1}{2}\sqrt{\frac{A_1^2 G_{v_2}^2 + A_2^2 G_{v_1}^2 + 2A_1 A_2 G_{v_1} G_{v_2}}{A_1 A_2 G_{v_1} G_{v_2}}}$$

$$= \frac{1}{2}\sqrt{\frac{A_1}{A_2}\frac{G_{v_2}}{G_{v_1}} + \frac{A_2}{A_1}\frac{G_{v_1}}{G_{v_2}} + 2}$$

Defining $a = A_1 G_{v_2}/A_2 G_{v_1}$, the equation becomes

$$\zeta = \frac{1}{2}\sqrt{a + \frac{1}{a} + 2}$$

It can be seen by inspection that ζ will be 1 for $a = 1$, and greater than 1 for any other value of a.

7-4 Vibration Caused by Rotating Unbalance

One type of vibrating system that deserves special study is the single-degree-of-freedom mass-spring-damper system that is caused to vibrate by a

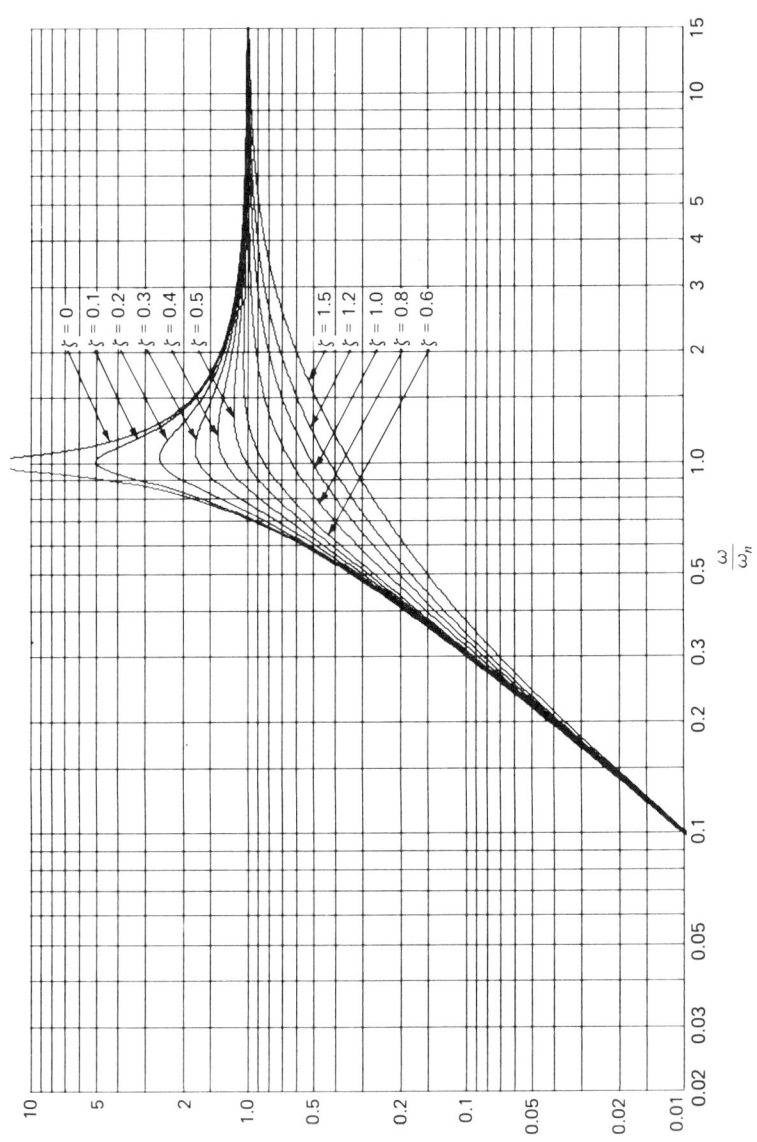

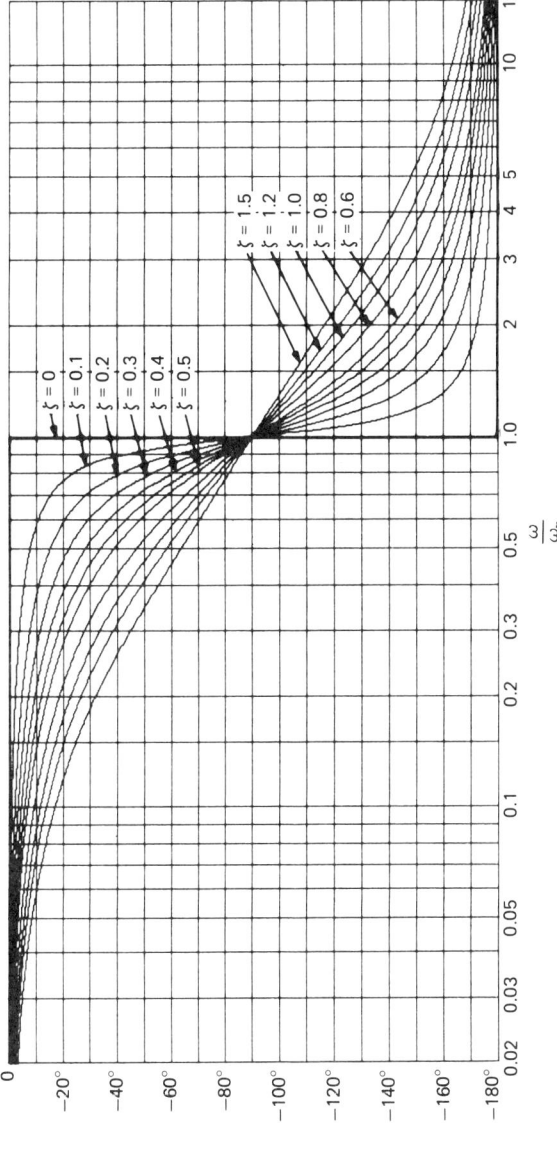

FIG 7-9. Vibration characteristics of single-degree-of-freedom mass-spring-damper systems excited by a rotating unbalance within the mass (the system of Fig. 7-4, described by Eq. 7-4).

rotating unbalance in the system (Fig. 7-4). This is a very common type of system, but, since the excitation amplitude varies with frequency, the frequency response curves of Fig. 7-5 are not as convenient to use as a set of curves developed specifically for this case.

In reality, a vibrating system of this type will often have more than one degree of freedom, so the rotating unbalance might cause several different modes of vibration to occur simultaneously. For our analysis, however, we will assume the system is constrained as in Fig. 7-4 to have only a single degree of freedom, and thus only the vertical component of the centrifugal force can cause vibration. From the equation that describes the system (Eq. 7-4, repeated here for convenience)

$$M\frac{d^2y}{dt^2} + c\frac{dy}{dt} + ky = mr\omega^2 \sin \omega t$$

we see that the amplitude of the sinusoidal forcing function is proportional to the square of its frequency (i.e., to the square of the speed of rotation of the unbalanced rotor).

If Eq. 7-4 is put into the general form of Eq. 7-5, then the amplitude ratio (From Eq. 7-7) is found to be

$$\frac{Y}{K_1} = \frac{Yk}{mr\omega^2} = \frac{1}{\sqrt{\left[1 - (\omega/\omega_n)^2\right]^2 + 4\zeta^2(\omega/\omega_n)^2}} \quad (7\text{-}13)$$

The right-hand side of Eq. 7-13 is given by the curves of Fig. 7-5.

When studying a system that vibrates because of a rotating unbalance, however, we are normally interested in knowing how the actual amplitude of vibration varies with frequency, and we see from Eq. 7-13 that Fig. 7-5 does not give this information directly. By multiplying both sides of Eq. 7-13 by the term $(\omega/\omega_n)^2$, and noting that $\omega_n^2 = k/M$, we obtain the equation

$$\frac{MY}{mr} = \frac{(\omega/\omega_n)^2}{\sqrt{\left[1 - (\omega/\omega_n)^2\right]^2 + 4\zeta^2(\omega/\omega_n)^2}} \quad (7\text{-}14)$$

Plotting the right-hand side of Eq. 7-14 yields the set of normalized curves, Fig. 7-9, that give us the desired type of information directly. Note that M is the total mass of the system (including the rotating unbalance mass), Y is its maximum displacement, when vibrating, from the position of static equilibrium, and m is the rotating unbalance mass, which is located at a radius r from its center of rotation. It is interesting to see that the amplitude of vibration is essentially the same at all frequencies appreciably above the resonance frequency. Note that the phase angle is given by Eq. 7-8, so the curves of ϕ are the same as those in Fig. 7-5. Since normalized coordinates have been used, Fig. 7-9 can be used to facilitate the analysis of vibrating systems of this type with any combination of parameters.

In practice, a system with vibration from rotating unbalance might be an electric motor or similar device mounted to the floor by motor mounts having flexibility and damping. If such a motor always runs at the same speed, then it should be relatively simple to minimize the vibration problem by proper selection of component values so that operation at the required speed is either well below or well above the resonance frequency. If, however, the motor must be run at several different speeds, or cycled over a fairly wide range of speeds, then the situation is more complicated and much judgement is required in parameter selection so that the overall vibration problem is minimized. In all cases, of course, the problem is helped by reducing the amount of rotor unbalance.

7-5 Displacement Input Acting Through a Dashpot and Spring in Parallel

A mass caused to vibrate by a sinusoidal motion input acting through a spring and damper in parallel (as illustrated in Fig. 7-3) is an important special case to consider. The differential equation describing the motion of the mass (Eq. 7-3, repeated here for convenience),

$$M \frac{d^2y}{dt^2} + c \frac{dy}{dt} + ky = cX\omega \cos \omega t + kX \sin \omega t$$

has a forcing function that consists of two parts: a sine term and a cosine term. The equation can be solved by using the principle of *superposition* and the equations and/or curves developed in Section 7-3. Such an approach would be reasonable when solving for one specific input, but inconvenient when considering the total frequency-response characteristics.

In order to maintain a unified approach in solution techniques, the steady-state solution to Eq. 7-3 presented below is based on the method presented in Chapter 3, even though a somewhat less laborious solution would result from the complex-number technique given in Appendix D.

Equation 7-3 is first divided through by k. Using the expressions for ω_n and ζ (Eqs. 6-5 and 6-15), it can then be written

$$\frac{1}{\omega_n^2} \frac{d^2y}{dt^2} + \frac{2\zeta}{\omega_n} \frac{dy}{dt} + y = X\left(2\zeta \frac{\omega}{\omega_n} \cos \omega t + \sin \omega t\right) \quad (7\text{-}15)$$

Rewriting the forcing function in the form of a single harmonic function based on Eq. 3-7, the equation becomes

$$\frac{1}{\omega_n^2} \frac{d^2y}{dt^2} + \frac{2\zeta}{\omega_n} \frac{dy}{dt} + y = K \sin(\omega t + \psi) \quad (7\text{-}16)$$

with

$$K = X\sqrt{4\zeta^2\left(\frac{\omega}{\omega_n}\right)^2 + 1} \qquad (7\text{-}17)$$

$$\psi = \tan^{-1}\left(2\zeta\frac{\omega}{\omega_n}\right) \qquad (7\text{-}18)$$

By assuming

$$y_{ss} = A\sin(\omega t + \psi) + B\cos(\omega t + \psi)$$

the method of solution follows the same pattern as the solution of Eq. 7-5, producing the result

$$y_{ss} = \frac{\left[1-(\omega/\omega_n)^2\right]K}{\left[1-(\omega/\omega_n)^2\right]^2 + 4\zeta^2(\omega/\omega_n)^2}\sin(\omega t + \psi)$$

$$- \frac{2\zeta(\omega/\omega_n)K}{\left[1-(\omega/\omega_n)^2\right]^2 + 4\zeta^2(\omega/\omega_n)^2}\cos(\omega t + \psi) \qquad (7\text{-}19)$$

This may be changed to the alternative form (again using Eq. 3-7)

$$y_{ss} = \left[\frac{\left[1-(\omega/\omega_n)^2\right]^2 K^2}{\left\{\left[1-(\omega/\omega_n)^2\right]^2 + 4\zeta^2(\omega/\omega_n)^2\right\}^2}\right.$$

$$\left. + \frac{4\zeta^2(\omega/\omega_n)^2 K^2}{\left\{\left[1-(\omega/\omega_n)^2\right]^2 + 4\zeta^2(\omega/\omega_n)^2\right\}^2}\right]^{1/2}\sin(\omega t + \psi + \gamma) \qquad (7\text{-}20)$$

with

$$\gamma = \tan^{-1}\left[\frac{-2\zeta(\omega/\omega_n)}{1-(\omega/\omega_n)^2}\right] \qquad (7\text{-}21)$$

Equation 7-20 may be reduced, by algebraic manipulation, to the form

$$y_{ss} = \frac{K}{\sqrt{\left[1-(\omega/\omega_n)^2\right]^2 + 4\zeta^2(\omega/\omega_n)^2}}\sin(\omega t + \psi + \gamma) \qquad (7\text{-}22)$$

Substituting for K from Eq. 7-17 produces

$$y_{ss} = X\sqrt{\frac{4\zeta^2(\omega/\omega_n)^2 + 1}{\left[1-(\omega/\omega_n)^2\right]^2 + 4\zeta^2(\omega/\omega_n)^2}}\sin(\omega t + \psi + \gamma) \qquad (7\text{-}23)$$

7-5 DISPLACEMENT INPUT ACTING THROUGH A DASHPOT AND SPRING IN PARALLEL

The solution may now be put into the desired final form

$$y_{ss} = Y \sin(\omega t + \phi) \tag{7-24}$$

Comparison of Eqs. 7-23 and 7-24 shows the dimensionless amplitude ratio to be

$$\frac{Y}{X} = \sqrt{\frac{4\zeta^2(\omega/\omega_n)^2 + 1}{[1 - (\omega/\omega_n)^2]^2 + 4\zeta^2(\omega/\omega_n)^2}} \tag{7-25}$$

and the phase angle between the output y and the input x to be

$$\phi = \psi + \gamma$$

To obtain a more usable expression for the phase angle ϕ, we use the trigonometric identity

$$\tan \phi = \tan(\psi + \gamma) = \frac{\tan \psi + \tan \gamma}{1 - \tan \psi \tan \gamma}$$

Combining with Eqs. 7-18 and 7-21, this becomes

$$\tan \phi = \frac{2\zeta(\omega/\omega_n) - \dfrac{2\zeta(\omega/\omega_n)}{1 - (\omega/\omega_n)^2}}{1 + \dfrac{[2\zeta(\omega/\omega_n)]^2}{1 - (\omega/\omega_n)^2}} \tag{7-26}$$

Multiplying numerator and denominator by $1 - (\omega/\omega_n)^2$ reduces the equation to the form

$$\tan \phi = \frac{-2\zeta(\omega/\omega_n)^3}{1 - (\omega/\omega_n)^2 + 4\zeta^2(\omega/\omega_n)^2} \tag{7-27}$$

or

$$\phi = \tan^{-1}\left[\frac{-2\zeta(\omega/\omega_n)^3}{1 - (\omega/\omega_n)^2 + 4\zeta^2(\omega/\omega_n)^2}\right] \tag{7-28}$$

Figure 7-10 is the family of frequency-response curves for the system, corresponding to Eqs. 7-25 and 7-28. It can be used for directly determining amplitude and phase relationships for the system of Fig. 7-3, avoiding the necessity of solving the equations for each individual problem.

ROTATIONAL SYSTEMS

An input can also be applied to a flywheel through a torsional spring and rotary dashpot acting in parallel, as illustrated in Fig. 7-11. Since this is an

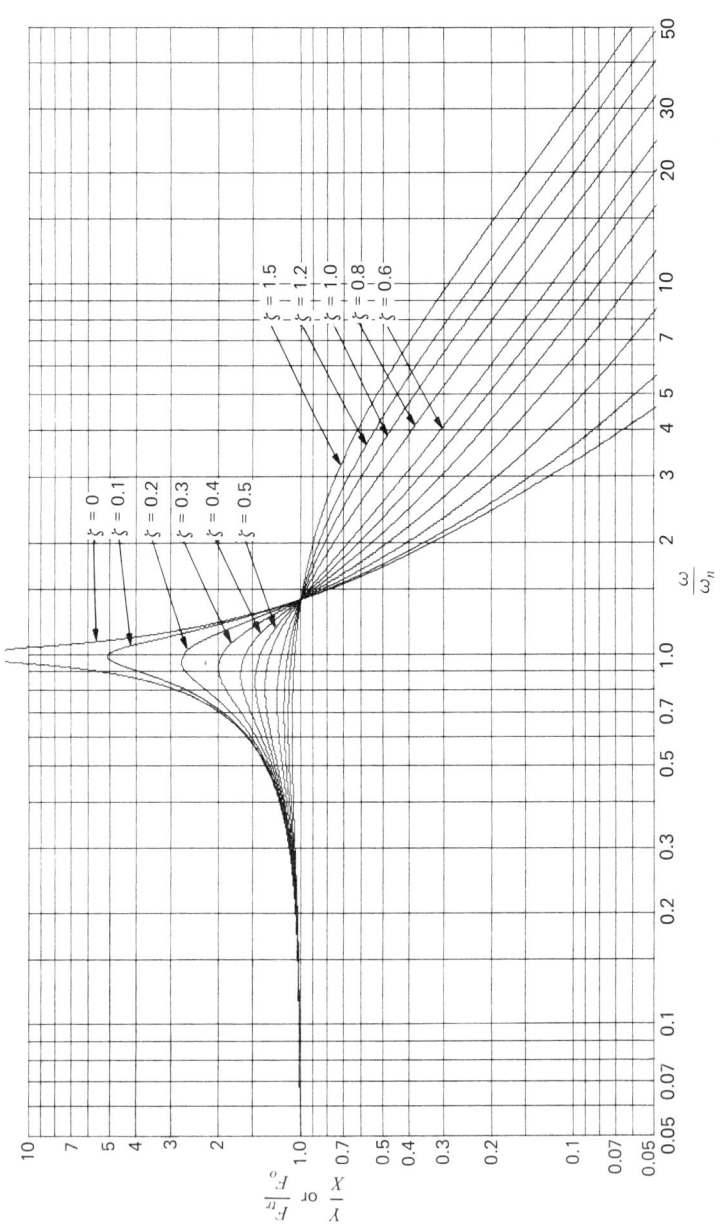

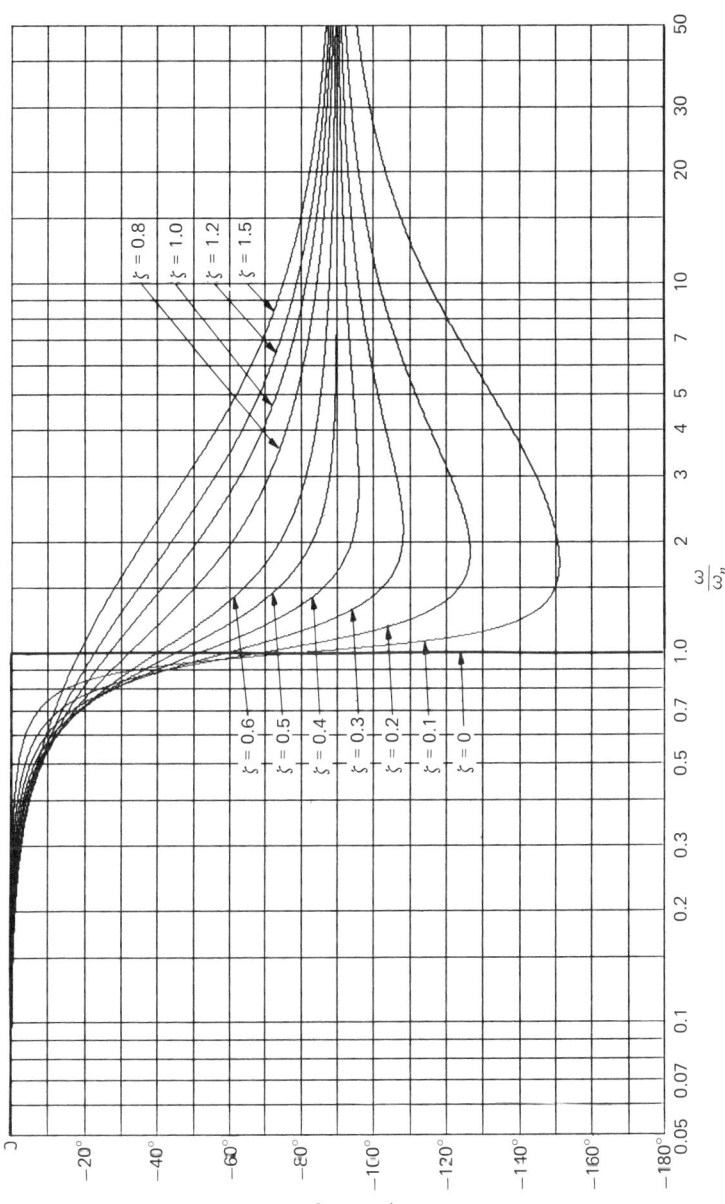

FIG 7-10. Curves having two uses:
(a) Frequency-response curves for a mass caused to vibrate by a displacement input acting through a spring and dashpot in parallel (as in Fig. 7-3). The equation of motion for the system is given by Eq. 7-15.
(b) Transmissibility curves for the system of Fig. 7-12, giving the ratio of the peak force (F_{tr}) transmitted to the surrounding structure to the force F_0 that would be transmitted if the mounting were rigid.

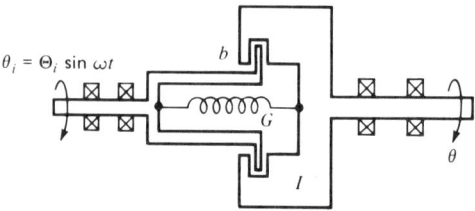

FIG 7-11. A rotational system with an angular displacement input acting through a torsional spring and rotary dashpot in parallel.

exact analog of the system of Fig. 7-3, it is described by the same equation of motion (Eq. 7-15, with the translational displacements y and X replaced by the corresponding rotational displacements θ and Θ_i, and with ω_n and ζ defined for rotational parameters by Eqs. 6-8 and 6-22). The frequency-response curves of Fig. 7-10 are therefore equally valid for rotational or translational systems.

7-6 Transmissibility

In Section 7-4 the vibration characteristics of a mass-spring-damper system excited by a rotating unbalance were studied. One important aspect of such a system that has not yet been covered, however, is the force transmitted by the spring and damper elements to the surrounding structure. Without adequate isolation between a vibrating machine and the room in which it is located, serious problems of noise, building vibration, and even structural damage may occur.

Figure 7-12 illustrates a machine with a pair of counterrotating rotors of equal unbalance (so that horizontal components of the centrifugal force vectors cancel), isolated from the floor by springs and dashpots with a net spring rate k and net damping coefficient c. Since this is equivalent to the system of Fig. 7-4, it is described by Eq. 7-4,

$$M \frac{d^2y}{dt^2} + c \frac{dy}{dt} + ky = mr\omega^2 \sin \omega t$$

Considering the forcing function in the equation of motion, it can be said that the vibrating force in the machine is

$$F = F_0 \sin \omega t \tag{7-29}$$

with

$$F_0 = mr\omega^2 \tag{7-30}$$

7-6 TRANSMISSIBILITY

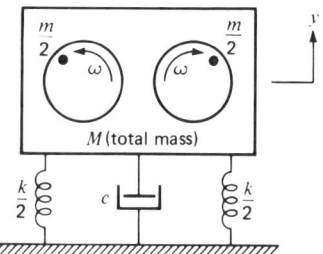

FIG 7-12. A machine with rotating unbalance isolated from the floor by springs and dashpots.

(A reciprocating unbalance, with harmonic motion of the reciprocating element, would also produce a vibrating force represented by Eq. 7-29, but Eq. 7-30 would not be applicable.)

The force transmitted to the floor due to machine vibration would be given by Eq. 7-29 if the machine were mounted rigidly to the floor. When mounting is as in Fig. 7-12, the force transmitted is that caused by spring compression and dashpot velocity,

$$f_{tr} = ky + c\frac{dy}{dt} \quad (7\text{-}31)$$

Since the motion of the mass is given by the equation

$$y = Y \sin(\omega t + \phi)$$

Eq. 7-31 may be written

$$f_{tr} = kY \sin(\omega t + \phi) + cY\omega \cos(\omega t + \phi) \quad (7\text{-}32)$$

This may be put into the alternative form (based on Eq. 3-7)

$$f_{tr} = F_{tr} \sin(\omega t + \phi + \lambda) \quad (7\text{-}33)$$

where

$$F_{tr} = Y\sqrt{k^2 + (c\omega)^2} \quad (7\text{-}34)$$

$$\lambda = \tan^{-1}\left(\frac{c\omega}{k}\right) \quad (7\text{-}35)$$

The *transmissibility* (TR) is defined as the *peak force* transmitted to the surrounding structure (the floor, in the case of Fig. 7-12) through the spring and damper support divided by the peak force that would be transmitted if the mounting were rigid,

$$TR = \frac{F_{tr}}{F_0} \quad (7\text{-}36)$$

Substitution for F_{tr} and F_0 from Eqs. 7-34 and 7-30 yields

$$TR = \frac{Y\sqrt{k^2 + (c\omega)^2}}{mr\omega^2} \qquad (7\text{-}37)$$

From Eq. 7-13, which applies to the system under discussion, is obtained

$$mr\omega^2 = Yk\sqrt{\left[1 - \left(\frac{\omega}{\omega_n}\right)^2\right]^2 + 4\zeta^2\left(\frac{\omega}{\omega_n}\right)^2}$$

Substitution of this expression into Eq. 7-37 yields

$$TR = \frac{\sqrt{k^2 + (c\omega)^2}}{k\sqrt{\left[1 - (\omega/\omega_n)^2\right]^2 + 4\zeta^2(\omega/\omega_n)^2}} \qquad (7\text{-}38)$$

Dividing both numerator and denominator by k and using the relationship

$$\frac{c^2\omega^2}{k^2} = 4\zeta^2\left(\frac{\omega}{\omega_n}\right)^2$$

the final answer is obtained:

$$TR = \sqrt{\frac{1 + 4\zeta^2(\omega/\omega_n)^2}{\left[1 - (\omega/\omega_n)^2\right]^2 + 4\zeta^2(\omega/\omega_n)^2}} \qquad (7\text{-}39)$$

Note that the expression for transmissibility (Eq. 7-39) is identical to that for the amplitude ratio of the system excited by a displacement acting through a spring and dashpot in parallel (Eq. 7-25). Figure 7-10 may therefore also be used for quickly determining the transmissibility of a linear system corresponding to Fig. 7-12 for which mechanical parameters are known.

Although the phase angle between the transmitted and exciting forces can be shown to be given by Eq. 7-28 (i.e., to be the same as the phase angle between the output and input of the system of Fig. 7-3), it is not considered significant enough to include in the definition of transmissibility, and it is normally ignored.

The curves of Fig. 7-10 show some very important characteristics of transmissibility. For all values of ζ, transmissibility is found to be 1.0 at $\omega/\omega_n = 0$ and $\omega/\omega_n = \sqrt{2}$. Between these two frequencies, transmissibility is in all cases greater than 1 (i.e., the vibration transmitted to the surroundings is greater than it would be with rigid mounting). For $\omega/\omega_n > \sqrt{2}$, transmissibility is less than 1 for all values of ζ. In order to isolate a vibrating machine successfully from the surrounding structure, the frequency of excitation must therefore be greater than $\sqrt{2}$ times the natural frequency of the system, and should preferably be at least 3 or 4 times the natural frequency. Best results are obtained with zero damping. Damping, however, reduces the transmissibility with $\omega/\omega_n < \sqrt{2}$, so that it may be advantageous to add damping if (1) the excitation frequency varies over a wide range during operation of the machine, with $\omega/\omega_n < \sqrt{2}$ an appreciable amount of time, or (2) the machine accelerates and/or decelerates slowly through the critical speed range when turned on or off. (For the second situation a good alternative solution might be to introduce damping when needed during the starting or stopping of the machine, but to eliminate it during the steady-state operation.)

The amount of damping that should be used will vary for each specific application. The value chosen represents a compromise, since low-frequency isolation characteristics can be improved only at the expense of the high-frequency characteristics.

The concept of transmissibility may also be used for a rotational system that is connected to the surrounding structure through torsional springs and rotary dashpots acting in parallel.

7-7 Systems with Simultaneous Inputs at Two or More Frequencies

A vibrating system will sometimes be excited by two or more periodic functions occuring simultaneously. Multiple inputs often occur because the device producing the excitation has not only a fundamental frequency but higher harmonics as well. In Fig. 7-13a a simple mass-spring-damper

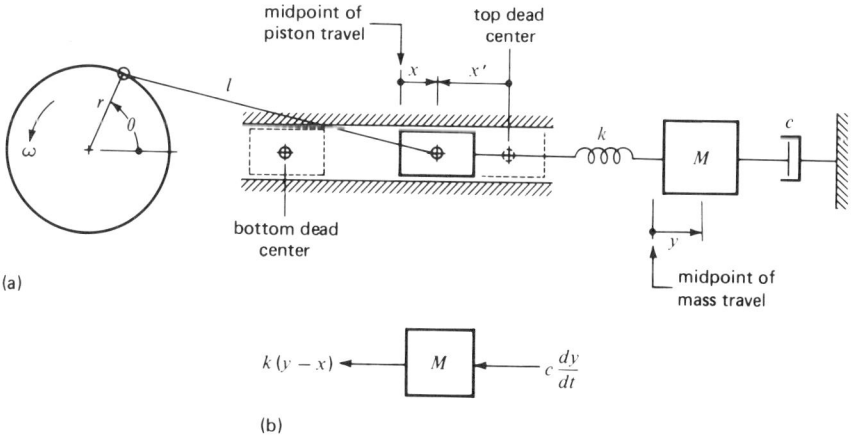

FIG 7-13. (a) System with motion excitation from a piston-crank mechanism.
(b) Free-body diagram of the mass M.

system has motion excitation caused by a piston actuated through a crankshaft and connecting rod. With the crankshaft rotating at the angular velocity ω, the displacement x of the piston from its midpoint is given by the approximate equation[3]

$$x = r \cos \omega t - \frac{r^2}{4l}(1 - \cos 2\omega t) \tag{7-40}$$

The equation of motion of the system is (note the free-body diagram of Fig. 7-13b).

$$M\frac{d^2y}{dt^2} + c\frac{dy}{dt} + k(y - x) = 0$$

or,

$$M\frac{d^2y}{dt^2} + c\frac{dy}{dt} + ky = -k\frac{r^2}{4l} + kr \cos \omega t + k\frac{r^2}{4l} \cos 2\omega t \tag{7-41}$$

The input to the system therefore has a constant and two harmonic components: one at the frequency ω and the other at the frequency 2ω.

For determining the response of a linear system with two or more inputs, the principle of *superposition* may be used; that is, the response is equal to *the algebraic sum of the responses that would occur if the inputs took place one at a time*. For the system of Fig. 7-13a the output will have three superimposed components: one a constant, the second a harmonic function at the frequency ω, and the third a harmonic function at 2ω. The amplitudes and phase angles of each of the two harmonic output components can be determined by using the frequency-response curves of Fig. 7-5. Note that a lightly damped system with simultaneous inputs at different frequencies may have large amplitudes of vibration if *any one* of these inputs is at or near the system's resonance frequency.

Example 7-3. A single-cylinder piston air compressor is driven by an electric motor through a V-belt as shown in Fig. 7-14a. The compres-

---

[3]The approximate relationship derived in most kinematic textbooks for piston displacement x' away from the position of top dead center is

$$x' \approx r\left(1 - \cos \theta + \frac{\sin^2 \theta}{2l/r}\right)$$

By use of the trigonometric identity

$$\sin^2 \theta = \tfrac{1}{2}(-\cos 2\theta + 1)$$

the equation becomes

$$x' = r\left(1 + \frac{r}{4l} - \cos \theta - \frac{r}{4l} \cos 2\theta\right)$$

By assuming that $t = 0$ when $\theta = 0$ (i.e., $\theta = \omega t$) and noting that $x = r - x'$, Eq. 7-40 is obtained.

7-7 SYSTEMS WITH SIMULTANEOUS INPUTS AT TWO OR MORE FREQUENCIES

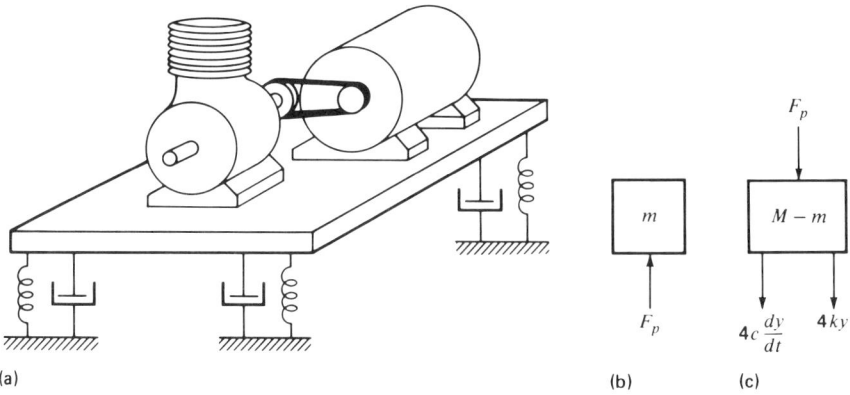

FIG 7-14. (a) A piston air compressor driven by an electric motor.
(b) Modified free-body diagram of the piston (F_p is the vertical component of the force on the piston from the connecting rod).
(c) Modified free-body diagram of the main mass.

sor has a stroke of 3.0 in., a connecting rod length of 4.0 in., and an effective piston weight of 1.0 lb. The crankshaft is balanced, and the effective weight of the connecting rod is negligible. (By including part of the connecting rod in the effective piston weight and taking part of its weight into account when balancing the crankshaft, this is a reasonable approximation.) The electric motor is very well balanced. The compressor operates at 400 r/min. The total compressor-motor package, weighing 150 lb, is set on four springs and four dashpots. Determine the response of the system to the compressor unbalance, considering motion in the vertical direction only. Assume that $k = 680$ lb/in. for each spring and $c = 3.25$ lb · s/in. for each dashpot.

Solution: For developing the system equation, we define M as the total mass of the unit and m as the piston mass. The vertical position of the piston is obtained by using Eq. 7-40 and noting that the piston position with respect to the platform must be superimposed upon the platform position,

$$y_p = y + r \cos \omega t - \frac{r^2}{4l}(1 - \cos 2\omega t)$$

The force necessary to give it this motion is, from Newton's second law (see Fig. 7-14b),

$$F_p = m \frac{d^2}{dt^2}\left[y + r \cos \omega t - \frac{r^2}{4l}(1 - \cos 2\omega t)\right]$$

Applying Newton's second law to the main mass, with the reaction to F_p as one of the forces on the mass (see Fig. 7-14c), yields

$$(M - m)\frac{d^2y}{dt^2} + 4c\frac{dy}{dt} + 4ky$$

$$= -m\frac{d^2}{dt^2}\left[y + r\cos\omega t - \frac{r^2}{4l}(1 - \cos 2\omega t)\right]$$

which reduces to

$$M\frac{d^2y}{dt^2} + 4c\frac{dy}{dt} + 4ky = mr\omega^2 \cos\omega t + m\frac{r^2}{l}\omega^2 \cos 2\omega t$$

Since the system equation has two forcing functions, the total steady-state solution will be found by superposition. Although the form of the equation is such that Fig. 7-9 is applicable to the first forcing function and could be made applicable to the second with some manipulation, we will use Fig. 7-5 in our solution.

The two forcing functions are cosine terms, whereas the forcing function of Eq. 7-5, upon which Fig. 7-5 is based, is a sine term. This causes no problem in using Fig. 7-5, however, since the equations and curves could just as readily have been developed on the basis of a cosine forcing function, with the same results in terms of amplitude ratio and phase angle. Or, alternatively, θ could have been defined for this problem in a manner resulting in sine forcing functions.

With the given system parameters, the equation is

$$\frac{150}{386}\frac{d^2y}{dt^2} + 4(3.25)\frac{dy}{dt} + 4(680)y = \frac{(1)(1.5)(41.85^2)}{386}\cos 41.85t$$

$$+ \frac{(1)(1.5^2)(41.85^2)}{(386)(4)}\cos 83.7t$$

By setting the coefficient of the zeroth derivative equal to 1 to match the general form of Eq. 7-5, this becomes

$$0.000143\frac{d^2y}{dt^2} + 0.00478\frac{dy}{dt} + y = 0.00250 \cos 41.85t$$

$$+ 0.000938 \cos 83.7t$$

The solution will have the form

$$y = Y_1 \cos(41.85t + \phi_1) + Y_2 \cos(83.7t + \phi_2)$$

with Y_1, Y_2, ϕ_1, and ϕ_2 to be determined by the use of Fig. 7-5 (or, if greater accuracy is required, from Eqs. 7-7 and 7-8). Solving for the undamped natural frequency and damping ratio,

$$\frac{1}{\omega_n^2} = 0.000143$$

$$\omega_n = 83.7 \text{ rad/s}$$

$$\frac{2\zeta}{\omega_n} = 0.00478$$

$$\zeta = 0.20$$

For the first forcing function,

$$\frac{\omega}{\omega_n} = \frac{41.8}{83.7} = 0.50$$

The amplitude ratio and phase angle are obtained from the curve as follows:

$$\frac{Y_1}{K_1} = 1.3$$

$$Y_1 = 1.3 K_1 = 1.3(0.00250) = 0.00325 \text{ in.}$$

$$\phi_1 = -15° = -0.262 \text{ rad}$$

For the second forcing function (noting that the frequency of excitation is twice that of the compressor)

$$\frac{\omega}{\omega_n} = \frac{83.7}{83.7} = 1.0$$

The amplitude ratio and phase angle are in this case

$$\frac{Y_2}{K_1} = 2.5$$

$$Y_2 = 2.5 K_1 = 2.5(0.000938) = 0.00235 \text{ in.}$$

$$\phi_2 = -90° = -1.571 \text{ rad}$$

The total system response, by superposition, is therefore

$$y = 0.00325 \cos(41.8t - 0.262) + 0.00235 \cos(83.7t - 1.571) \quad (Ans.)$$

The response of the system and the two components of which it consists are illustrated in Fig. 7-15. For this particular example, the vibration amplitude caused by the first harmonic is about 72 percent of that caused by the fundamental frequency.

7-8 Conclusion

Forced vibration is that occurring in a system from a continuing periodic input that may be in the form of either a force or a displacement. Virtually all vibration problems in machinery lie in the category of forced vibrations.

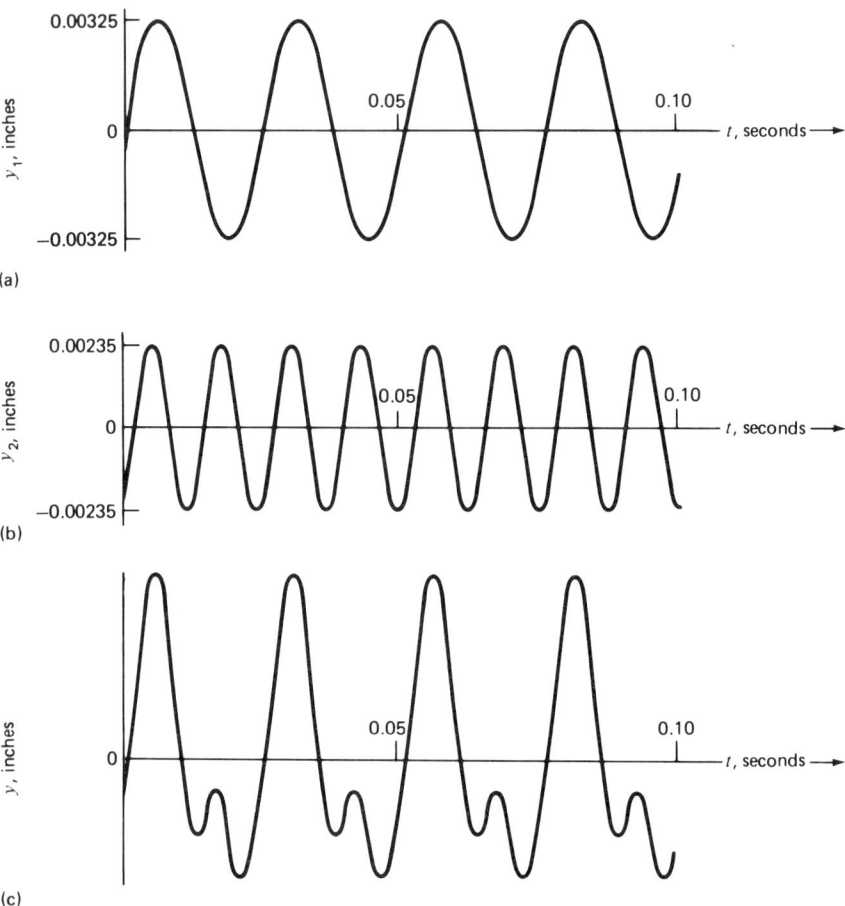

FIG 7-15. Vibration characteristics of the piston compressor system of Example 7-3.
(a) Component of vibration at the fundamental frequency of the compressor.
(b) Component of vibration at the first harmonic of the compressor.
(c) Total response of system, which is the algebraic sum of (a) and (b).

7-8 CONCLUSION

In this chapter the responses of single-degree-of-freedom vibrating systems to several different types of harmonic inputs have been investigated. In solving the differential equations of motion for the responses we have included only the steady-state solutions (the particular integrals). The transient (complementary) solution has been omitted in each case since it will die out rather quickly in any real system with damping.

Response characteristics of the systems investigated have been presented not only in equation form but also as plots of amplitude ratio and phase angle as functions of frequency. These are families of curves for each type of input and have been presented in dimensionless form for maximum utility. The use of these curves in the study of forced vibrations can save much labor. They can be used for both translational and rotational vibrating systems as well as for a wide range of other second-order dynamic systems.

Large output amplitudes may occur in forced vibration when the input frequency is at or near the resonance frequency of the system. For systems with little damping, resonance occurs at $\omega/\omega_n \approx 1$, but it takes place at a somewhat lower frequency for systems with an appreciable amount of damping, as can be seen in Figs. 7-5 and 7-10. If there is a sufficiently high damping ratio (the exact value depending upon the type of input), then a condition of resonance will not occur at any speed.

In order to avoid forced vibration problems in the design or construction of mechanical equipment, it is important to make sure that no periodic inputs occur near the system's resonance frequency. One should avoid attaching a device to a vibrating machine if the frequency of vibration is near the resonance frequency of that device. This is usually a straightforward precaution if operation occurs at one speed only. If the speed of operation (and therefore the frequency of vibration) varies, as is true, for example, with an internal combustion engine in a vehicle, the resonance frequency of any attached device should lie outside the total range of vibration frequencies that will occur.

The frequency of vibration of a machine is normally equal to its speed of operation. It is important to realize, however, that significant vibrations will often occur at frequencies that are *harmonics* of the speed of operation. Failure to consider harmonics will sometimes result in resonance problems.

A machine with an unbalanced rotating component or a reciprocating component will often transmit undesirable vibration forces to the floor (or other surrounding structure). The magnitude of the transmitted forces can be greatly reduced by mounting the machine on springs and dashpots. Such an isolation system must be properly designed, however, or else it might actually make the situation worse.

Problems

7-1. A system is excited by the given harmonic input force. Assuming steady-state vibration,
(a) determine the peak-to-peak amplitude of the mass.
(b) determine y and dy/dt for the instant of time at which the input force F is maximum and positive.

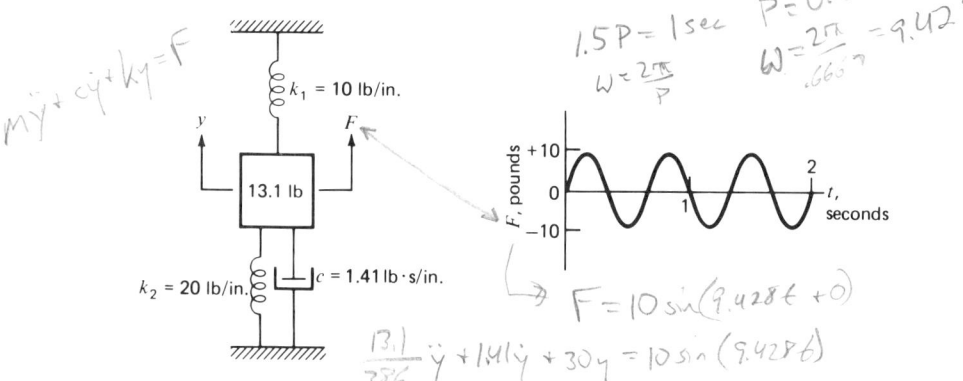

7-2. The system shown is subjected to a sinusoidal input at the end of the torsion bar with a maximum simple amplitude of 4°.
(a) If at an input frequency of $600/\pi$ cycles per minute the peak-to-peak amplitude of the disk is observed to be 1.6°, determine the damping coefficient b.
(b) What will be the maximum angular velocity of the disk (i.e., $d\theta_o/dt$) if the input frequency is changed to $300/\pi$ cycles per minute?
(c) Determine the peak value of angular twist between the two ends of the shaft at $300/\pi$ cycles per minute.

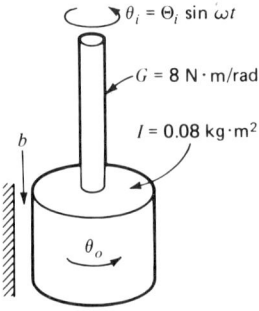

PROBLEMS

7-3. The total weight of the machine illustrated is 100 lb. It is supported by four springs and four dashpots as shown. The motor operates between 89 and 178 Hz. Determine the maximum permissible motor unbalance if the peak-to-peak amplitude is not to exceed 0.010 in. Assume that $k = 25$ lb/in. for each spring and $c = 0.20$ lb · s/in. for each dashpot.

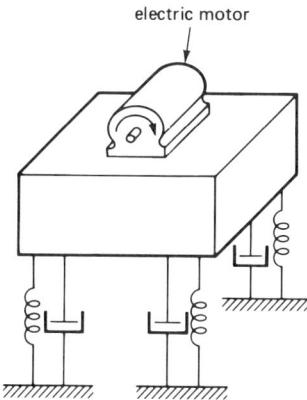

7-4. The system is excited by the given harmonic motion.
(a) Write the differential equation that describes the system.
(b) Determine the steady-state peak-to-peak amplitude of θ_o.
(c) What would be the ratio Θ_o/Θ_i if the input frequency were very low?

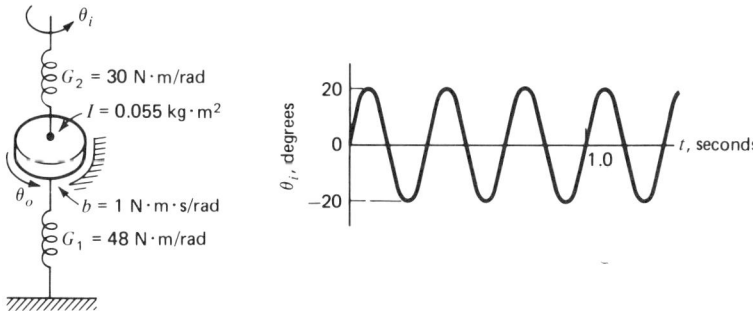

7-5. Write the equation of motion for the system in the figure, where x is the input and y the response. Determine the system response.

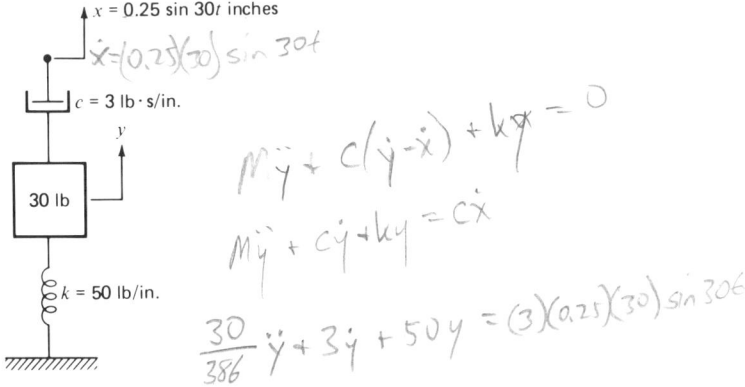

$$m\ddot{y} + c(\dot{y}-\dot{x}) + ky = 0$$

$$m\ddot{y} + c\dot{y} + ky = c\dot{x}$$

$$\frac{30}{386}\ddot{y} + 3\dot{y} + 50y = (3)(0.25)(30)\sin 30t$$

7-6. The system shown has two inputs, x_1 and x_2. Write the equation of motion and determine the response y. Use the principle of superposition.

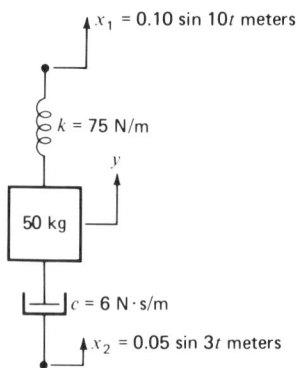

$c(\dot{x}-\dot{y}) = -c(\dot{y}-\dot{x})$

7.5

$-ky$

$\Sigma F = m\ddot{y} = -c(\dot{y}-\dot{x})$
$\qquad -ky$

$\boxed{m\ddot{y} + c\dot{y} + ky = c\dot{x}}$

7-7. Assume that the beam shown is made of steel ($E = 30 \times 10^6$ lb/in.^{2}) and has negligible mass compared to the lumped mass M.
  (a) Determine the undamped natural frequency ω_n.
  (b) If the input is $F = 30 \sin 20t$ pounds, the steady-state vibration of the mass has the form
  $$y_{ss} = Y \sin(\omega t + \phi)$$
  Determine Y, ω, and ϕ.

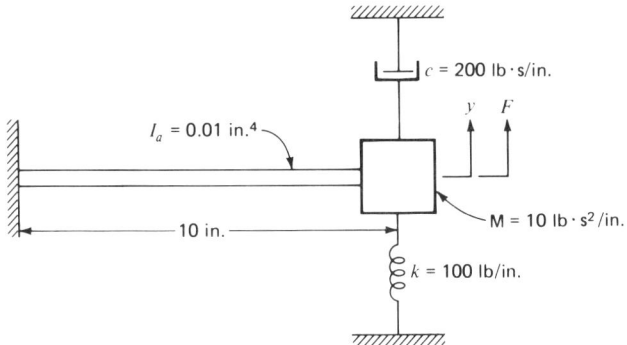

7-8. (a) Determine the oscillation amplitude and frequency of the mass in the figure between $t = t_1$ and $t = t_2$. State any assumptions necessary.
  (b) What is the oscillation frequency from $t = t_2$ on?
  (c) Sketch the vibration of the mass (y vs. t) from $t = t_1$ until the movement ceases.

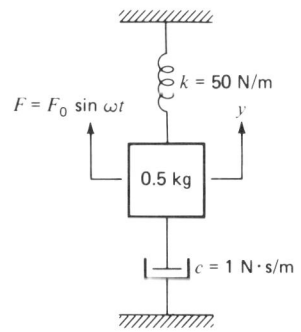

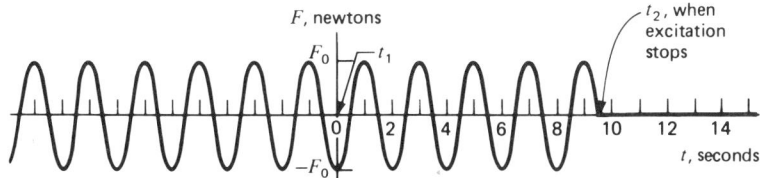

7-9. A sinusoidal vertical motion of 2 inches peak-to-peak amplitude is applied at different frequencies to the top of the spring illustrated. Determine the maximum peak-to-peak amplitude (inches) that the mass could have, and the frequency (hertz) at which this would occur.

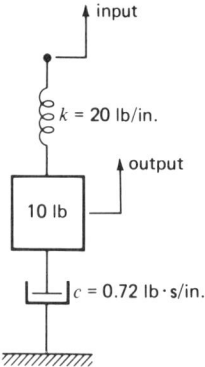

7-10. The input to the system shown is $0.10 \sin 30t$ (radians).
  (a) Determine the value of b that limits the output amplitude to 0.16 rad peak-to-peak. (Give units.)
  (b) Determine the peak *angular velocity* of the disk with the given input and your calculated value of b.

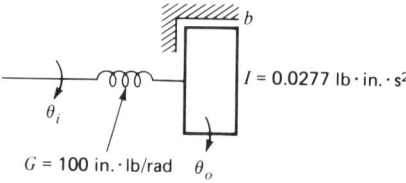

7-11. The system shown is excited by the given harmonic motion.
  (a) Write the differential equation that describes the system.
  (b) Determine the steady-state peak-to-peak amplitude of the mass.
  (c) Determine the maximum stretch in the springs over the cycle.
  (d) What would be the amplitude ratio Y/X if the input frequency were very low?

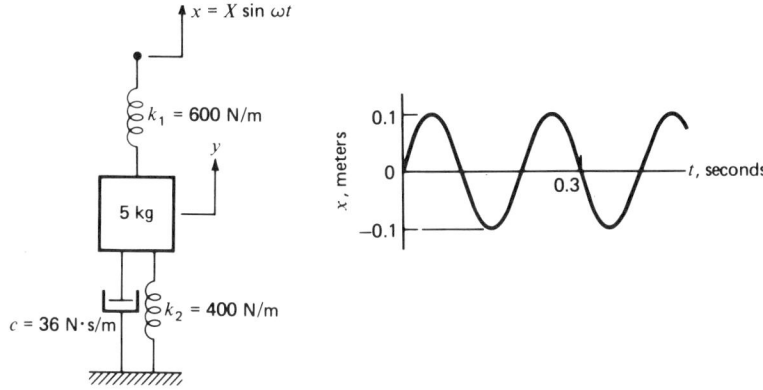

7-12. The machine in the illustration has a total weight of 10 lb, with an unbalance weight of 1.0 lb rotating at a radius of 2.0 in. Operation occurs with the unbalance weight rotating at speeds between 18 and 60 rad/s. Determine the minimum value of the damping coefficient c needed to keep the peak-to-peak amplitude of vibration below 0.60 in.

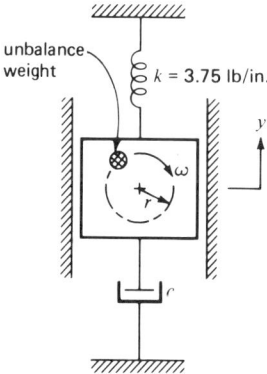

7-13. A mass supported by a spring and dashpot contains a crank-piston machine as shown. The crankshaft is balanced and the weight of the connecting rod can be considered negligible. Determine the response of the mass for three different speeds of the crank-piston machine: (a) 100 r/min; (b) 500 r/min; (c) 1000 r/min. Consider both the fundamental frequency of excitation and the first harmonic.

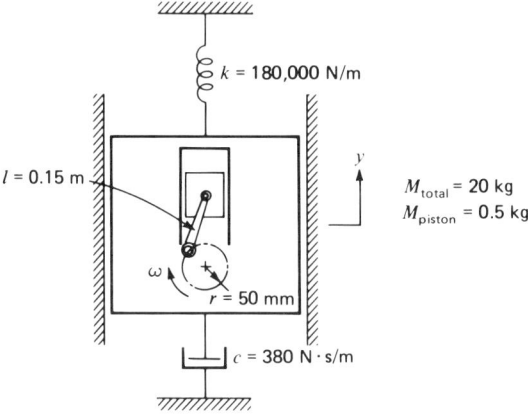

7-14. Using the loop analysis technique, write the differential equation describing each electric circuit shown. Determine for steady-state operation the amplitude of the harmonically varying current and the phase angle of the current with respect to the input voltage. Assume for part (a) that transients will die out.

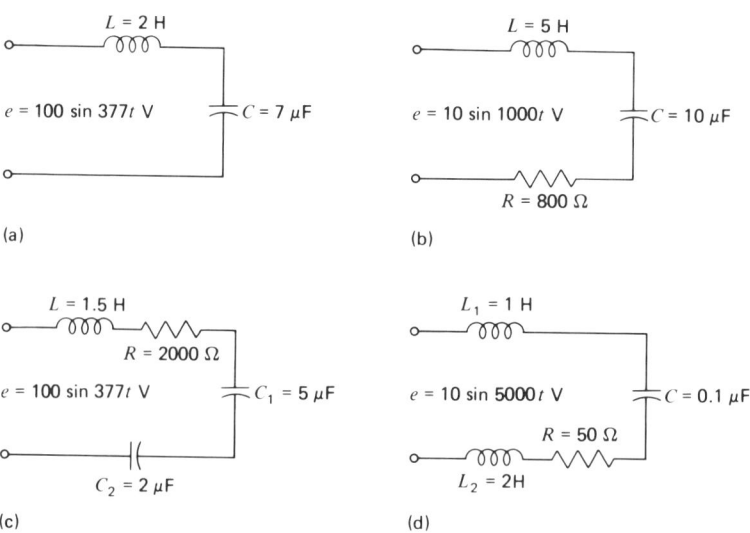

7-15. For the electric circuits of Prob. 7-14, determine the maximum voltage drop occurring each cycle across each of the elements.

7-16. Using the node analysis technique obtain the differential equation describing each electric circuit shown. Determine the amplitude of the harmonically varying voltage e_1, and the phase angle of e_1 with respect to the input voltage e.

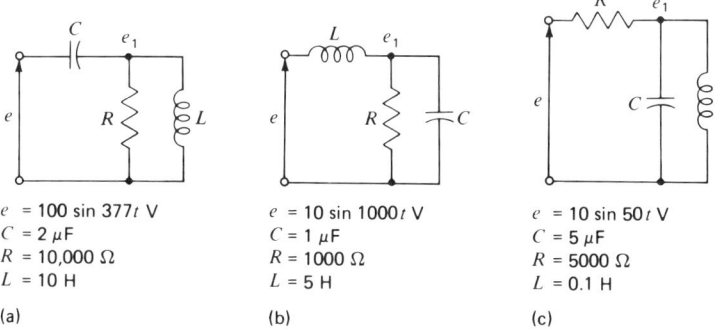

(a)
$e = 100 \sin 377t$ V
$C = 2\ \mu F$
$R = 10,000\ \Omega$
$L = 10$ H

(b)
$e = 10 \sin 1000t$ V
$C = 1\ \mu F$
$R = 1000\ \Omega$
$L = 5$ H

(c)
$e = 10 \sin 50t$ V
$C = 5\ \mu F$
$R = 5000\ \Omega$
$L = 0.1$ H

7-17. For the electric circuits of Prob. 7-16, determine the maximum current flowing each cycle through each of the elements.

7-18. (a) For the given two-tank liquid-level system with linear outlet valves, determine ω_n, ω_d, and ζ. Consider δh_2 as the output.

(b) If the input flow varies harmonically as represented by the equation

$$q_i = 20 + 3 \sin \omega t\ (\text{m}^3/\text{s})$$

determine the minimum permissible value of ω if h_2 is not to vary more than ± 0.25 meter.

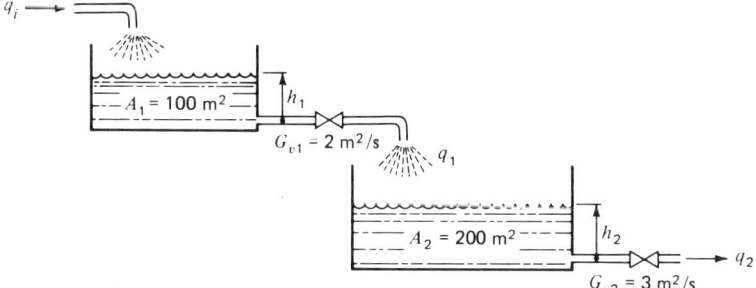

7-19. Repeat Prob. 7-18 for the case in which the two outlet valves are replaced by nonlinear ones having the flow-head characteristics $q_1 = 10\sqrt{h_1}$ and $q_2 = 15\sqrt{h_2}$, with q in m³/s and h in meters. The other system parameters and the input q_i remain the same.

7-20. For the given system with steady-state vibration, determine the peak-to-peak amplitude of the mass and the phase angle of the output y with respect to the input x.

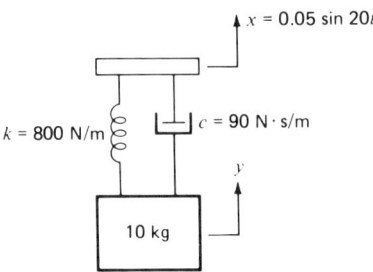

7-21. For the rotational system with the given input, determine the allowable range of values for the damping coefficient b if the output amplitude is not to exceed ± 0.05 rad.

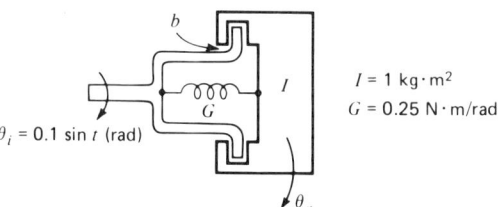

7-22. Write the differential equation for the given system. Determine the required spring constant k if the amplitude ratio (Y/X) is to be independent of the damping coefficient.

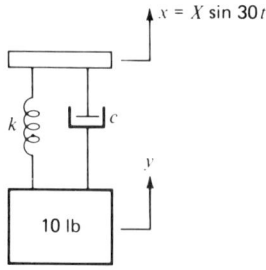

7-23. Write the differential equation for the given system, and determine for steady-state vibration the amplitude ratio and phase angle of θ_o with respect to θ_i. Assume the following values: $I = 2$ kg · m², $G = 10$ N · m/rad, and $b = 2$ N · m · s/rad.

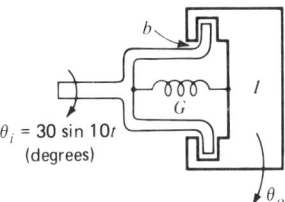

7-24. A machine weighing a total of 20 lb is mounted on four identical springs and dashpots as shown to reduce the vibration transmitted to the floor. The dc electric motor has an unbalance of 10 oz · in. and a speed that varies from 0 to 5000 r/min. Assume that $k = 500$ lb/in. for each spring.
  (a) Determine the required damping coefficients if the transmissibility is not to exceed 2.25 for any motor speed.
  (b) With the damping found in (a), determine the peak vibration force transmitted to the floor for $\omega/\omega_n = 1$, $\sqrt{2}$, 2, and 4.

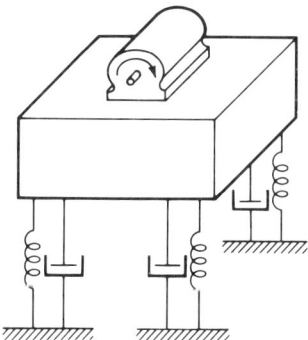

7-25. A refrigeration machine with a total mass of 900 kg is mounted on four springs, with no damping. The constant-speed compressor gives a harmonically varying unbalance force, in newtons, that can be represented by the equation $F = 200 \sin 160t$. Determine the proper spring constant (or an acceptable range of spring constants) if the peak vibration force transmitted to the floor is not to exceed 40 newtons.

More Complex Single-Degree-of-Freedom Systems

8-1 Introduction

In previous chapters a number of different types of dynamic systems were described and modeled mathematically. These were all relatively simple single-degree-of-freedom systems that could be represented by a single differential equation. There are other systems, however, that appear to be significantly more complex but are still single-degree-of-freedom systems that can be modeled by the same general equations given in Chapter 7.

8-2 Determining the Degrees of Freedom of a Mechanical System

For the two-mass system of Fig. 8-1, knowledge of y_1 and its derivatives will give us only part of the answer; it is necessary to also know about y_2 in order to understand the vibration characteristics of the system fully. Since

8-2 DETERMINING THE DEGREES OF FREEDOM OF A MECHANICAL SYSTEM

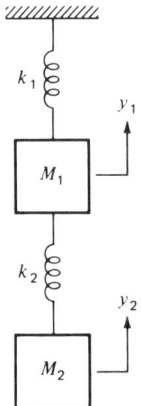

FIG 8-1. A two-mass system having two degrees of freedom.

there is no direct kinematic relationship between y_1 and y_2, the system has two degrees of freedom.

Figure 8-2 illustrates another type of two-mass system. A little thought should convince the reader that this system is in a different category than that of Fig. 8-1. Although it is necessary to know both y_1 and y_2 to understand how the system is behaving, the mechanical linkage gives a direct kinematic relationship between these two variables, and thus knowledge of either one allows the other to be directly determined. This particular system therefore has only a single degree of freedom.

Many vibrating systems contain rotational and translational components coupled together. Such systems may have a single degree of freedom

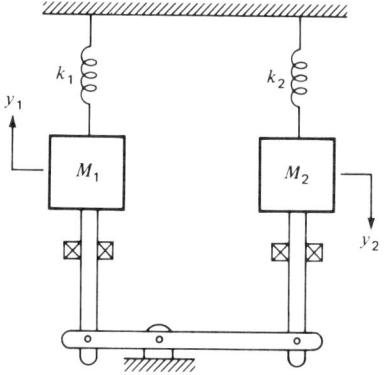

FIG 8-2. A two-mass system having a single degree of freedom.

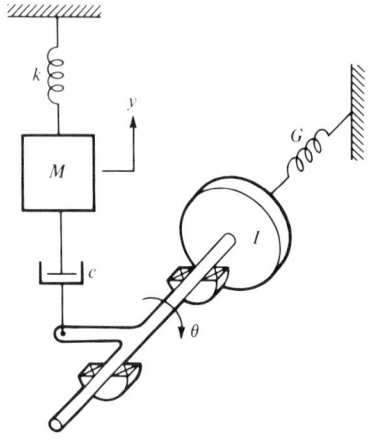

FIG 8-3. A combination rotational and translational vibrating system having two degrees of freedom.

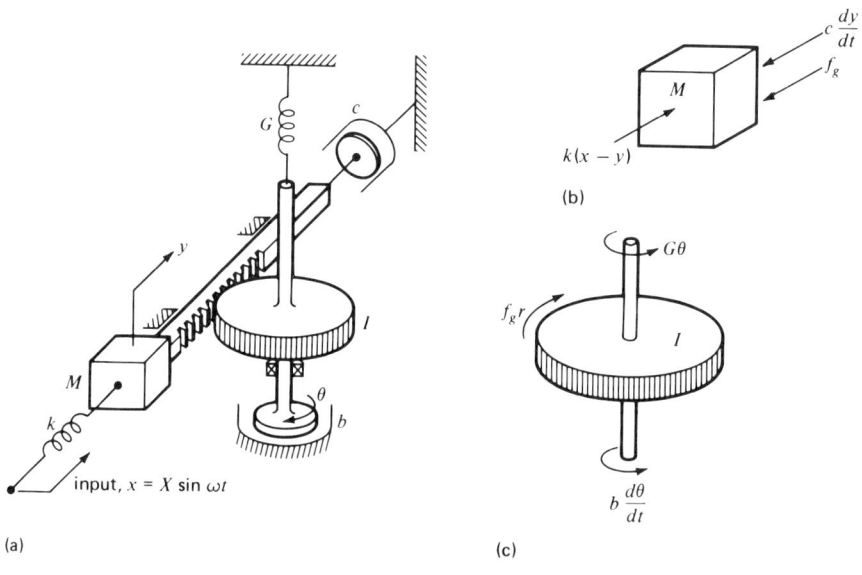

FIG 8-4. (a) A combination rotational and translational vibrating system having a single degree of freedom.
(b) Modified free-body diagram of translating mass, showing all force components acting in the y direction.
(c) Modified free-body diagram of rotational subsystem, showing all torques acting about the axis of rotation.

or contain two or more degrees of freedom, depending upon the exact manner in which they are coupled. The system of Fig. 8-3 is a two-degree-of-freedom system, whereas the somewhat similar system illustrated in Fig. 8-4a has only a single degree of freedom. The difference between these two systems is in the manner of coupling between the rotating flywheel and the translating mass. In Fig. 8-3 there is no positive type of linkage between the two inertias since they are joined through a dashpot. In Fig. 8-4a, on the other hand, the mass and flywheel are coupled through a rack and pinion, so the angular position of the flywheel has a specific kinematic relationship to the position of the mass regardless of the other details of vibratory motion.

8-3 Method of Analysis

If, in studying the motion of a system like that of Fig. 8-4a, we focus our attention upon the translational components, it is often convenient to think of the rotational inertia I as having the same effect on the system as if the mass M were increased by some certain value. Likewise, a proper increase in the spring constant k would allow for the effect of G, and an equivalent translational damping coefficient could include the effect of b. Conversely, the system might be considered to be equivalent in vibration characteristics to a purely rotational system with a single moment of inertia, a single torsional spring, and a single torsional damper, the three components having the proper equivalent values.

The way which is probably best for analyzing such a system—that is, for developing and solving its equation of motion—can be summarized by the following steps:

1. Consider individually each part of the system that can be considered as a rotational or translational subsystem and to which Newton's second law of motion can be applied in a straightforward manner. The equation of motion should be written for each of these subsystems with the aid of *modified free-body diagrams* (diagrams showing only those force components or torque components acting on the inertia elements that will affect their vibratory motion).
2. Combine the equations into a single equation of motion for the total system by using the known relationships existing at each interface.
3. Manipulate the resulting equation, if possible, into one of the general forms given in Chapter 7. This procedure will facilitate the evaluation of the system's dynamic characteristics by allowing the use of the corresponding nondimensional frequency-response diagrams or equations.

MORE COMPLEX SINGLE-DEGREE-OF-FREEDOM SYSTEMS

The following examples will illustrate the application of the foregoing steps to single-degree-of-freedom systems containing two or more subsystems.

Example 8-1.

(a) Develop the differential equation describing the vibratory motion of the system of Fig. 8-4a, putting it in a form that will facilitate the use of the frequency-response curves developed in Chapter 7 to study the system's characteristics.

(b) Explain how the maximum value of the tangential gear force (F_g) could be obtained for any given set of system parameters and input.

Solution:

(a) The system can be seen to consist of two subsystems: (1) a translating mass acted upon by a spring force, dashpot force, and a force at the gear-rack interface; (2) a rotating flywheel acted upon by a torsional spring, a rotary dashpot, and the torque produced by the gear-rack interface force.

By applying Newton's second law of motion to the mass M, with the aid of the modified free-body diagram of Fig. 8-4b, we obtain the equation

$$M\frac{d^2y}{dt^2} + c\frac{dy}{dt} + k(y - x) = -f_g \tag{8-1}$$

where f_g is the tangential component of the force between the gear and pinion at the pitch point. (Since the mass is constrained to move only in the y direction, we can ignore force components that do not act in this direction. By the same token, since the axis of I is fixed by bearings, only the force components on the gear that causes a torque about its axis—that is, the tangential component f_g—need be considered when analyzing the rotational subsystem.) It should be noted that f_g will vary with time in some manner that is not readily apparent.

Newton's second law applied to the rotational subsystem (with the aid of the modified free-body diagram of Fig. 8-4c) results in the equation

$$I\frac{d^2\theta}{dt^2} + b\frac{d\theta}{dt} + G\theta = f_g r \tag{8-2}$$

Note that the tangential force f_g acting on the gear is the reaction of that acting on the rack, and therefore it has the same magnitude but acts in the opposite direction.

8-3 METHOD OF ANALYSIS

The two subsystem equations can be combined by the elimination of f_g to obtain the single equation describing the total system, which is

$$M\frac{d^2y}{dt^2} + c\frac{dy}{dt} + k(y - x) + \frac{I}{r}\frac{d^2\theta}{dt^2} + \frac{b}{r}\frac{d\theta}{dt} + \frac{G}{r}\theta = 0 \quad (8\text{-}3)$$

The equation should now be written in terms of a single coordinate, which can be either y or θ depending upon which form is most convenient for the particular analysis being done. The relationship between y and θ is purely kinematic, being simply

$$y = r\theta \quad (8\text{-}4)$$

if we make the logical choice of defining our coordinates so that $y = 0$ when $\theta = 0$. It follows that the first and second derivatives are related by the equations

$$\frac{dy}{dt} = r\frac{d\theta}{dt} \quad (8\text{-}5)$$

$$\frac{d^2y}{dt^2} = r\frac{d^2\theta}{dt^2} \quad (8\text{-}6)$$

Assuming that we are somewhat more interested in y than θ, the equation of motion can now, by combining Eqs. 8-3, 8-4, 8-5, and 8-6, be written as

$$\frac{Mr^2 + I}{kr^2 + G}\frac{d^2y}{dt^2} + \frac{cr^2 + b}{kr^2 + G}\frac{dy}{dt} + y = \frac{kr^2 X}{kr^2 + G}\sin \omega t \quad (8\text{-}7)$$

(*Ans.*)

This is the equation of a single-degree-of-freedom vibrating system. It has been put into a form that matches that of Eq. 7-5. The frequency-response curves of Fig. 7-5 can therefore be directly applied to determine its vibration characteristics for any combination of system parameters. Note that the undamped natural frequency is

$$\omega_n = \sqrt{\frac{kr^2 + G}{Mr^2 + I}} \quad (8\text{-}8)$$

(b) The gear force f_g has disappeared in the development of the overall equation of motion. It may sometimes be important in systems of this type to know force f_g for stress considerations. This force can be readily determined once we know the vibration amplitude for a given input and a given set of system parameters.

With a sinusoidal input, the total system (which includes the rotational subsystem) vibrates with harmonic motion; that is,

$$\theta = \Theta \sin(\omega t + \phi) \quad (8\text{-}9)$$

Considering the rotational subsystem alone, the single input to it ($f_g r$) must be harmonic in order for Eq. 8-9 to be valid. The *subsystem* equation of motion (Eq. 8-2) may therefore be written[1]

$$I\frac{d^2\theta}{dt^2} + b\frac{d\theta}{dt} + G\theta = F_g r \sin \omega t \qquad (8\text{-}10)$$

where F_g is the *peak* value of the tangential gear force f_g. Manipulation of Eq. 8-10 to the general form of Eq. 7-5 produces

$$\frac{I}{G}\frac{d^2\theta}{dt^2} + \frac{b}{G}\frac{d\theta}{dt} + \theta = \frac{F_g r}{G} \sin \omega t \qquad (8\text{-}11)$$

The natural frequency and damping ratio of the rotational subsystem are

$$\omega_n = \sqrt{\frac{G}{I}} \quad \text{and} \quad \zeta = \frac{b}{2\sqrt{IG}}$$

For a given input frequency, ω/ω_n is therefore available, and the amplitude ratio (equal to $\Theta/(F_g r/G)$ in this case) may be obtained directly from the frequency-response curves of Fig. 7-5. If Θ is also known, the peak value of the tangential gear force, F_g, is readily determined. (Θ can be obtained by considering the total system, as outlined in part (a) of the example, using Eqs. 8-4 and 8-7).

Example 8-2. An electric motor is used to cause an antenna dish to scan back and forth, covering a given angle of view. The drive transmission consists of a two-stage gear train, as illustrated in Fig. 8-5, with gear diameters as shown. The moments of inertia are

$I_1 = 4.0$ lb $\cdot$ in.$\cdot$s^2 (motor armature and pinion)

$I_2 = 6.0$ lb $\cdot$ in.$\cdot$s^2 (countershaft with two gears)

$I_3 = 8000$ lb $\cdot$ in.$\cdot$s^2 (antenna and drive gear)

A helical torsional spring with stiffness $G = 1000$ in.$\cdot$lb/rad is placed between the antenna and ground as shown to ensure that the antenna will always point in a particular direction when the electric motor is turned off. The three shafts are much stiffer torsionally than the helical torsional spring. Damping is present in each of the three shafts, but the values of the damping coefficients cannot be readily determined by calculation.

---

[1]To be strictly precise, there should be a phase angle in the forcing function of Eq. 8-10, since $x(t)$ and $f_g(t)$ are not normally in phase. This phase angle can be ignored, however, since we are interested only in the maximum value of the gear force.

8-3 METHOD OF ANALYSIS

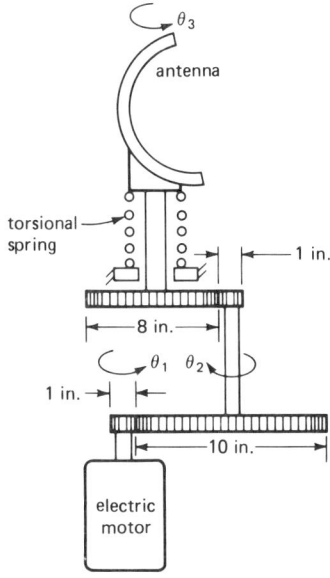

FIG 8-5. An antenna drive system.

(a) Write the differential equation describing the system (with the electric motor turned off), and determine the undamped natural frequency of rotational vibration.

(b) If it were found desirable to increase the undamped natural frequency to a value of 0.20 rad/s, would it be feasible to do this by removing weight from the antenna structure?

(c) If the system were found to have too little damping and we wanted to increase the damping ratio a given amount with the smallest possible size of rotary dashpot, on which shaft should it be placed?

Solution:

(a) The vibrating system is composed of three rotational subsystems connected through gear interfaces. With the aid of the three modified free-body diagrams given in Fig. 8-6, we can write the three subsystem equations of motion as

$$I_1 \frac{d^2\theta_1}{dt^2} + b_1 \frac{d\theta_1}{dt} = -f_{g_1} r_1 \tag{8-12}$$

$$I_2 \frac{d^2\theta_2}{dt^2} + b_2 \frac{d\theta_2}{dt} = f_{g_1} r_2 - f_{g_2} r_3 \tag{8-13}$$

$$I_3 \frac{d^2\theta_3}{dt^2} + b_3 \frac{d\theta_3}{dt} + G\theta_3 = f_{g_2} r_4 \tag{8-14}$$

In writing these equations, the assumed directions of the gear forces are somewhat arbitrary except that the action and reaction at each of the two gear interfaces must be shown to act in opposite directions.

Equations 8-12 and 8-14 are manipulated into the forms

$$\frac{r_2}{r_1} I_1 \frac{d^2\theta_1}{dt^2} + \frac{r_2}{r_1} b_1 \frac{d\theta_1}{dt} = -f_{g_1} r_2 \quad (8\text{-}15)$$

$$\frac{r_3}{r_4} I_3 \frac{d^2\theta_3}{dt^2} + \frac{r_3}{r_4} b_3 \frac{d\theta_3}{dt} + \frac{r_3}{r_4} G\theta_3 = f_{g_2} r_3 \quad (8\text{-}16)$$

Equations 8-15 and 8-16 are now substituted into Eq. 8-13 to obtain the single system equation

$$I_2 \frac{d^2\theta_2}{dt^2} + b_2 \frac{d\theta_2}{dt} + \frac{r_2}{r_1} I_1 \frac{d^2\theta_1}{dt^2} + \frac{r_2}{r_1} b_1 \frac{d\theta_1}{dt} + \frac{r_3}{r_4} I_3 \frac{d^2\theta_3}{dt^2}$$
$$+ \frac{r_3}{r_4} b_3 \frac{d\theta_3}{dt} + \frac{r_3}{r_4} G\theta_3 = 0 \quad (8\text{-}17)$$

At this stage of development the differential equation still has three dependent variables: θ_1, θ_2, and θ_3. To write the equation in terms of a single dependent variable it is necessary to use the kinematic relationships pertaining to the gear train, which are

$$\theta_1 = \frac{r_4 r_2}{r_3 r_1} \theta_3 \quad (8\text{-}18)$$

$$\theta_2 = \frac{r_4}{r_3} \theta_3 \quad (8\text{-}19)$$

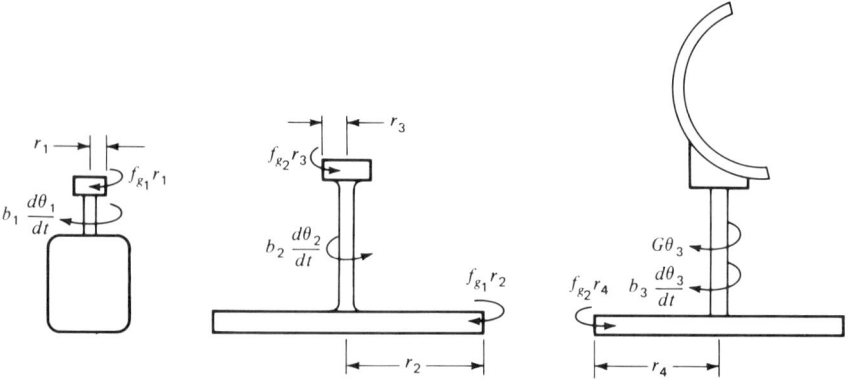

FIG 8-6. Modified free-body diagrams for the components of the antenna drive system.

8-3 METHOD OF ANALYSIS

and the corresponding derivative relationships

$$\frac{d\theta_1}{dt} = \frac{r_4 r_2}{r_3 r_1} \frac{d\theta_3}{dt} \qquad \frac{d^2\theta_1}{dt^2} = \frac{r_4 r_2}{r_3 r_1} \frac{d^2\theta_3}{dt^2} \qquad (8\text{-}20)$$

$$\frac{d\theta_2}{dt} = \frac{r_4}{r_3} \frac{d\theta_3}{dt} \qquad \frac{d^2\theta_2}{dt^2} = \frac{r_4}{r_3} \frac{d^2\theta_3}{dt^2} \qquad (8\text{-}21)$$

Using the above, the system equation can be put into the form

$$\frac{I_3 + \left(\frac{r_4}{r_3}\right)^2 I_2 + \left(\frac{r_2 r_4}{r_1 r_3}\right)^2 I_1}{G} \frac{d^2\theta_3}{dt^2}$$

$$+ \frac{b_3 + \left(\frac{r_4}{r_3}\right)^2 b_2 + \left(\frac{r_2 r_4}{r_1 r_3}\right)^2 b_1}{G} \frac{d\theta_3}{dt} + \theta_3 = 0 \qquad (8\text{-}22)$$

Since Eq. 8-22 has the same form as Eq. 7-5, we know that the undamped natural frequency is

$$\omega_n = \sqrt{\frac{G}{I_3 + (r_4/r_3)^2 I_2 + (r_2 r_4/r_1 r_3)^2 I_1}} \qquad (8\text{-}23)$$

Substituting the known values of the moments of inertia and the torsional spring rate into Eq. 8-23, we obtain

$$\omega_n = \sqrt{\frac{1000}{8000 + (8^2)(6) + (80^2)(4)}} = \sqrt{\frac{1000}{33{,}984}}$$

$$= 0.1716 \text{ rad/s} \qquad (Ans.)$$

(b) It is not possible to increase ω_n to 0.20 rad/s merely by removing weight from the antenna. Even if the moment of inertia of the antenna were reduced to zero, the natural frequency of the system would be

$$\omega_n = \sqrt{\frac{1000}{25{,}984}} = 0.196 \text{ rad/s}$$

in the absence of any other changes. Note that this is true even though the moment of inertia of the motor and the intermediate shaft are only a small fraction of that of the antenna.

(c) Comparing the equation of motion of the system (Eq. 8-22) with the general form of Eq. 7-5, it is seen that

$$\frac{2\zeta}{\omega_n} = \frac{b_3 + (r_4/r_3)^2 b_2 + (r_2 r_4/r_1 r_3)^2 b_1}{G} \qquad (8\text{-}24)$$

With the given gear ratios, we find that

$$\left(\frac{r_4}{r_3}\right)^2 = 64$$

and

$$\left(\frac{r_2 r_4}{r_1 r_3}\right)^2 = 6400$$

It is therefore clear that a given rotary dashpot would have the most effect in increasing ζ if it were placed on the motor shaft, with its damping coefficient appearing as b_1 in Eq. 8-24.

8-4 Equivalent Inertia, Damping, and Spring Rate

Example 8-2 illustrates several significant points that warrant further discussion. First of all, the relative significance of the moment of inertia of each of the three rotational subsystems should be noted. If we relate them all to the antenna, then we find, for example, that the increase in the effective moment of inertia of the antenna due to the motor armature is equal to the moment of inertia of the armature times the *square of the total gear reduction between the two*. In this particular example the inertia of the motor armature predominates even though it is only 0.0005 that of the antenna. If we were to relate everything to the motor shaft, then the component of the effective moment of inertia there due to the antenna would equal I_3 *divided* by the square of the total gear ratio between the two. It is apparent that a given moment of inertia is more significant if it is on a faster rotating shaft.

When the dynamic characteristics of a geared rotational system are being analyzed[2] the moment of inertia of any shaft may be replaced by an equivalent moment of inertia at any other shaft in the system. If the speed ratio N of the two shafts is given by the equation

$$\omega_2 = \frac{\omega_1}{N} \qquad (8\text{-}25)$$

---

[2]This discussion also applies, of course, to rotational systems in which the shafts are connected by means of belts or chains.

8-4 EQUIVALENT INERTIA, DAMPING, AND SPRING RATE

the *equivalent* moment of inertia at shaft 1 (I_{equiv}) corresponding to the *true* moment of inertia at shaft 2 (I_2) is

$$I_{equiv} = \frac{I_2}{N^2} \tag{8-26}$$

Similar statements can be made with regard to damping. A given rotary dashpot will produce more damping torque (and therefore be more effective) if it is put on a faster rotating shaft. With the speed relationship of Eq. 8-25, the equivalent damping coefficient at shaft 1 (b_{equiv}) corresponding to the true damping coefficient at shaft 2 (b_2) is

$$b_{equiv} = \frac{b_2}{N^2} \tag{8-27}$$

If the system contains torsional springs, the equivalent spring rate at shaft 1 (G_{equiv}) corresponding to the true rate of the spring(s) attached to shaft 2 (G_2) is

$$G_{equiv} = \frac{G_2}{N^2} \tag{8-28}$$

Example 8-3. The rotary tiller in Fig. 8-7 is driven by an internal combustion engine acting through a belt drive and a worm gear. Assume the following values: for the engine shaft, $I = 0.135$ lb · in.·s² and $D_{pulley} = 3.5$ in.; for the intermediate shaft, $I = 0.047$ lb · in.·s² and $D_{pulley} = 5.0$ in., with a single-thread worm; for the tine shaft, $I = 1.56$ lb · in.·s², with a 22-tooth worm gear.

(a) Determine the effective moment of inertia at the engine shaft.

(b) Determine the expression for the effective damping coefficient at the engine shaft if we label the damping coefficients b_1, b_2, and b_3 for the engine, intermediate, and tine shafts, respectively.

Solution: The speed ratio between the engine and intermediate shafts is

$$N = \frac{\omega_{eng}}{\omega_{int}} = \frac{5.0}{3.5} = 1.429$$

The speed ratio between the engine and tine shafts is

$$N = \frac{\omega_{eng}}{\omega_{int}} \cdot \frac{\omega_{int}}{\omega_{tine}} = \frac{5.0}{3.5} \cdot \frac{22}{1} = 31.43$$

(a) The effective moment of inertia at the engine shaft is therefore (from Eq. 8-26)

$$I_{equiv} = 0.135 + \frac{0.047}{1.429^2} + \frac{1.56}{31.43^2}$$

$$= 0.1596 \text{ lb} \cdot \text{in.} \cdot \text{s}^2 \qquad (Ans.)$$

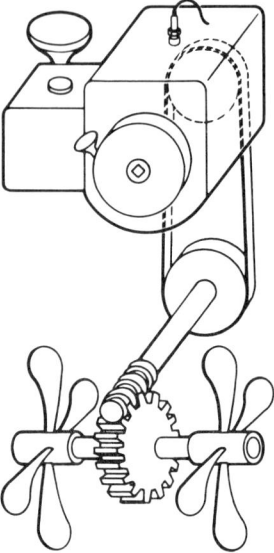

FIG 8-7. A rotary tiller.

(b) The effective damping ratio is obtained in a similar manner:

$$b_{\text{eff}} = b_1 + \frac{b_2}{1.429^2} + \frac{b_3}{31.43^2} \qquad (Ans.)$$

8-5 Conclusion

Mechanical vibrating systems may be very complex, with many components, and still have only a single degree of freedom. Regardless of the complexity, a system will have a single degree of freedom if the motions of all the inertia elements have a direct kinematic relationship so that knowledge of the motion of any one will allow the motion of all others to be directly determined.

A complex single-degree-of-freedom mechanical system will, just as the simple ones, have a single undamped natural frequency ω_n and a single damping ratio ζ. The frequency-response characteristics may be obtained from the dimensionless curves of Chapter 7, and the free-vibration characteristics may be obtained from Fig. 6-6. It is often convenient for analytical purposes to consider an equivalent system in which all components have been lumped into a single inertia, a single damper, and a single spring.

In developing the equations of motion for systems of the type discussed in this chapter it is relatively easy to make mistakes in sign. Such mistakes can be readily spotted in the final equation of motion, however. The inertia terms must all be additive since inertia in any subsystem can only increase the effective inertia term in the equation of the total system. Similarly, all damping terms, which denote a loss of energy, must also be additive. With real springs (having positive spring rates), the spring rate terms will be additive.

For an additional check on the validity of a solution, note that all additive terms must have the same units. In Eq. 8-8, for example, the two spring terms kr^2 and G will each have the units $N \cdot m/rad$ (or $in. \cdot lb/rad$), and the two inertia terms Mr^2 and I will each have the units $kg \cdot m^2$ (or $lb \cdot in. \cdot s^2$).

Problems

8-1. Develop the differential equation of motion for the system of Example 8-1 which has θ as the dependent variable.

8-2. For Example 8-1 determine the maximum gear force F_g that occurs during the vibration, following the procedure outlined in that example. Use the system parameters $M = 0.5$ lb $\cdot$ s²/in., $k = 30$ lb/in., $c = 2$ lb $\cdot$ s/in., $I = 1.0$ lb $\cdot$ in.·s², $G = 46$ in.·lb/rad, $b = 7$ in.·lb · s/rad, $r = 3$ in., $x = 1.5 \sin 7.58t$.

8-3. The differential equation for small motions of the system shown is determined to be (with the given sign convention for y and θ)

$$10\frac{d^2y}{dt^2} + 6y - 16\frac{d^2\theta}{dt^2} - 8\frac{d\theta}{dt} - 4\theta = 0$$

Determine the undamped natural frequency.

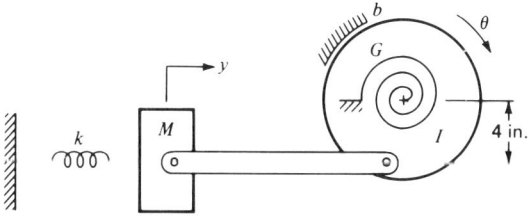

8-4. A differential equation for small motions of the system shown in the figure was derived to be

$$M\frac{d^2y}{dt^2} + c\frac{dy}{dt} + ky + \frac{I}{r}\frac{d^2\theta}{dt^2} + \frac{G}{r^2}y = f(t)$$

Determine the damped and undamped natural frequencies, and the damping ratio. Assume the following values: $M = 50$ kg, $I = 0.6$ kg $\cdot$ m^2, $k = 1200$ N/m, $G = 2$ N $\cdot$ m/rad, $c = 0.25$ N $\cdot$ s/m, and $r = 0.15$ m.

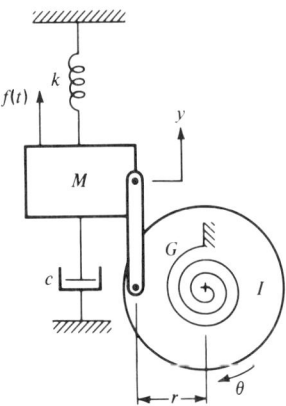

8-5. Write the applicable equation of motion and solve for the undamped natural frequency of the system shown. Assume the following values: $I_1 = 2.0$ lb $\cdot$ in.$\cdot$s^2, $I_2 = 0.25$ lb $\cdot$ in.$\cdot$s^2, $G_1 = 500$ in.$\cdot$lb/rad, $G_2 = 200$ in.$\cdot$lb/rad, $r_1 = 6$ in., and $r_2 = 4$ in.

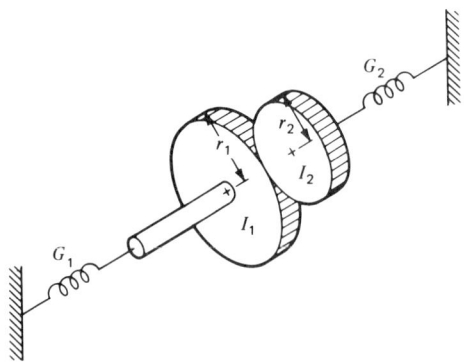

PROBLEMS

8-6. Develop the equation of motion for the system shown, and determine its natural frequency in hertz. Use the following parameters: $I_1 = 0.05$ kg · m², $I_2 = 0$, $I_3 = 0.08$ kg · m², $G_1 = G_2 = G_3 = 10$ N · m/rad, $r_1 = 0.10$ m, $r_2 = 0.03$ m, and $r_3 = 0.12$ m.

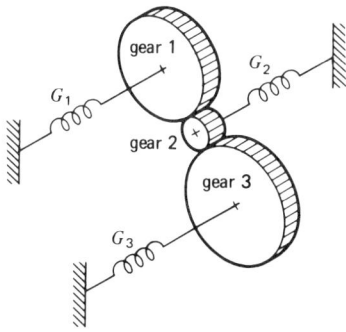

8-7. Write the equation of motion for the given system with y as the dependent variable. Obtain the expression for the undamped natural frequency. What restriction is necessary in order to treat it as a linear system?

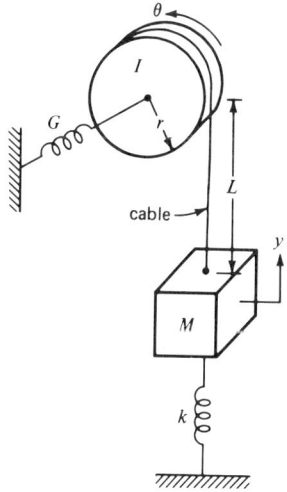

8-8. Determine the undamped natural frequency of the given system in hertz. Assume that the lever rod has negligible mass and is very stiff.

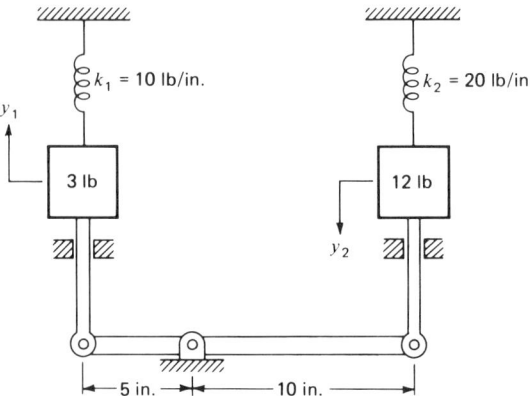

8-9. Develop the differential equation that describes the system illustrated, assuming that the cable is kept under tension at all times. Determine the expression for the undamped natural frequency.

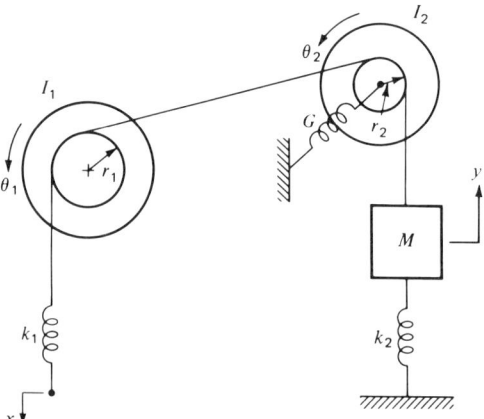

PROBLEMS

8-10. (a) Determine the differential equation of motion for the three-mass translational system shown, with y_1 as the sole dependent variable appearing in the equation. Assume the lever rods to be massless and very stiff.

(b) Develop a set of rules for translational lever systems of this type, similar to those obtained for geared rotational systems (Eqs. 8-26, 8-27, 8-28), that allow the definition of equivalent parameters (mass, spring rate, and damping coefficient) at M_1 for all elements of the system.

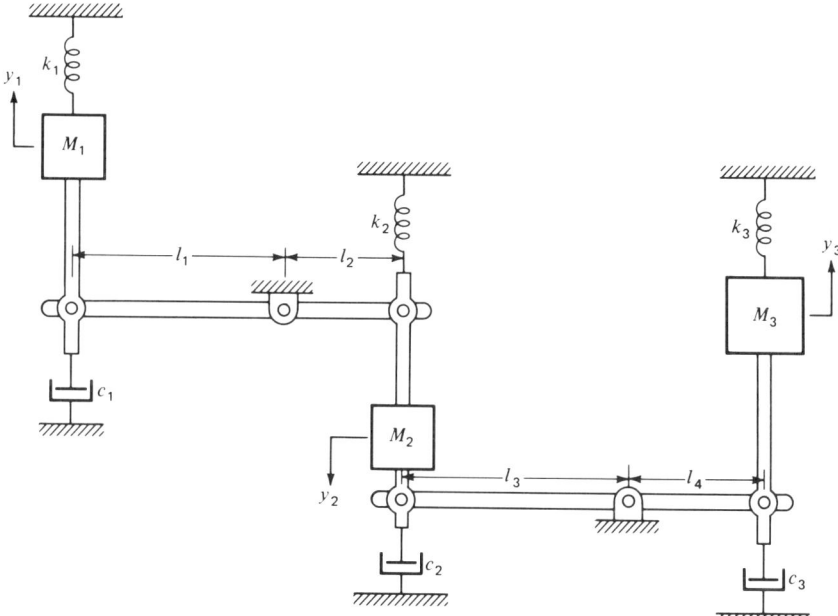

8-11. Develop the differential equation describing the given system with θ_0 as the dependent variable. Determine the peak-to-peak amplitude of θ_0, and the phase angle of θ_0 with respect to θ_i under steady-state conditions. Assume the following values: $I_1 = 0.08$ kg · m², $I_2 + I_3 = 0.25$ kg · m², $I_4 = 0.06$ kg · m², $G_1 = 12$ N · m/rad, $G_2 = 35$ N · m/rad, $b = 1.0$ N · m · s/rad, $r_1 = 0.10$ m, $r_2 = 0.05$ m, $r_3 = 0.15$ m, and $r_4 = 0.08$ m.

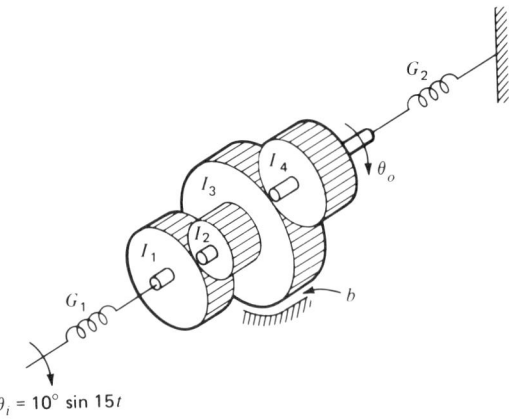

$\theta_i = 10°\sin 15t$

8-12. Assume the following parameters for the system shown: $I = 0.10$ kg · m², $r_1 = 0.08$ m, $r_2 = 0.14$ m, $M = 5$ kg, and $k = 800$ N/m.
(a) Write the differential equation of motion for the output y.
(b) Determine the maximum amplitude of y and its phase angle with respect to x for steady-state vibration with $x = 0.2\sin 3t$ (meters).
(c) For an input amplitude fixed at 0.2 meter, determine the frequency range over which a linear analysis is not possible because both cables do not remain in tension at all times.

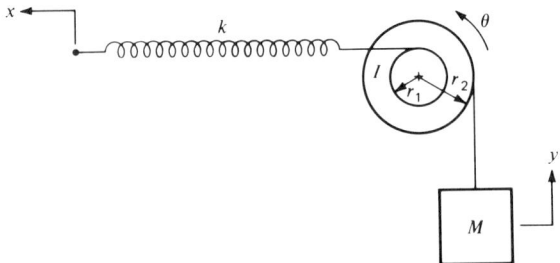

8-13. Write the differential equation of motion for the given system with θ as the dependent variable. Determine the amplitude of θ and its phase angle with respect to the input x under steady-state conditions. Use the following parameters: $k = 50$ lb/in., $c = 10$ lb $\cdot$ s/in., $r = 3$ in., $M = 1.0$ lb $\cdot$ s^2/in., and $I = 8$ lb $\cdot$ in.$\cdot$s^2.

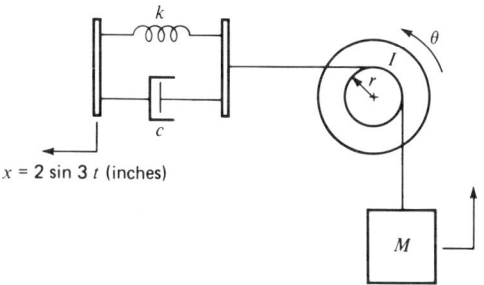

$x = 2 \sin 3t$ (inches)

Vibrating Systems with More Than One Degree of Freedom

9-1 Introduction

The number of degrees of freedom of a vibrating system is equal to the number of independent coordinates that must be specified to completely describe its vibratory motion for any set of conditions. We would therefore expect the vibration characteristics of a multi-degree-of-freedom system to be considerably more complex than those of a single-degree-of-freedom system, and this is indeed the case. In many cases, however, it is relatively simple to *write* the equations of motion to describe a system having more than one degree of freedom. The *solution* of these equations, on the other hand, may be quite difficult.

9-2 Writing the Equations of Motion

To write the differential equations describing a multi-degree-of-freedom mechanical vibrating system, Newton's second law of motion should be

9-2 WRITING THE EQUATIONS OF MOTION

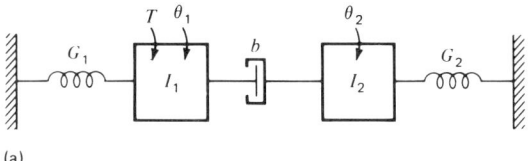
(a)

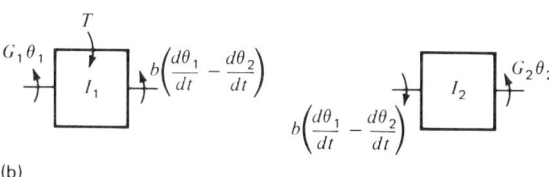
(b)

FIG 9-1. (a) Rotational two-degree-of-freedom vibrating system.
(b) Free-body diagrams of the two flywheels.

---

applied in turn to each inertia element of the system.[1] The result is a set of simultaneous ordinary linear differential equations, the number of equations in this set being equal to the number of degrees of freedom of the system.

Example 9-1. Write the differential equations of motion describing the two-degree-of-freedom rotational system illustrated in Fig. 9-1a.

Solution: Applying Newton's second law of motion to the flywheel on the left with the aid of the free-body diagram of Fig. 9-1b, we obtain

$$T - G_1\theta_1 - b\left[\frac{d\theta_1}{dt} - \frac{d\theta_2}{dt}\right] = I_1 \frac{d^2\theta_1}{dt^2}$$

In like manner, we obtain for the other flywheel

$$- G_2\theta_2 + b\left[\frac{d\theta_1}{dt} - \frac{d\theta_2}{dt}\right] = I_2 \frac{d^2\theta_2}{dt^2}$$

---

[1] In some instances it is necessary to apply Newton's second law to a point in the system that is massless. Example 9-3 illustrates such a case.

The system is therefore described by the pair of simultaneous differential equations

$$I_1 \frac{d^2\theta_1}{dt^2} + b\left[\frac{d\theta_1}{dt} - \frac{d\theta_2}{dt}\right] + G\theta_1 = T$$

$$I_2 \frac{d^2\theta_2}{dt^2} + b\left[\frac{d\theta_2}{dt} - \frac{d\theta_1}{dt}\right] + G\theta_2 = 0 \qquad (Ans.)$$

The two dependent variables θ_1 and θ_2 are not related by a simple constant (if they were, this would be a single-degree-of-freedom system). Two independent equations are therefore necessary in order to solve for θ_1 and θ_2.

Example 9-2. Write the differential equations of motion that describe the three-mass translational system of Fig. 9-2.

Solution: Applying Newton's second law to each of the three masses in turn with the aid of the free-body diagrams of Fig. 9-2b, we obtain

$$M_1 \frac{d^2y_1}{dt^2} + c_1\left[\frac{dy_1}{dt} - \frac{dy_3}{dt}\right] + k_1 y_1 + k_2(y_1 - y_2) = 0$$

$$M_2 \frac{d^2y_2}{dt^2} + c_2\left[\frac{dy_2}{dt} - \frac{dy_3}{dt}\right] + k_2(y_2 - y_1) + k_3(y_2 - y_3) = 0$$

$$M_3 \frac{d^2y_3}{dt^2} + c_2\left[\frac{dy_3}{dt} - \frac{dy_2}{dt}\right] + c_1\left[\frac{dy_3}{dt} - \frac{dy_1}{dt}\right]$$

$$+ k_3(y_3 - y_2) + k_4 y_3 = 0 \qquad (Ans.)$$

Since this is a three-degree-of-freedom system, three indepedent differential equations are necessary to completely describe it.

Note that in multi-degree-of-freedom mechanical vibrating systems, springs and dampers may be connected between any two of the masses and/or between any mass and ground. The dynamic interactions of such a system are extremely complex, and we are not very likely to visualize intuitively how the system will vibrate under a given set of conditions. The writing of the applicable dynamic equations is still relatively simple, however, no matter how the spring and damper elements are connected, if the following basic principles are observed:

1. The change in force from a spring is equal to the product of the spring constant k and its *net change in length*. (Or, in the case of a torsional spring, the change in torque equals the product of the torsional spring constant G and the *net change in angle of relative rotation* between the two ends.)

9-2 WRITING THE EQUATIONS OF MOTION

2. The force from a viscous damper is equal to the product of the damping coefficient c and the *velocity difference* between the two parts of the damper. (Or, *torque* equals the product of the torsional coefficient b and the *difference in angular velocity* between the two parts.)

To help avoid sign problems, it is advisable always to use a consistent sign convention for all parts of a given system. Note that in Fig. 9-1 the clockwise direction of rotation (when viewing the system from the left) has been chosen as positive for all parameters associated with *both* flywheels, and in Fig. 9-2 the upward direction is considered positive for all parameters. Although the final result is independent of the sign convention used, a logical and consistent choice will make things much easier.

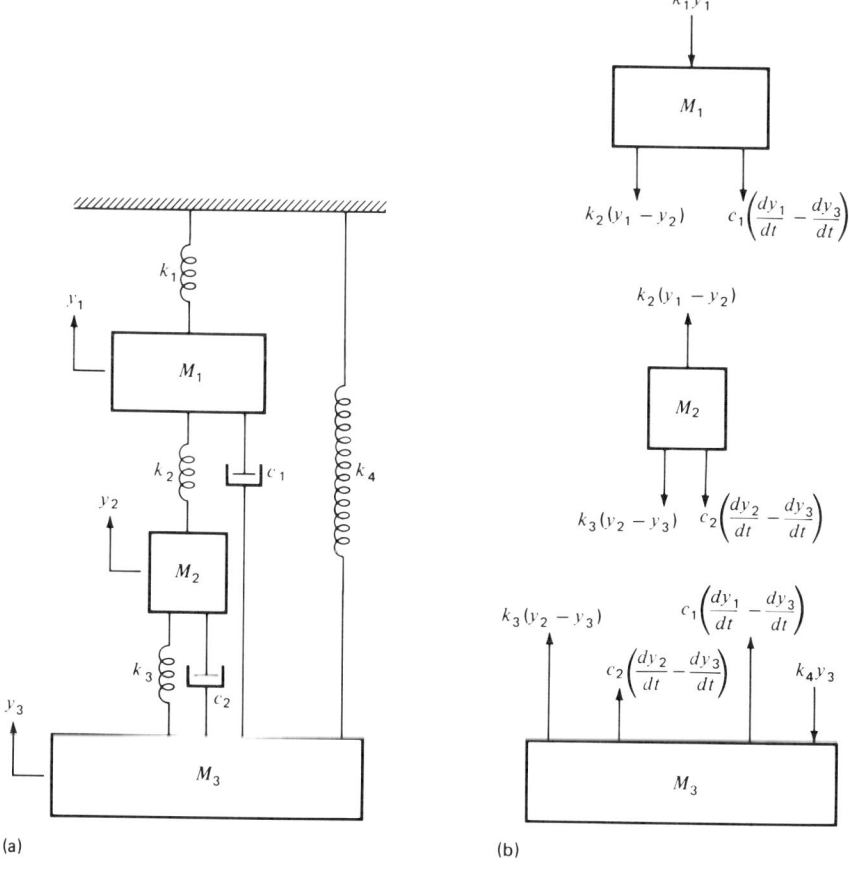

FIG 9-2. (a) Translational three-degree-of-freedom vibrating system.
(b) Free-body diagrams of the three masses.

Example 9-3. Write the differential equations of motion for the vibrating system of Fig. 9-3.

Solution: There is no equivalent component that can replace the spring and dashpot in series (unlike the case of the two springs in series, which can be replaced by a single equivalent spring). With no direct kinematic relationship between y_1 and y_2, the system has two degrees of freedom, and Newton's second law must be applied not only to the mass M but also to the massless link between the spring and dashpot. The resultant equations of motion are

$$M\frac{d^2y_1}{dt^2} + (k_1 + k_2)y_1 = k_2 y_2$$

$$c\frac{dy_2}{dt} + k_2 y_2 = k_2 y_1 \qquad (Ans.)$$

It becomes increasingly laborious to obtain closed-form mathematical solutions to the differential equations of motion as the number of degrees of freedom increases. One way to study the vibration characteristics of multi-degree-of-freedom systems is to simulate them on an analog computer. The analog computer is ideally suited for studying sets of simultaneous differential equations (nonlinear as well as linear) and allows a wide range of forcing functions and initial conditions to be investigated. The motions of the inertia elements can be plotted side by side as a function of time on a multichannel strip chart recorder. Such information can greatly facilitate the evaluation and design of relatively complex systems. An introduction to analog computer techniques is given in Section 17-4.

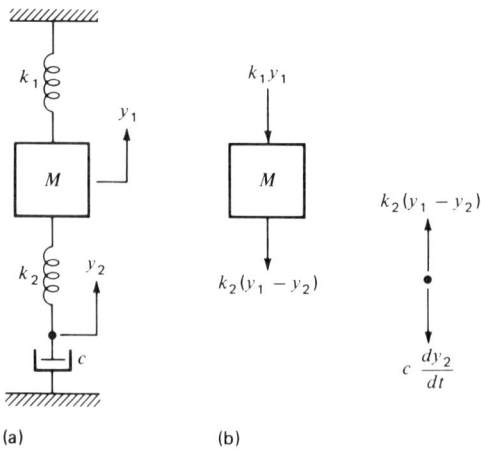

FIG 9-3. (a) A single-mass system with two degrees of freedom.
(b) Free-body diagrams of the mass M and the massless link between the spring and dashpot.

9-3 Two-Mass System Without Damping

One type of two-degree-of-freedom vibrating system that is relatively common is illustrated in Fig. 9-4. This is a translational system having two masses, two springs, but no damping. Although all real systems will have some damping, it is often small enough so that the theoretical performance of the corresponding undamped system is essentially the same in most respects as that of the real system. Very useful data can therefore be obtained by analyzing the undamped system, and the absence of damping terms appreciably simplifies the analysis. The equations of motion for the system of Fig. 9-4 are readily obtained as

$$M_1 \frac{d^2 y_1}{dt^2} + k_1 y_1 + k_2(y_1 - y_2) = F_1 \tag{9-1}$$

$$M_2 \frac{d^2 y_2}{dt^2} + k_2(y_2 - y_1) = 0 \tag{9-2}$$

or, in operator form,

$$\left[M_1 p^2 + (k_1 + k_2) \right] y_1 - k_2 y_2 = F_1 \tag{9-3}$$

$$y_1 = \left[\frac{M_2}{k_2} p^2 + 1 \right] y_2 \tag{9-4}$$

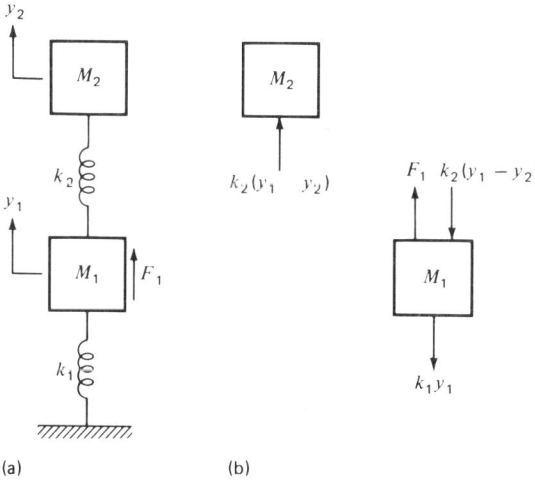

FIG 9-4. (a) Simple two-mass system (the vibration absorber).
(b) Free-body diagrams of the two masses.

FREE VIBRATION

We will consider free vibration first, setting $F_1 = 0$. With this simplification, Eqs. 9-3 and 9-4 can be combined to produce a single fourth-order equation in y_1:

$$\{M_1 M_2 p^4 + [(k_1 + k_2) M_2 + k_2 M_1] p^2 + k_1 k_2\} y_1 = 0 \quad (9\text{-}5)$$

The characteristic equation in this case has only three terms and can be considered as a quadratic equation in p^2. Solving for p^2 by means of the quadratic equation yields

$$p^2 = \frac{-[(k_1 + k_2) M_2 + k_2 M_1]}{2 M_1 M_2}$$

$$\pm \frac{\sqrt{[(k_1 + k_2) M_2 + k_2 M_1]^2 - 4 k_1 k_2 M_1 M_2}}{2 M_1 M_2} \quad (9\text{-}6)$$

With real springs and masses (i.e., with positive values for all k's and M's in the above equation), the term under the radical can be shown to be always positive. Since the term outside of the radical can be seen to be greater than the radical, the two values of p^2 will both be real negative numbers. We determine the four roots by taking the square roots of the two values of p^2 obtained from Eq. 9-6; they are all pure imaginary numbers and may be written

$$p_1, p_2 = \pm j\omega_{n_1} \quad (9\text{-}7)$$

$$p_3, p_4 = \pm j\omega_{n_2} \quad (9\text{-}8)$$

with

$$\omega_{n_1} = \left[\frac{(k_1 + k_2) M_2 + k_2 M_1}{2 M_1 M_2} - \frac{\sqrt{[(k_1 + k_2) M_2 + k_2 M_1]^2 - 4 k_1 k_2 M_1 M_2}}{2 M_1 M_2} \right]^{1/2} \quad (9\text{-}9)$$

$$\omega_{n_2} = \left[\frac{(k_1 + k_2) M_2 + k_2 M_1}{2 M_1 M_2} + \frac{\sqrt{[(k_1 + k_2) M_2 + k_2 M_1]^2 - 4 k_1 k_2 M_1 M_2}}{2 M_1 M_2} \right]^{1/2} \quad (9\text{-}10)$$

9-3 TWO-MASS SYSTEM WITHOUT DAMPING

The final solution for y_1 may now be obtained in the manner presented in Chapter 3, and is found to be

$$y_1 = (C_1 \sin \omega_{n_1} t + C_2 \cos \omega_{n_1} t) + (C_3 \sin \omega_{n_2} t + C_4 \cos \omega_{n_2} t) \quad (9\text{-}11)$$

If we combine Eqs. 9-3 and 9-4 in a manner that gives us a fourth-order equation in y_2, we find that the resultant characteristic equation is *identical* to that corresponding to Eq. 9-5. The solution for y_2 can therefore be written

$$y_2 = (C_5 \sin \omega_{n_1} t + C_6 \cos \omega_{n_1} t) + (C_7 \sin \omega_{n_2} t + C_8 \cos \omega_{n_2} t) \quad (9\text{-}12)$$

with the four constants depending on initial conditions but also being related to the constants of Eq. 9-11. Note that we have defined ω_{n_1} as the *lower* of the two natural frequencies.

From Eqs. 9-11 and 9-12 we can see one very significant difference between this system and an undamped single-degree-of-freedom system. In this system we do not necessarily find the two masses vibrating with pure harmonic motion. The form that free vibration takes is dependent upon the initial conditions, and in the general case can be seen to be the *superposition* of harmonic motion at the two natural frequencies ω_{n_1} and ω_{n_2}.

With certain combinations of initial conditions, however, free vibration will occur with both of the two masses vibrating with simple harmonic motion at the *same frequency* (either ω_{n_1} or ω_{n_2}). If they are vibrating at the lower natural frequency ω_{n_1}, they will be in phase with each other; if they are vibrating at ω_{n_2}, they will be 180° out of phase. These are called the *principal modes* of vibration. A two-degree-of-freedom system has two principal modes corresponding to its two natural frequencies.

With one set of initial conditions that will produce free vibration in the first principal mode, Eqs. 9-11 and 9-12 can be reduced to the form[2]

$$y_1 = C_1 \sin \omega_{n_1} t \quad (9\text{-}13)$$

$$y_2 = C_5 \sin \omega_{n_1} t \quad (9\text{-}14)$$

---

[2]An alternative set of initial conditions will produce vibration in the first principal mode that is represented by the equations

$$y_1 = C_2 \cos \omega_{n_1} t$$
$$y_2 = C_6 \cos \omega_{n_1} t$$

Other sets of initial conditions can produce first-principal-mode vibration that is represented by the equations

$$y_1 = C_1 \sin \omega_{n_1} t + C_2 \cos \omega_{n_1} t$$
$$y_2 = C_5 \sin \omega_{n_1} t + C_6 \cos \omega_{n_1} t$$

with the restriction that $C_1/C_5 = C_2/C_6$. The ratio of the amplitudes of the two masses (given by Eq. 9-15 or Eq. 9-16) is independent of the initial conditions, however, as long as they do produce first-principal-mode vibration.

Substitution into Eq. 9-1 (with $F_1 = 0$) produces the ratio of vibration amplitudes:

$$\frac{C_1}{C_5} = \frac{k_2}{-M_1 \omega_{n_1}^2 + k_1 + k_2} \quad (9\text{-}15)$$

Substitution of Eqs. 9-13 and 9-14 into Eq. 9-2 produces an alternative expression for the ratio of amplitudes,

$$\frac{C_1}{C_5} = -\frac{M_2}{k_2} \omega_{n_1}^2 + 1 \quad (9\text{-}16)$$

Either Eq. 9-15 or 9-16 can be used to determine the relative amplitudes of the two masses when free vibration occurs in the first principal mode. The answer will be the same regardless of which one is used; the two equations can in fact be shown to be equivalent by substitution of Eq. 9-9 for ω_{n_1} into each. The fact that the ratio C_1/C_5 is positive for any combination of values for the masses and spring constants shows that the two masses do indeed move in phase in the first principal mode.

Free vibration in the second principal mode, with one appropriate set of initial conditions, can be represented by the equations

$$y_1 = C_3 \sin \omega_{n_2} t \quad (9\text{-}17)$$

$$y_2 = C_7 \sin \omega_{n_2} t \quad (9\text{-}18)$$

which give the two equivalent equations for relative vibration amplitudes:

$$\frac{C_3}{C_7} = \frac{k_2}{-M_1 \omega_{n_2}^2 + k_1 + k_2} \quad (9\text{-}19)$$

$$\frac{C_3}{C_7} = -\frac{M_2}{k_2} \omega_{n_2}^2 + 1 \quad (9\text{-}20)$$

As in the case of the first principal mode, these two equations will produce the same answer. We find, however, that the ratio C_3/C_7 will always be *negative*, indicating that the two masses vibrate 180° out of phase with each other in the second principal mode of free vibration.

FORCED VIBRATION

The two-degree-of-freedom system of Fig. 9-4 might be excited by the sinusoidal forcing function $F_1 = F \sin \omega t$. Such a system with no damping would never reach a state of *steady-state* vibration corresponding to the *particular integral*, since the *complementary* solution would not die out. But all real systems do, of course, have some degree of damping. It is therefore logical to solve for the *particular integral* of the undamped system as a

9-3 TWO-MASS SYSTEM WITHOUT DAMPING

close approximation to the steady-state response of the corresponding real system with a small amount of damping.

To solve for the steady-state solution, we again use the method presented in Chapter 3. With $F_1 = F \sin \omega t$, Eqs. 9-3 and 9-4 can be combined into a single fourth-order equation in y_1:

$$\left[(M_1 p^2 + k_1 + k_2)(M_2 p^2 + k_2) - k_2^2\right] y_1 = (M_2 p^2 + k_2) F \sin \omega t \quad (9\text{-}21)$$

(It should be noted that the characteristic equation of Eq. 9-21 is the same as that of Eq. 9-5. It has been written in factored form to facilitate the development of Eq. 9-23.) Dividing Eq. 9-21 by $k_1 k_2$, we obtain

$$\left[\left(\frac{M_1}{k_1} p^2 + 1 + \frac{k_2}{k_1}\right)\left(\frac{M_2}{k_2} p^2 + 1\right) - \frac{k_2}{k_1}\right] y_1 = \left(\frac{M_2}{k_2} p^2 + 1\right)\frac{F}{k_1} \sin \omega t$$

$$(9\text{-}22)$$

It is very informative at this point to rewrite these two equations in terms of the following system parameters:

$\omega_{11} = \sqrt{\dfrac{k_1}{M_1}}$, the natural frequency associated with M_1 and k_1 alone

$\omega_{22} = \sqrt{\dfrac{k_2}{M_2}}$, the natural frequency associated with M_2 and k_2 alone

$Y_{st} = \dfrac{F}{k_1}$, static deflection of M_1 that would occur if a steady force F were applied

With these definitions, Eq. 9-22 can be rewritten

$$\left[\left(\frac{p^2}{\omega_{11}^2} + 1 + \frac{k_2}{k_1}\right)\left(\frac{p^2}{\omega_{22}^2} + 1\right) - \frac{k_2}{k_1}\right] y_1 = \left(\frac{p^2}{\omega_{22}^2} + 1\right) Y_{st} \sin \omega t \quad (9\text{-}23)$$

We assume the solution

$$y_1 = Y_1 \sin \omega t + Y_1' \cos \omega t \quad (9\text{-}24)$$

Substituting Eq. 9-24 and the required derivatives

$$p^2 y_1 = -Y_1 \omega^2 \sin \omega t - Y_1' \omega^2 \cos \omega t$$

$$p^4 y_1 = Y_1 \omega^4 \sin \omega t + Y_1' \omega^4 \cos \omega t$$

$$p^2 (Y_{st} \sin \omega t) = -Y_{st} \omega^2 \sin \omega t$$

into Eq. 9-23, we obtain

$$\left\{\left[-\left(\frac{\omega}{\omega_{11}}\right)^2 + 1 + \frac{k_2}{k_1}\right]\left[-\left(\frac{\omega}{\omega_{22}}\right)^2 + 1\right] - \frac{k_2}{k_1}\right\}[Y_1 \sin \omega t$$

$$+ Y_1' \cos \omega t] = \left[-\left(\frac{\omega}{\omega_{22}}\right)^2 + 1\right]Y_{st} \sin \omega t \qquad (9\text{-}25)$$

This equation must be valid for any value of ωt; that is, it must be valid when $\sin \omega t = 0$ and also when $\cos \omega t = 0$. Taking these two cases, we can therefore write the pair of equations

$$\left\{\left[-\left(\frac{\omega}{\omega_{11}}\right)^2 + 1 + \frac{k_2}{k_1}\right]\left[-\left(\frac{\omega}{\omega_{22}}\right)^2 + 1\right] - \frac{k_2}{k_1}\right\}Y_1$$

$$= \left[-\left(\frac{\omega}{\omega_{22}}\right)^2 + 1\right]Y_{st} \qquad (9\text{-}26)$$

$$\left\{\left[-\left(\frac{\omega}{\omega_{11}}\right)^2 + 1 + \frac{k_2}{k_1}\right]\left[-\left(\frac{\omega}{\omega_{22}}\right)^2 + 1\right] - \frac{k_2}{k_1}\right\}Y_1' = 0 \quad (9\text{-}27)$$

By solving Eq. 9-27 we find $Y_1' = 0$. Solving Eq. 9-26 for the dimensionless ratio Y_1/Y_{st} yields

$$\frac{Y_1}{Y_{st}} = \frac{1 - (\omega/\omega_{22})^2}{\left[1 - (\omega/\omega_{22})^2\right]\left[-(\omega/\omega_{11})^2 + (k_2/k_1) + 1\right] - (k_2/k_1)} \qquad (9\text{-}28)$$

Since the assumed solution (Eq. 9-24) reduces to $y_1 = Y_1 \sin \omega t$, the vibration of M_1 will be either in phase (if Y_1 is positive) or 180° out of phase (if Y_1 is negative) with the forcing function $[-(\omega/\omega_{22})^2 + 1]Y_{st} \sin \omega t$.

In a similar manner we find that the second mass will also, for any value of ω, be either in phase or 180° out of phase with the forcing function; that is,

$$y_2 = Y_2 \sin \omega t \qquad (9\text{-}29)$$

The amplitude ratio for M_2 is found to be

$$\frac{Y_2}{Y_{st}} = \frac{1}{\left[1 - (\omega/\omega_{22})^2\right]\left[-(\omega/\omega_{11})^2 + (k_2/k_1) + 1\right] - (k_2/k_1)} \qquad (9\text{-}30)$$

Some very interesting observations can be made from Eqs. 9-28 and 9-30. The amplitude Y_1 will obviously be zero at $\omega = \omega_{22}$—that is, when the frequency of the forcing function is equal to the natural frequency

9-3 TWO-MASS SYSTEM WITHOUT DAMPING

associated with M_2 and k_2. Amplitude Y_1 will, on the other hand, become infinite at the frequency that causes the denominator of Eq. 9-28 to be zero. Amplitude Y_2 cannot be zero for any finite input amplitude, but it does become infinite when the denominator of Eq. 9-30 is zero. The denominators of Eqs. 9-28 and 9-30 are identical, so we find infinite amplitude occurring at the same frequencies for the two masses. To determine these resonance frequencies, we set the denominator equal to zero,

$$\left[1 - \left(\frac{\omega}{\omega_{22}}\right)^2\right]\left[-\left(\frac{\omega}{\omega_{11}}\right)^2 + \frac{k_2}{k_1} + 1\right] - \frac{k_2}{k_1} = 0 \qquad (9\text{-}31.)$$

The resonance frequencies can now be obtained by solving for ω. Equation 9-31, however, can be shown by proper manipulation to be equivalent to Eq. 9-5. *The two resonance frequencies of this undamped two-mass system under forced vibration are therefore the same as the natural frequencies of free vibration, corresponding to the two principal modes, and are given by Eqs. 9-9 and 9-10.*

Frequency-response characteristics of M_1 for a typical system of this type are given in Fig. 9-5. Although plotted in nondimensional form, the curve is valid only for a system with $\omega_{11} = \omega_{22}$ and $M_1/M_2 = 5$. Although

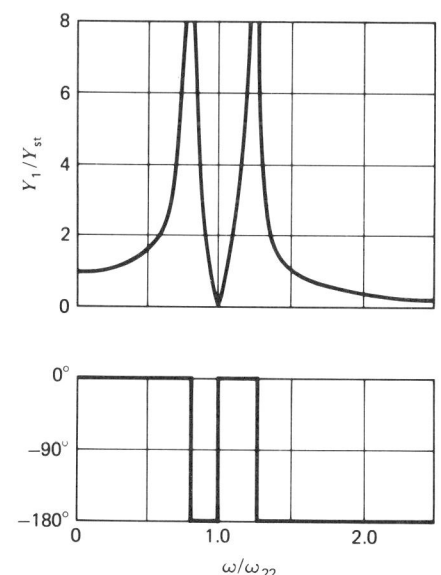

FIG 9-5. Frequency response characteristics of M_1 in Fig. 9-4, with $F_1 = F \sin \omega t$. Valid for $M_1 = 5M_2$ and $\omega_{11} = \omega_{22}$.

a different curve would be required for any other set of system parameters, certain characteristics remain the same. The condition of zero amplitude at $\omega = \omega_{22}$ always lies between the two resonance peaks at ω_{n_1} and ω_{n_2}, and the abrupt phase angle changes for M_1 always occur at ω_{n_1}, ω_{22}, and ω_{n_2}.

THE VIBRATION ABSORBER

Because of the unusual vibration characteristics of the undamped two-mass system of Fig. 9-4, as illustrated by the frequency-response curve of Fig. 9-5, simple one-mass systems are sometimes changed to two-mass systems in order to reduce vibration problems. Mass M_1 would typically be a machine attached to the floor through a flexible mounting system having a net spring rate k_1. A *vibration absorber* for this machine would then consist of a smaller mass M_2 and a spring with stiffness k_2 attached directly to M_1. As can be seen from Fig. 9-5, this technique has the potential of greatly reducing the amplitude of vibration of M_1 if the absorber is "tuned" to the frequency of the undesired vibration. (The device can theoretically eliminate the vibration of M_1 completely, but the effect of damping in any real system will cause a certain amount of vibration amplitude to remain even when the excitation is exactly at the tuned frequency.)

With improper design, however, the use of such a device could backfire, since the single natural frequency of the original system has been replaced by two natural frequencies, and operation at either one could cause serious problems. Vibration absorbers are primarily useful, therefore, for machinery operating at a single fixed speed, although devices have been built that can be continuously "tuned" by changing the value of k_2 to maintain maximum effectiveness as the speed of operation is varied [2].

9-4 Two-Mass System with Damping

If the amount of damping present in a two-mass system is small, then the solution obtained for the undamped equivalent system is normally adequate. In free vibration the lightly damped system would differ from the undamped case in that the amplitudes would decrease with time, but otherwise the vibration characteristics would be essentially the same. With forced vibration of a lightly damped two-mass system, the frequency-response characteristics would be almost the same as for the equivalent undamped system, the main difference being found at and near the resonance frequencies where amplitudes would be lower. The values of the resonance frequencies are virtually unchanged by a small amount of damping, and the amplitude ratio curves lie very close to the undamped ones at any appreciable distance away from the resonance frequencies.

9-4 TWO-MASS SYSTEM WITH DAMPING

If damping is considerable, however, it must be included in the analysis to obtain accurate results. Let us consider the system illustrated in Fig. 9-6. Note that this is the same system as analyzed previously except that dashpots have been added in parallel with the two springs.

FREE VIBRATION

With free vibration the two equations of motion for the system of Fig. 9-6a can be written (note the free-body diagrams of Fig. 9-6b)

$$M_1 \frac{d^2 y_1}{dt^2} + c_1 \frac{dy_1}{dt} + c_2 \left(\frac{dy_1}{dt} - \frac{dy_2}{dt} \right) + k_1 y_1 + k_2 (y_1 - y_2) = 0 \quad (9\text{-}32)$$

$$M_2 \frac{d^2 y_2}{dt^2} + c_2 \left(\frac{dy_2}{dt} - \frac{dy_1}{dt} \right) + k_2 (y_2 - y_1) = 0 \quad (9\text{-}33)$$

Putting these into operator form, we have

$$\left[M_1 p^2 + (c_1 + c_2) p + k_1 + k_2 \right] y_1 = \left[c_2 p + k_2 \right] y_2 \quad (9\text{-}34)$$

$$\left[M_2 p^2 + c_2 p + k_2 \right] y_2 = \left[c_2 p + k_2 \right] y_1 \quad (9\text{-}35)$$

Eqs. 9-34 and 9-35 can be combined into a single equation for y_1; the

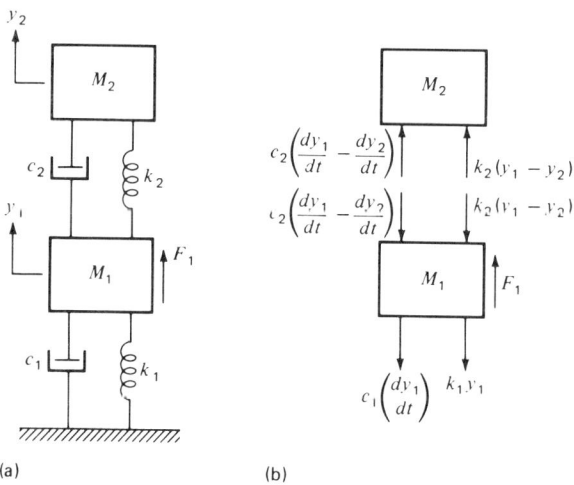

FIG 9-6. (a) Two-mass vibrating system with damping.
(b) Free-body diagrams of the two masses.

result is

$$[(M_1 p^2 + (c_1 + c_2)p + k_1 + k_2)(M_2 p^2 + c_2 p + k_2)$$
$$- (c_2 p + k_2)^2] y_1 = 0 \quad (9\text{-}36)$$

If we combine Eqs. 9-34 and 9-35 in a manner that produces a single expression for y_2, we will find that the characteristic equation is the same. The characteristic equation in this case (and that for all other two-degree-of-freedom two-mass systems) will have the form

$$Ap^4 + Bp^3 + Cp^2 + Dp + E = 0 \quad (9\text{-}37)$$

It will have four roots, but these are not so readily obtained as in the case of zero damping. The factoring can be done, however, using one of the methods in Appendix C or by using an appropriate digital computer program. It is most convenient not to factor the characteristic equation completely, but to leave it as the product of two quadratic terms, putting it in the general form

$$\left[\left(\frac{1}{\omega_{n_1}}\right)^2 p^2 + \frac{2\zeta_1}{\omega_{n_1}} p + 1\right]\left[\left(\frac{1}{\omega_{n_2}}\right)^2 p^2 + \frac{2\zeta_2}{\omega_{n_2}} p + 1\right] = 0 \quad (9\text{-}38)$$

Frequencies ω_{n_1} and ω_{n_2} are the undamped natural frequencies of the system, that is, the natural frequencies of the corresponding undamped system. If ζ_1 and ζ_2 are both less than 1.0, free vibration may have components at the *two damped natural frequencies*

$$\begin{aligned} \omega_{d_1} &= \omega_{n_1}\sqrt{1 - \zeta_1^2} \\ \omega_{d_2} &= \omega_{n_2}\sqrt{1 - \zeta_2^2} \end{aligned} \quad (9\text{-}39)$$

The number of cycles that each component persists in free vibration depends upon the value of its damping ratio.

If one of the two damping ratios is greater than 1.0, then there can be no free-vibration component at the corresponding natural frequency. If both ζ's are greater than 1.0, there can be no free vibration at all, and initial conditions on the two masses (displacement or velocity) will merely cause them to move in an asymptotic manner to their positions of static equilibrium.

FORCED VIBRATION

Forced vibration of a damped two-mass system will also, as in the undamped case, cause both of the masses to vibrate at the excitation frequency. Although we can derive exact equations for the motions of the two masses, it is also possible to determine some forced-vibration characteristics by consideration of the characteristic equation alone. Large

9-4 TWO-MASS SYSTEM WITH DAMPING

output amplitudes will occur when the forcing frequency is in the vicinity of ω_{n_1} if ζ_1 is low, or in the vicinity of ω_{n_2} if ζ_2 is low. The amplitudes of motion depend upon the values of ζ_1 and ζ_2, increasing in each case with a decrease in ζ.

Example 9-4. The electric circuit of Fig. 9-7 has been analyzed and found, for a certain set of component parameters, to be represented by the equation

$$1.2346 \times 10^{-4} \frac{d^4 i_2}{dt^4} + 1.8815 \times 10^{-3} \frac{d^3 i_2}{dt^3} + 0.032667 \frac{d^2 i_2}{dt^2}$$
$$+ 0.073333 \frac{di_2}{dt} + i_2 = 0.200 \omega E \cos \omega t - 0.410 \omega^2 E \sin \omega t$$

Determine the damped and undamped natural frequencies of the system, and the values of ζ_1 and ζ_2.

Solution: The characteristic equation of the system is

$$1.2346 \times 10^{-4} p^4 + 1.8815 \times 10^{-3} p^3 + 0.032667 p^2$$
$$+ 0.073333 p + 1 = 0$$

It can be factored (by means of a "canned" digital computer program or by one of the hand methods outlined in Appendix C) into the form

$$(0.02778 p^2 + 0.006667 p + 1)(0.004444 p^2 + 0.06667 p + 1) = 0$$

Comparing the above to Eq. 9-38, a generalized factored form of the characteristic equation of a fourth-order dynamic system, we find that

$$\frac{1}{\omega_{n_1}^2} = 0.02778 \qquad \frac{2\zeta_1}{\omega_{n_1}} = 0.006667$$

$$\frac{1}{\omega_{n_2}^2} = 0.004444 \qquad \frac{2\zeta_2}{\omega_{n_2}} = 0.06667$$

from which the undamped natural frequencies and damping ratios are

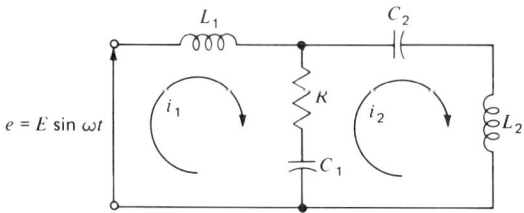

FIG 9-7. Electric circuit for Example 9-4.

found

$$\omega_{n_1} = 6.000 \text{ rad/s} \quad \zeta_1 = 0.0200$$
$$\omega_{n_2} = 15.00 \text{ rad/s} \quad \zeta_2 = 0.500 \quad (Ans.)$$

Finally, the damped natural frequencies are determined by applying Eq. 9-39,

$$\omega_{d_1} = \omega_{n_1}\sqrt{1 - \zeta_1^2} = 5.999 \text{ rad/s}$$
$$\omega_{d_2} = \omega_{n_2}\sqrt{1 - \zeta_2^2} = 12.99 \text{ rad/s} \quad (Ans.)$$

9-5 Vibrating Systems with More Than Two Degrees of Freedom

The damped and undamped natural frequencies and associated damping ratios can be found for vibrating systems with three or more inertia elements by a straightforward extension of the methods that have been presented for two-mass systems, although there may be considerable labor involved in factoring the higher-order characteristic equations unless a digital computer program is used. The number of natural frequencies in each case equals the number of inertia elements, provided that there is no direct kinematic relationship between any of them.

9-6 Conclusion

The equations of motion for vibrating systems having more than one degree of freedom are normally easy to obtain by simply applying Newton's second law of motion to each inertia element (and to each massless link between a spring and dashpot that has no direct kinematic relationship to any inertia element or other such massless link). Although it may not be an easy task to solve the set of simultaneous differential equations obtained, the method of solution presented in this chapter is a straightforward extension of the basic principles given in Chapter 3.

A vibrating system with two or more masses having no direct kinematic relationship with one another will have the number of natural frequencies equal to the number of masses. With each natural frequency there is an associated damping ratio. If a multi-degree-of-freedom system has a periodic input, resonance can occur if the frequency of that input is at or near one of the system's natural frequencies, provided that the damping ratio associated with that natural frequency is appreciably less than 1.0.

PROBLEMS

The vibration absorber is a device that makes use of the unique characteristics of a two-mass system for greatly reducing unwanted vibrations in a machine. It is useful, however, only for those systems in which the frequency of the input does not vary appreciably.

Problems

9-1. Show that Eqs. 9-5 and 9-31 are equivalent.

9-2. Obtain the set of differential equations that describes each of the given systems. (Do not combine the separate equations.)

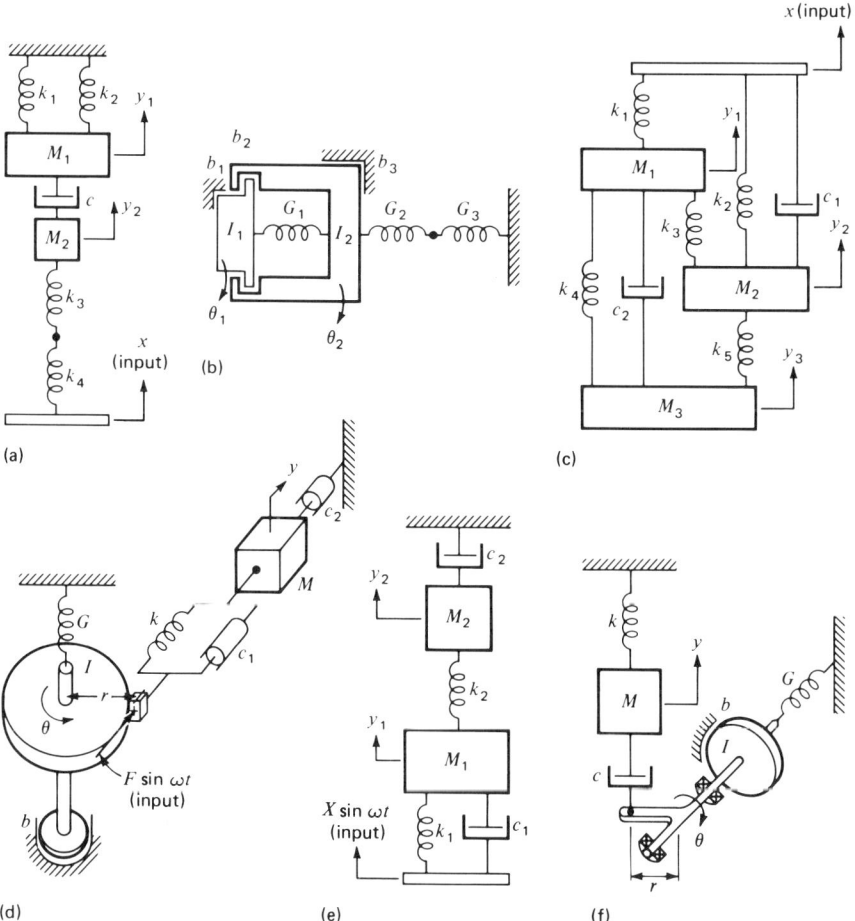

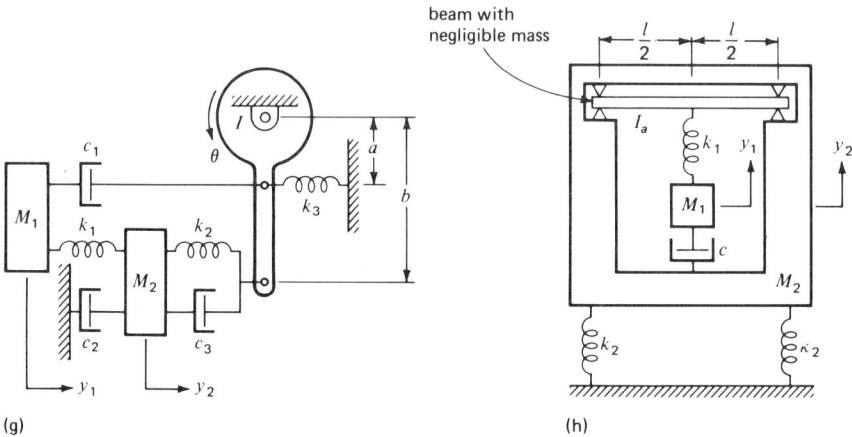

(g) (h)

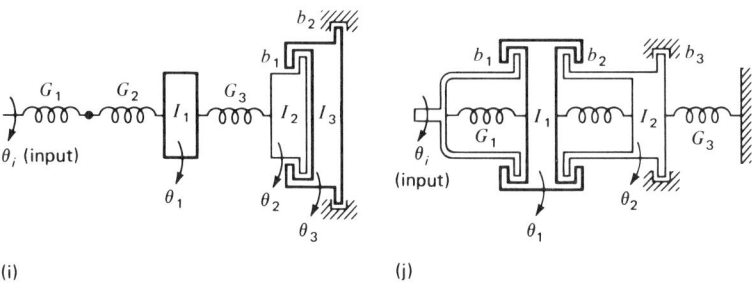

(i) (j)

9-3. Write the pair of differential equations that describe the given system; then combine them to obtain a single equation that relates y_1 to the input x.

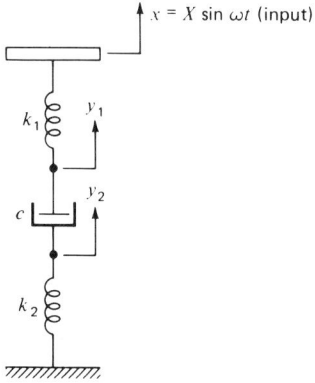

$x = X \sin \omega t$ (input)

9-4. Derive the differential equations of motion for the given system in terms of the two dependent variables θ_1 and θ_2. y should not appear in the final equations.

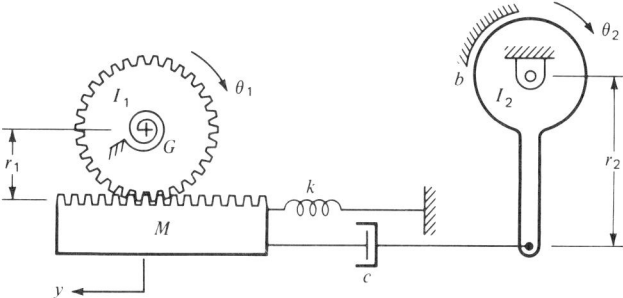

9-5. Assuming negligible damping and a flexible steel beam spring of negligible mass,
  (a) write the pair of differential equations that describes the system shown; then combine to obtain a single equation that relates y_1 to the input F.
  (b) determine the damped and undamped natural frequencies of the system, and the corresponding damping ratios.

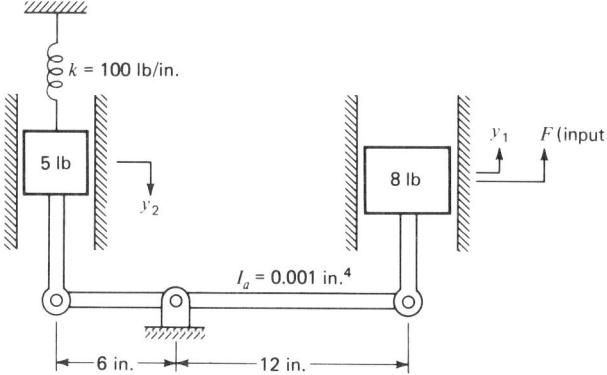

9-6. A pulley system has a compliant cable with parameters as given in the figure. Assuming that damping is negligible, that operation occurs so that the cable is always in tension, and that the mass of the cable is negligible in comparison with the other inertia elements,
(a) determine the natural frequencies of vibration.
(b) determine the steady-state amplitudes of vibration of θ and y for the input $x = 5 \sin 100t$ (in.).
(c) determine the frequency of input that will cause the flywheel to remain motionless.

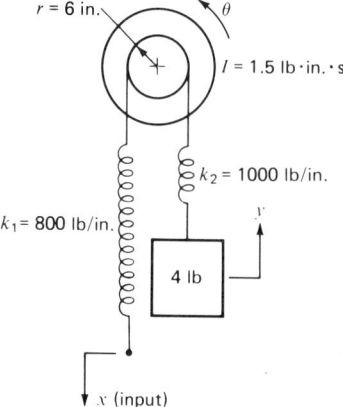

9-7. Assuming that damping is negligible, that operation occurs so that the cable is always in tension, and that the mass of the cable is negligible, write the differential equations that describe the given pulley system with a compliant cable. (Do not combine the separate equations.)

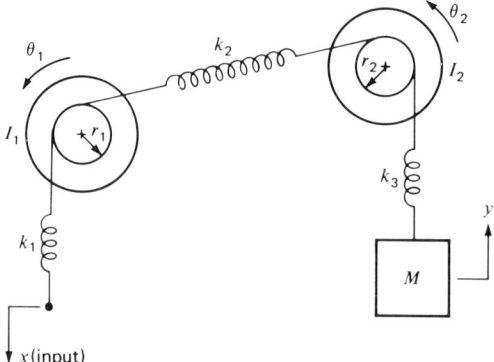

PROBLEMS

9-8. The two-mass system shown has been analyzed and found to be represented, with the forcing function x equal to zero, by the equation (in factored form)

$$(p^2 + 1.8p + 9{,}500)(p^2 + 95p + 35{,}000)y_1 = 0$$

(a) Determine the damped and undamped natural frequencies and the corresponding damping ratios.

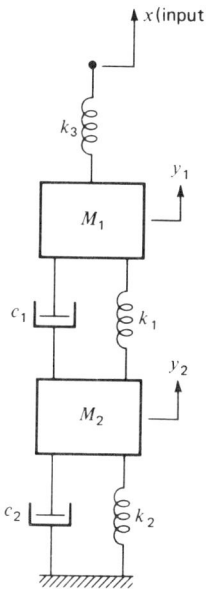

9-9. For the given system derive a single differential equation for θ_1 in terms of I_1, I_2, G, and the input. (θ_2 should not appear in the equation.)

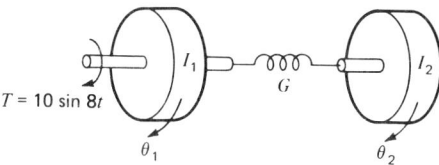

9-10. Write the pair of differential equations that describe the given system; then combine them into a single equation with y as the dependent variable.

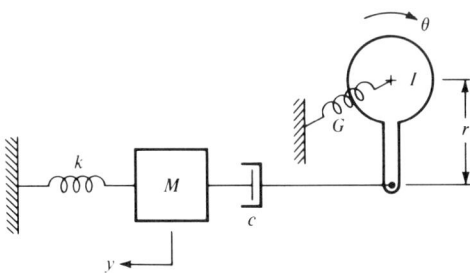

Distributed Parameter Systems

10-1 Introduction

The terms *lumped parameter system* and *distributed parameter system* are used to distinguish between two basically different types of model used for mechanical vibrating systems. In a lumped parameter model, the springs and dashpots have zero mass, and each mass is concentrated at a point (or, in a rotational system, the moment of inertia is concentrated in a plane). In a distributed parameter model, on the other hand, the mass is distributed in some manner throughout the components, and the compliance (springiness) of material composing the mass acts as the spring. If the material has hysteresis, then damping is also distributed throughout the system components. A distributed parameter system has an infinite number of degrees of freedom, since knowing the motion of one specific point does not permit one to directly determine the motion of other points.

Although all real vibrating systems are in reality distributed parameter systems, a great number of them may be approximated very accurately by a lumped parameter model having a small number of degrees of freedom. Such an approximation is obviously most accurate for systems in which the springs have much more compliance than the rest of the structure and in which the mass of these springs is small relative to the total mass.

Although this chapter considers only mechanical vibrating systems, the concepts of lumped and distributed parameter models can also be applied to other types of dynamic systems. For the precise analysis of some electric systems, for example, it is necessary to consider the resistance, capacitance, and inductance that are distributed throughout the system components and wiring.

Some very common vibrating devices or systems cannot, however, be approximated accurately by lumped parameter models unless the chosen model has a great number of degrees of freedom. Consider, for example, the lateral vibration characteristics of the cantilever beam of Fig. 10-1a. We might try to approximate this device by a massless cantilever spring having the same stiffness as the true beam and a single lumped mass at the end having the same value as the total mass of the true beam, as illustrated in Fig. 10-1b. The vibration characteristics calculated for such a model would be quite different, however, from those of the actual beam. The actual beam would have an infinite number of natural frequencies, whereas the crude model would indicate a single natural frequency that coincided with none. The two-mass two-spring approximation of Fig. 10-1c would give results somewhat nearer to reality, but still far from satisfactory. A lumped parameter model having many point masses connected together by very short massless beam springs, Fig. 10-1d, could be used to get accurate answers, however. Although a closed-form mathematical

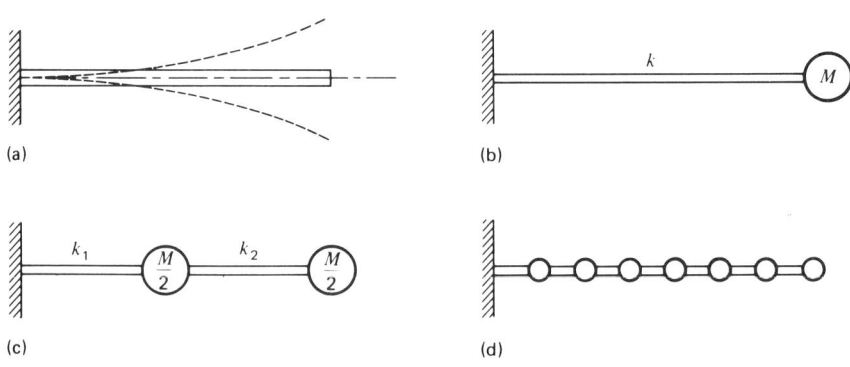

FIG 10-1. Models of a cantilever beam in transverse vibration.
    (a) Vibrating cantilever beam (distributed parameters).
    (b) Crude single-mass lumped parameter model of the beam.
    (c) Two-mass model.
    (d) Multimass model.

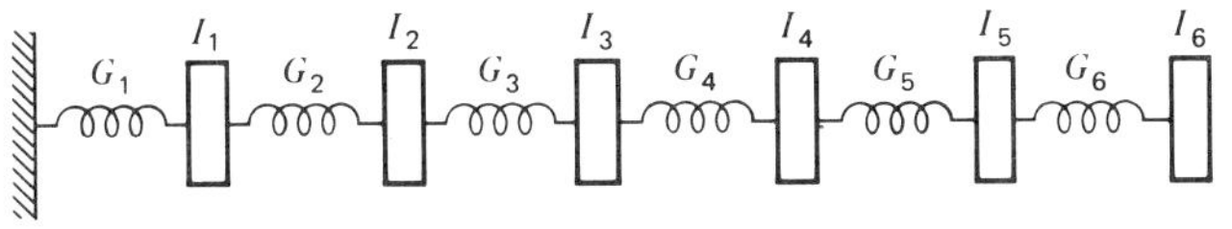

10-2 RIGOROUS ANALYSIS OF A DISTRIBUTED PARAMETER SYSTEM

FIG 10-2. Lumped parameter model of a cantilever beam for the study of rotational vibration.

solution for the vibration characteristics of such a model with many degrees of freedom is difficult to obtain, techniques have been developed based on trial-and-error methods that are very satisfactory [13, 15]. Such analytical methods are particularly easy to implement if a digital computer can be used to eliminate the drudgery of the many repetitious calculations they require.

The above discussion is also valid for rotational vibrating systems. The cantilever beam of Fig. 10-1a will also vibrate rotationally if some type of angular excitation is applied or if it is set into free vibration with a torsional twist as an initial condition. The rotational vibration characteristics of this cantilever beam can be accurately found by analyzing the lumped parameter rotational model (Fig. 10-2) consisting of many disks, or flywheels, connected together by short, massless torsional springs [4]. As in the case of lateral vibration, a digital computer program is required to make this approach practical.

A turbine blade is a good example of a distributed parameter element for which it is very important to know the vibration characteristics. A turbine blade is basically a cantilever beam having a fairly complex cross-sectional shape that is usually tapered toward the tip. It is important to know the natural frequencies of turbine blades because excitation at a resonance frequency will often cause rapid failure through fatigue. Such blades can vibrate both laterally and rotationally, and there are some vibration modes in which torsion and bending are coupled. Although vibration characteristics of turbine blades are invariably checked experimentally before a turbine design is completed, digital computer programs based on very sophisticated lumped parameter models are commonly used to give accurate predictions.

10-2 Rigorous Analysis of a Distributed Parameter System

There are a number of distributed parameter systems that can be analyzed mathematically without resorting to approximate numerical solutions, and it is instructive to consider an example of this type of analysis. If the cantilever beam of Fig. 10-3a is made of a homogeneous material and has

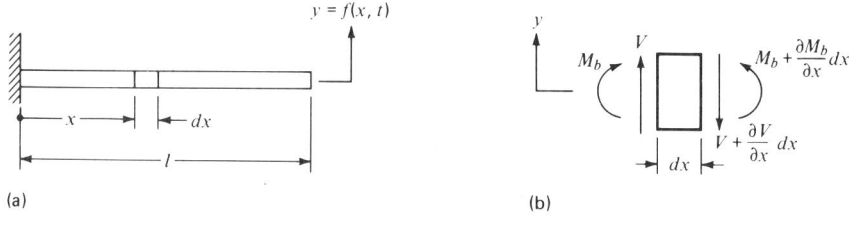

FIG 10-3. (a) A cantilever beam.
(b) Free-body diagram of an infinitesimal element of the beam.

a *uniform cross section* over its total length, then the following equation from the field of *strength of materials* [19] can be used to describe its bending characteristics:

$$M_b = -EI_a \frac{d^2y}{dx^2} \qquad (10\text{-}1)$$

with M_b = *bending moment* at any distance x along the beam, E = Young's modulus, and I_a = *area moment of inertia* of the uniform cross section. (Equation 10-1 applies to any beam in bending: cantilever, simply supported, etc.). A single differentiation of Eq. 10-1 with respect to the distance x produces the equation for shear force V,

$$\frac{dM_b}{dx} = V = -EI_a \frac{d^3y}{dx^3} \qquad (10\text{-}2)$$

A second differentiation yields an expression for the *load q per unit length*,

$$\frac{dV}{dx} = q = -EI_a \frac{d^4y}{dx^4} \qquad (10\text{-}3)$$

To proceed from Eq. 10-3 to an analysis of the vibratory motion of the beam we make the assumption that it is in a state of free vibration with harmonic motion (and negligible damping) at one of its natural frequencies ω_n. The load per unit length acting at each point along the beam and causing it to bend is, from Newton's second law, equal to the product of the *mass per unit length* λ and the lateral acceleration at that point; that is,

$$q = \lambda \frac{d^2y}{dt^2} \qquad (10\text{-}4)$$

A free-body diagram of an element of the beam of length dx is given by Fig. 10-3b. The difference in shear force between the left and right faces is what is required to give the element the acceleration corresponding to its vibratory motion. The force of gravity that may be acting on the beam can be ignored since it is normally much less than the inertia loading and merely causes the beam, if horizontal, to "droop" a small amount.

10-2 RIGOROUS ANALYSIS OF A DISTRIBUTED PARAMETER SYSTEM

Equations 10-3 and 10-4 are now combined, the result being the partial differential equation[1]

$$EI_a \frac{\partial^4 y}{\partial x^4} = -\lambda \frac{\partial^2 y}{\partial t^2} \qquad (10\text{-}5)$$

With harmonic motion at a natural frequency, each mass particle of the beam will have motion described by the equation

$$y = y_m(x)\sin \omega_n t \qquad (10\text{-}6)$$

with $y_m(x)$ the amplitude of vibration.[2]

By taking the second derivative of Eq. 10-6 with respect to time,

$$\frac{\partial^2 y}{\partial t^2} = -y_m \omega_n^2 \sin \omega_n t$$

and substituting it into Eq. 10-5, we obtain the equation that describes the vibration of the beam. At this point, however, for the sake of simplicity, we drop the $\sin \omega_n t$ term, so the resulting equation applies only to the maximum amplitude condition of each vibration cycle—that is, when $\sin \omega_n t = 1$. The result,

$$EI_a \frac{d^4 y_m}{dx^4} = \lambda y_m \omega_n^2 \qquad (10\text{-}7)$$

is a fourth-order differential equation, which may also be written as

$$\frac{d^4 y_m}{dx^4} - a^4 y_m = 0 \qquad (10\text{-}8)$$

with

$$a = \sqrt[4]{\frac{\lambda \omega_n^2}{EI_a}} \qquad (10\text{-}9)$$

Note that Eq. 10-8 can be applied to any beam of uniform cross section by use of the proper boundary conditions, since it was derived without regard to boundary conditions. For the cantilever beam presently under consideration, the four boundary conditions are

1. $y_m = 0$ at $x = 0$
2. $dy_m/dx = 0$ at $x = 0$
3. $d^2 y_m/dx^2 = 0$ at $x = l$ (10-10)
4. $d^3 y_m/dx^3 = 0$ at $x = l$

---

[1] The relationships at this point in the derivation show y to be a function of two independent variables (x and t), so partial derivatives must be used.

[2] Note that the peak amplitude of the vibratory motion varies along the length of the beam; that is, y_m is a function of x, and therefore y is a function of both x and t.

The above can be explained by noting first of all that there must be zero deflection and zero slope at the point of attachment of the cantilever. At the free end, however, the displacement and slope are unknown, but the impossibility of creating a shear stress or bending moment at the tip tells us, based on Eqs. 10-1 and 10-2, that the second and third derivatives must be zero at the tip.

To solve Eq. 10-8 we take its characteristic equation

$$p^4 - a^4 = 0$$

and find the four roots in the following manner:

$$p^2 = \pm a^2$$
$$p_1 = +\sqrt{+a^2} = a$$
$$p_2 = -\sqrt{+a^2} = -a$$
$$p_3 = +\sqrt{-a^2} = ja$$
$$p_4 = -\sqrt{-a^2} = -ja$$

With these roots and the method of solution presented in Chapter 3, the general solution of Eq. 10-8 is found to be

$$y_m = C_1 e^{ax} + C_2 e^{-ax} + C_3 \sin ax + C_4 \cos ax \qquad (10\text{-}11)$$

The first three derivatives of Eq. 10-11 are

$$\frac{dy_m}{dx} = C_1 a e^{ax} - C_2 a e^{-ax} + C_3 a \cos ax - C_4 a \sin ax$$

$$\frac{d^2 y_m}{dx^2} = C_1 a^2 e^{ax} + C_2 a^2 e^{-ax} - C_3 a^2 \sin ax - C_4 a^2 \cos ax$$

$$\frac{d^3 y_m}{dx^3} = C_1 a^3 e^{ax} - C_2 a^3 e^{-ax} - C_3 a^3 \cos ax + C_4 a^3 \sin ax$$

Combining these with the boundary condition equations (Eqs. 10-10), we obtain

$$y_m|_{x=0} = 0 = C_1 + C_2 + 0 + C_4 \qquad (10\text{-}12)$$

$$\left.\frac{dy_m}{dx}\right|_{x=0} = 0 = C_1 a - C_2 a + C_3 a - 0 \qquad (10\text{-}13)$$

$$\left.\frac{d^2 y_m}{dx^2}\right|_{x=l} = 0 = C_1 a^2 e^{al} + C_2 a^2 e^{-al} - C_3 a^2 \sin al - C_4 a^2 \cos al \qquad (10\text{-}14)$$

$$\left.\frac{d^3 y_m}{dx^3}\right|_{x=l} = 0 = C_1 a^3 e^{al} - C_2 a^3 e^{-al} - C_3 a^3 \cos al + C_4 a^3 \sin al \qquad (10\text{-}15)$$

10-2 RIGOROUS ANALYSIS OF A DISTRIBUTED PARAMETER SYSTEM

To solve the set of four simultaneous linear equations (Eqs. 10-12 through 10-15), we use the mathematical theorem [9]: *A homogeneous system of n equations in the same number of unknowns has nontrivial solutions if, and only if, its coefficient determinant is zero.* The coefficient determinant,

$$\begin{vmatrix} 1 & 1 & 0 & 1 \\ 1 & -1 & 1 & 0 \\ e^{al} & e^{-al} & -\sin al & -\cos al \\ e^{al} & -e^{-al} & -\cos al & \sin al \end{vmatrix} \quad (10\text{-}16)$$

can be reduced to

$$2 + \cos al \, (e^{al} + e^{-al})$$

Setting this equal to zero and noticing that

$$\frac{e^{al} + e^{-al}}{2} = \cosh al$$

we obtain the solution

$$\cos al \, \cosh al = -1 \quad (10\text{-}17)$$

To obtain the natural frequencies of vibration we must first determine those values of al that will satisfy Eq. 10-17. These are the points of intersection of two curves: $\cos al$ and $-1/\cosh al$ plotted as ordinates against al as the abscissa. The first intersection occurs at $al = 1.875$ rad. Substituting this value into Eq. 10-9, we obtain

$$al = \sqrt[4]{\frac{\lambda \omega_n^2 l^4}{EI_a}} = 1.875 \quad (10\text{-}18)$$

Solving for ω_n, we get

$$\omega_{n_1} = 3.515 \sqrt{\frac{EI_a}{\lambda l^4}} \quad (10\text{-}19)$$

as the lowest natural frequency of the beam. In like manner, we find the higher natural frequencies to be

$$\omega_{n_2} = 22.04 \sqrt{\frac{EI_a}{\lambda l^4}} \quad (10\text{-}20)$$

$$\omega_{n_3} = 61.7 \sqrt{\frac{EI_a}{\lambda l^4}} \quad (10\text{-}21)$$

and so on. (There is an infinite number of values of al that satisfy Eq. 10-17. The beam therefore has an infinite number of natural frequencies, but normally only the first few have any practical significance.)

In solving for ω_n, $y_m(x)$ and its derivatives have disappeared. This makes no difference if knowledge of the natural frequencies is all that is

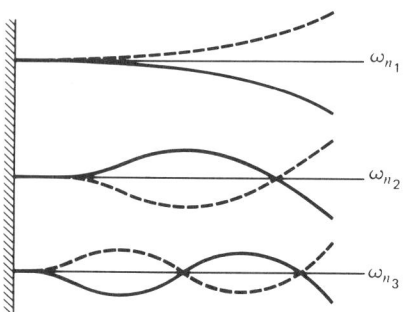

FIG 10-4. Vibration of a cantilever beam of uniform cross section.

required. If, however, it is also important to know the shape of the vibrating beam, this can be obtained by substitution of the value obtained for a (for the natural frequency of interest) back into the general solution (Eq. 10-11), rewriting it in the form

$$\frac{y_m(x)}{C_1} = e^{ax} + \frac{C_2}{C_1} e^{-ax} + \frac{C_3}{C_1} \sin ax + \frac{C_4}{C_1} \cos ax \quad (10\text{-}22)$$

The ratios C_2/C_1, C_3/C_1, and C_4/C_1 can be obtained by simultaneous solution of Eqs. 10-12 through 10-15, using the proper values of a and l.[3] This procedure will allow a shape function (giving *relative* amplitudes along the length of the beam) to be determined. Note that the *absolute* amplitude of vibration at any point depends also upon initial conditions for free vibration and upon the amplitude of the input for forced vibration at the natural frequency.

The shape that the cantilever beam takes when vibrating at each of the first three natural frequencies is illustrated by Fig. 10-4. Note that the fundamental natural frequency has a single node (point of zero vibration amplitude), the second mode has 2 nodes, and so forth.

Example 10-1.

(a) Determine the three lowest natural frequencies of a steel cantilever beam 20 in. long, with a thickness of 0.040 in. and a width of 1.00 in.

(b) Determine the shape function for the first natural frequency.

---

[3]The form of Eqs. 10-12 through 10-15 is such that one of the coefficients must be arbitrary [9]. The solution therefore produces the values of the other three coefficients with respect to the arbitrarily chosen one (C_1 in Eq. 10-22).

10-2 RIGOROUS ANALYSIS OF A DISTRIBUTED PARAMETER SYSTEM

Solution:
(a)
$$I_a = \frac{bh^3}{12} = \frac{(1.00)(0.040^3)}{12} = 5.333 \times 10^{-6} \text{ in.}^4$$

$$\lambda = \left(\frac{0.283}{386}\right)(1.00)(0.040) = 29.33 \times 10^{-6} \text{ lb} \cdot \text{s}^2/\text{in.}^2$$

From Eq. 10-19,

$$\omega_{n_1} = 3.515\sqrt{\frac{(30 \times 10^6)(5.333 \times 10^{-6})}{(29.33 \times 10^{-6})(20^4)}} = 20.52 \text{ rad/s}$$

$$f_{n_1} = \frac{\omega_{n_1}}{2\pi} = 3.266 \text{ Hz} \qquad (Ans.)$$

Similarly, from Eqs. 10-20 and 10-21,

$$\omega_{n_2} = 129 \text{ rad/s}$$
$$f_{n_2} = 20.5 \text{ Hz} \qquad (Ans.)$$

$$\omega_{n_3} = 360 \text{ rad/s}$$
$$f_{n_3} = 57.3 \text{ Hz} \qquad (Ans.)$$

(b)

$$a = \sqrt[4]{\frac{\lambda \omega_{n_1}^2}{EI_a}} = \sqrt[4]{\frac{(29.33 \times 10^{-6})(20.52)^2}{(30 \times 10^6)(5.333 \times 10^{-6})}} = 0.093742$$

Substituting $a = 0.093742$ and $l = 20$ into Eqs. 10-12 through 10-15 yields

$$C_1 + C_2 + C_4 = 0$$
$$C_1 - C_2 + C_3 = 0$$
$$6.5197 C_1 + 0.15338 C_2 - 0.95413 C_3 + 0.29937 C_4 = 0$$
$$6.5197 C_1 - 0.15338 C_2 + 0.29937 C_3 + 0.95413 C_4 = 0$$

Simultaneous solution of the four equations (with C_1 arbitrary) yields

$$C_2 = 6.52 C_1$$
$$C_3 = 5.52 C_1$$
$$C_4 = -7.52 C_1$$

Substitution into Eq. 10-22 provides the shape function

$$\frac{y_m(x)}{C_1} = e^{0.0937x} + 6.52 e^{-0.0937x} + 5.52 \sin 0.0937x$$

$$- 7.52 \cos 0.0937x \qquad (Ans.)$$

10-3 Beams with Other Boundary Conditions

The development of Eq. 10-8 was independent of boundary conditions. Equation 10-8 and its general solution (Eq. 10-11) therefore apply to any beam of uniform cross section. The three basic types of end conditions that may exist are: (1) fixed, (2) free, and (3) simply supported. The fixed and free conditions were found in the cantilever beam covered in the previous section. The simply supported end condition (as exists at both ends of the beam of Fig. 10-5) is represented by the equations

$$y_m = 0$$

$$\frac{d^2 y_m}{dx^2} = 0$$

with the slope dy_m/dx and the third derivative $d^3 y_m/dx^3$ variable.

Expressions for natural frequencies and shapes of lateral vibration can be determined for a beam of uniform cross section having any combination of end conditions by an analysis patterned along the lines of that used for the cantilever beam in Section 10-2. The steps of the analysis can be summarized as follows:

1. The boundary conditions at each end must be determined. These are then combined with the general solution (Eq. 10-11) and its first three derivatives to obtain a set of four simultaneous homogeneous equations (comparable to Eqs. 10-12 through 10-15).

2. Setting the coefficient determinant of the four simultaneous equations equal to zero and reducing it will yield the equation for the specific beam under consideration (comparable to Eq. 10-17), from which the natural frequencies of lateral vibration can be determined.

3. To determine the shape of vibration at any natural frequency, the

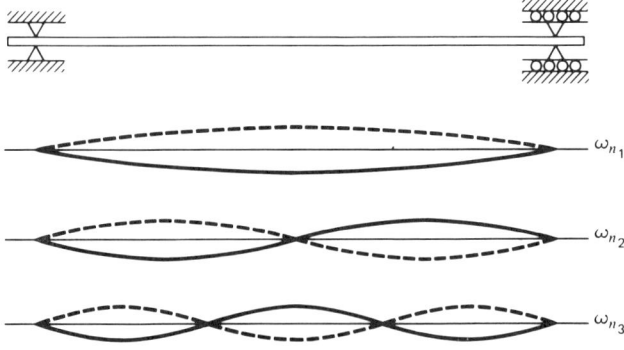

FIG 10-5. A simply supported beam of uniform cross section and its first three modes of free vibration.

value of a (Eq. 10-9) must first be determined for that particular ω_n. This value of a and the length l are then substituted into the four simultaneous equations obtained in step 1. By choosing an arbitrary value for one of the coefficients (e.g., setting $C_1 = 1$), the values of the other three can be determined. Substitution of the coefficients into Eq. 10-11 (or Eq. 10-22) will produce the equation that describes the shape of the vibration at the chosen natural frequency.

10-4 Conclusion

The beam of uniform cross section is a good example of a distributed parameter vibrating element for which the natural frequencies can be found through rigorous mathematical analysis. The same basic analytical approach can be followed for any set of end conditions. Vibration characteristics of many other objects (e.g., the lateral vibration of stretched wires, longitudinal vibration of rods, and transverse vibration of membranes and plates) can also be determined by analysis of a distributed parameter model [4, 20]. When this type of analysis is practical, the theoretical values obtained should be quite accurate.

One interesting distributed parameter problem is the surge of a helical spring. A helical spring has natural frequencies in the longitudinal direction. If one end has periodic motion at a low natural frequency, the coils will surge up and down. A surging spring may no longer supply the force for which it was designed and may fail prematurely from fatigue.

As the shape of a distributed parameter object becomes more complex, the rigorous analytical approach becomes less feasible. For such complex systems, approximate lumped parameter models and iterative analytical techniques (not covered in this introductory book) may advantageously be used [4, 13, 15]. Modern computer techniques make it possible to obtain quite accurate answers with sophisticated lumped parameter models.

Problems

10-1. Determine the first three natural frequencies of lateral vibration of the steel cantilever beam illustrated in the figure.

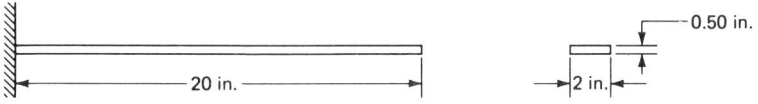

10-2. Determine the first three natural frequencies of the beam of Prob. 10-1 if
   (a) the material is changed to aluminum (density = 0.10 lb/in.^{3}, $E = 10 \times 10^6$ lb/in.^{2}).
   (b) the thickness is doubled.
   (c) the width is doubled.
   (d) the length is doubled.

10-3. (a) Determine the shape function for the first natural frequency of the beam of Prob. 10-1.
   (b) Using the function, draw the shape of the vibrating beam.

10-4. (a) Determine the shape functions for the second and third natural frequencies of the beam of Prob. 10-1.
   (b) Using the functions, draw the shape of the vibrating beam for both frequencies.

10-5. List the known and unknown boundary conditions for each of the given beams of uniform cross section.

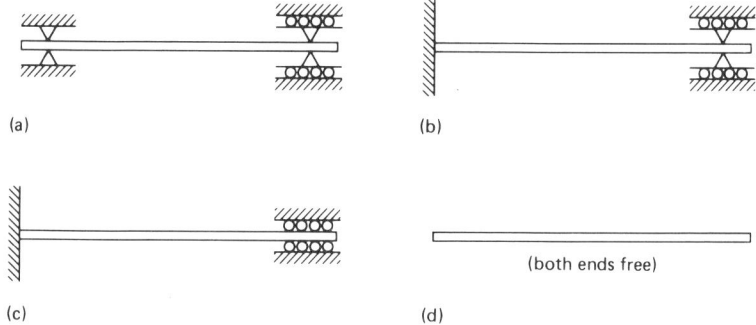

10-6. Determine expressions for the first three natural frequencies of lateral vibration of a simply supported beam of uniform cross section, that is, the beam illustrated for (a) of Prob. 10-5.

10-7. Compare ω_{n_1} of (a) a cantilever beam of uniform cross section with ω_n of (b) a crude lumped parameter model of the beam in which all the mass is at the end.

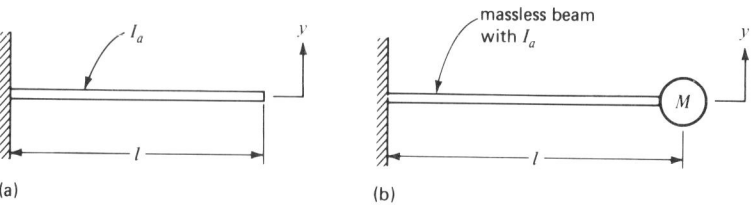

Critical Speeds of Rotors

11-1 Introduction

When a shaft or rotor[1] is run at certain speeds, the shaft may not remain straight, but may rotate in a bowed configuration. Speeds at which this phenomenon occurs are called *critical speeds*, and the resultant action of the shaft is commonly called *whirling* or *whipping*.

Critical-speed whirling can cause serious problems in machinery because of two different factors:

1. *Mechanical damage* may occur because of bending stress in the rotor shaft, mechanical interference between the rotor and its housing, or excessive bearing loads from centrifugal force and misalignment.
2. *Excessive vibration* may be induced in the machine and its surroundings because of the rotating centrifugal-force vector of the bowed shaft.

---

[1]A *rotor* is defined as the revolving part of a machine. In this book the term *rotor* is used to denote the total rotating assembly—the shaft as well as all attached disks.

11-2 Analysis of a Simple Lumped Parameter Rotor

In studying the theory of critical-speed whirling, it is convenient to analyze first a lumped parameter model with a single mass and no damping. The model rotor, shown in Fig. 11-1, is assumed to be mounted between a pair of simply-supported bearings (i.e., bearings that are self-aligning and offer no resistance to the bowing of the shaft). The bearings are, however, assumed to be infinitely stiff in the radial direction. The single disk of mass M can be at any position on the massless shaft. The center of gravity of the mass is offset a small distance ϵ (called the *eccentricity*) from the centerline of the shaft. The spring constant k of the shaft is for lateral bending and is defined as the lateral force required *at the mass* to displace it a given distance from its equilibrium position.

The rotor shown in Fig. 11-1 is mounted vertically, so the effect of gravity can be ignored in our analysis. (It can be shown, however, that the effect of gravity on such a rotor mounted horizontally does not affect the method of calculating the critical speed or its value.)

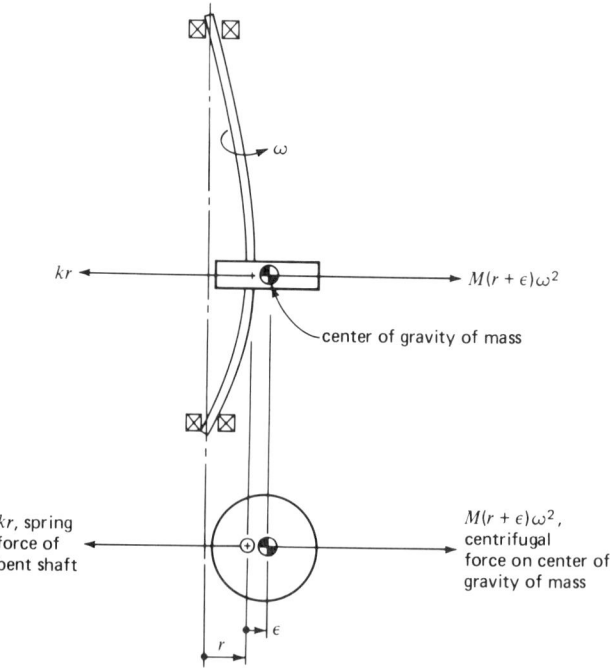

FIG 11-1. Single-mass lumped parameter model of whirling shaft.

11-2 ANALYSIS OF A SIMPLE LUMPED PARAMETER ROTOR

To analyze the dynamic action of the rotor, Newton's second law of motion is applied to the mass, with the assumptions that the centerline of the shaft is bowed a distance r from its equilibrium position, that the shaft is rotating at a constant speed ω, and that steady-state conditions are present. To keep the analysis simple, a coordinate system is chosen that is *fixed with respect to the rotor* so that the coordinate system itself rotates. The two perpendicular axes of the coordinate system are the centerline of the bearings and a radial line between that centerline and the center of gravity of the single mass. As the shaft whirls, its spring force must be of the proper value to give the radial acceleration of the mass corresponding to its circular path.[2] In order for the spring force to be aligned with the radial acceleration vector, a straight line perpendicular to the centerline of the bearings that passes through the center of gravity of the mass must also intersect the centerline of the bowed shaft. (See Fig. 11-1, and note that the straight line is one of the rotating coordinate axes).

By equating the spring force of the shaft to the product of the mass and the radial acceleration,

$$M(r + \epsilon)\omega^2 = kr \tag{11-1}$$

and solving for r, the radius of whirl, we obtain

$$r = \frac{\epsilon\omega^2}{k/M - \omega^2} \tag{11-2}$$

Equation 11-2 allows the amount of transverse shaft deflection to be calculated for any speed of rotation ω, and whirling can be said to exist for any speed at which r is large enough to produce an undesirable situation.

For very low speeds, Eq. 11-2 shows that r is insignificant. At $\omega = \sqrt{k/M}$, however, the denominator of Eq. 11-2 is zero, so r becomes infinite. The speed of rotation corresponding to a theoretically infinite whirl radius is called the *critical speed*, which is therefore given by the equation

$$\omega_{cr} = \sqrt{\frac{k}{M}} \tag{11-3}$$

with ω_{cr} in radians per second, and k and M with compatible spring and mass units.

---

[2]Another way to analyze the problem is to use the *principle of D'Alembert* (a French mathematician and philosopher of the eighteenth century). With this approach, an imaginary inertia force (centrifugal force for the rotor under consideration) is assumed to act on the mass. The mass is then assumed to be in static equilibrium; that is, the vector sum of the two forces on it must be zero. D'Alembert's principle is a convenient alternative method of analysis with the rotating coordinate system used here. It is also used in Chapter 12 for the analysis of dynamic balance problems. It could have been used in the earlier chapters for the analysis of mechanical vibrating systems, but the authors have chosen to use the direct application of Newton's second law of motion there.

At values of ω above the critical speed, the denominator of Eq. 11-2 becomes negative, so r and ϵ will have opposite signs. This means that the bow in the shaft will reverse direction as the speed of the rotor is raised above ω_{cr}, and the center of gravity of the mass will lie between the bearing centerline and the shaft centerline. At speeds much higher than the critical speed, Eq. 11-2 shows that $r \approx -\epsilon$, and thus the deflection of the shaft at very high speeds is the amount necessary to cause the center of gravity of the mass to lie almost exactly on the centerline of the bearings. Figure 11-2 is a dimensionless curve that gives the whirl amplitude (from Eq. 11-2) that would be found at any speed of rotation for the idealized single-mass model. It should be noted that Fig. 11-2 is based on the assumption of (1) a constant spring rate of the shaft regardless of the amount of deflection, (2) zero damping, and (3) steady-state operation; that is, the deflection at any given speed is that obtained after the rotor has been kept at that particular speed for a relatively long period of time. If a rotor is accelerated through the critical speed fairly quickly, the maximum whirl amplitude will be less than it would be if the rotor were allowed to remain at or near the critical speed for any appreciable period of time.

The lumped parameter model is actually a fairly close approximation of many real rotors. Even with precision balancing techniques, it is not possible to eliminate completely the eccentricity ϵ. Both journal bearings and antifriction bearings are often very stiff in comparison to the rotor shaft but offer little resistance to a small amount of shaft bowing. (Note that the critical speed is determined by the system characteristics when the shaft first starts to whirl. Therefore an increase of bearing resistance to

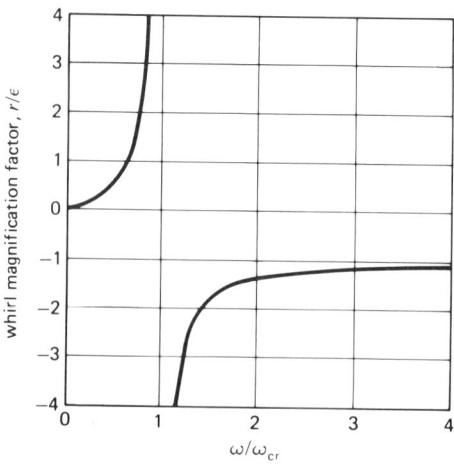

FIG 11-2. Whirl characteristics of an idealized single-mass rotor.

misalignment as the misalignment increases, while reducing the magnitude of whirl, will not affect the calculation of ω_{cr}.) Damping is normally very small for a whirling shaft, coming primarily from air resistance and the effect of bearing misalignment. A large disk mounted near the midpoint of the shaft, a fairly common situation, will make the mass of the shaft relatively insignificant. Calculations based upon Eq. 11-3 will, therefore, give fairly accurate results for a large number of actual rotor configurations.

The action of the rotor shaft while whirling at or near its critical speed warrants further discussion. Since the coordinate system rotates with the rotor, the fact that there is an equilibrium value of the ratio r/ϵ that corresponds to any speed ω means that the bend in the shaft stays fixed as it rotates. In other words, the shaft material that is in tension because of the bending remains in tension throughout the complete cycle of rotation. Critical speed whirling therefore does *not* cause fatigue of the shaft material, as one might tend to assume. Also, with no continual flexing of the shaft, there is no hysteresis to add to the system damping.

COMPARISON OF CIRCULAR VIBRATION AND WHIRLING

If the lateral vibration characteristics of the idealized single-mass rotor are studied, the natural frequency of lateral vibration will be found to be the same as the critical speed; that is,

$$\omega_n = \sqrt{\frac{k}{M}} = \omega_{cr} \qquad (11\text{-}4)$$

This is no mere coincidence, since the two phenomena are definitely related. If the mass is displaced laterally a small amount and then released, it will vibrate laterally at the natural frequency ω_n. If, by a proper choice of initial conditions, free lateral vibration can be induced in two perpendicular planes at once, as shown in Fig. 11-3, with the two sine waves of equal amplitude but 90° out of phase, the resultant path of the mass will be a circle about the bearing centerline[3] the same path it takes during whirling. The one significant difference between this situation and a whirling shaft is that with the circular *vibration* the shaft is not rotating, so the outer material is continuously alternating between tensile and compressive stress, a condition of fatigue. This is in contrast to critical-speed whirling in which the shaft becomes bent into a bowed configuration that it holds as it rotates.

---

[3]The two-dimensional vibration can be induced by hand in a very flexible rotor, after some practice, by pulling the mass down, then pushing it sideways as it is released to start it in its circular orbit. If the operation is not done perfectly, the resultant path of the mass will be elliptical.

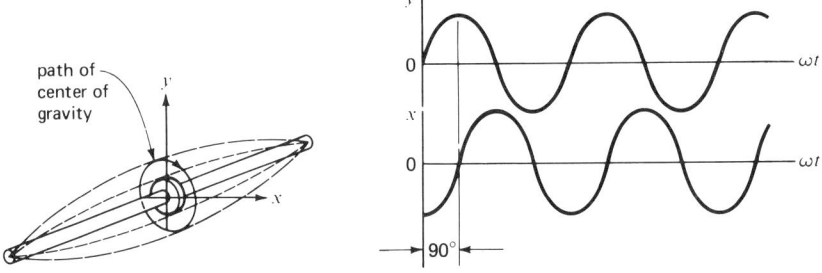

FIG 11-3. Lateral vibration of a single-mass rotor in two perpendicular planes, with 90° phase difference so that motion of the mass center of gravity is a circle about the bearing centerline.

11-3 Effect of Compliance in Bearings and Bearing Mounts

In the derivation of Eq. 11-3 it was assumed that the only compliance (springiness) in the system was that of the rotor shaft. In actuality, however, lateral compliance in any part of the system will have virtually the same effect in determining the critical speed. Additional lateral compliance may be present in the bearings themselves (compliance of the oil film of journal bearings or at the point contact between ball and race in a ball bearing) or it may be in the bearing housings or support structure. If this type of compliance is the same in all lateral directions, the effect on the critical speed is the same as the compliance of the rotor shaft.[4]

Figure 11-4 illustrates how a very stiff rotor mounted in a pair of compliant bearings will whirl at or near the critical speed. In this extreme case, the total rotor shaft moves in the circular orbit. Whirling action of any single-mass rotor will be somewhere between the extremes of Figs. 11-1 and 11-4. Since the compliant elements of the system act as springs in series, Eq. 2-8 can be used to determine the resultant net spring constant,

$$k = \frac{k_r k_b k_h}{k_r k_b + k_r k_h + k_b k_h} \tag{11-5}$$

where

k_r = spring rate of rotor shaft

k_b = effective spring rate at mass from bearings

k_h = effective spring rate at mass from bearing housings

---

[4]If the compliance is not uniform in all directions, the situation is somewhat more complex, with two critical speeds instead of one and an elliptical whirl orbit rather than a circular one [18].

11-3 EFFECT OF COMPLIANCE IN BEARINGS AND BEARING MOUNTS

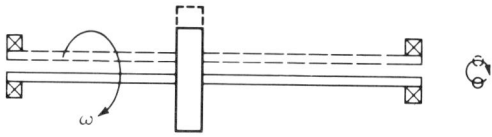

FIG 11-4. Whirl characteristics of a single-mass rotor with the shaft much stiffer than the bearings.

This net effective spring rate is then used in Eq. 11-3 for the general case in which the compliance of all three elements is significant.

Example 11-1. The steel rotor illustrated in Fig. 11-5a is supported by two ball bearings of different sizes. The bearings and bearing housings all have significant compliance, with effective spring constants (uniform in all radial directions) as given in Fig. 11-5a. Determine the critical speed of the rotor.

Solution: To solve for the critical speed, the net spring constant must be determined from Eq. 11-5. The spring constants k_r, k_b, and k_h are therefore required. To obtain the spring constant of the rotor shaft, the applicable beam equation from Appendix A (Case 3) is

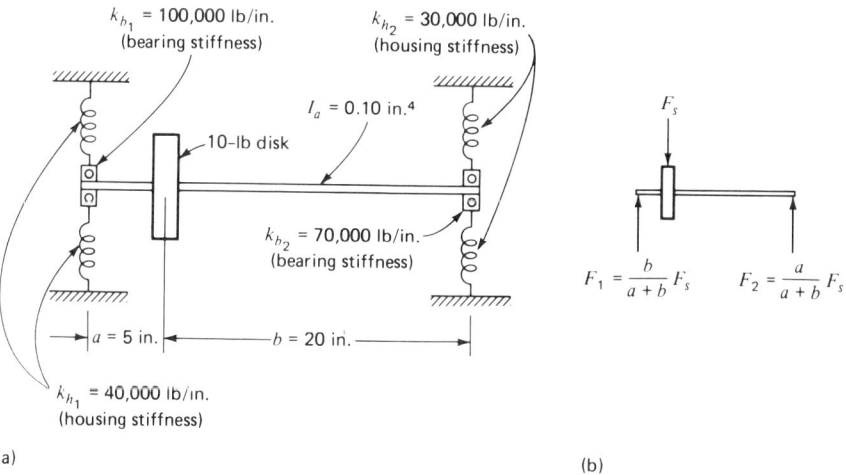

(a)   (b)

FIG 11-5. (a) A single-mass rotor with compliance in the rotor shaft, the bearings, and the bearing housings.
(b) Bearing reactions caused by a radial load applied at the mass.

applied, based on a force applied at the mass:

$$k_r = \frac{F_s}{y} = \frac{3EI_a l}{a^2 b^2} = \frac{3(30 \times 10^6)(0.10)(25)}{(5^2)(20^2)} = 22{,}500 \text{ lb/in.}$$

To obtain k_b we must determine the radial deflection of the mass from bearing compliance that results from a given force applied to the mass. The bearing reactions resulting from the application of the radial force F_s to the mass (see Fig. 11-5b) are found, from basic statics principles, to be

$$F_1 = \frac{b}{a+b} F_s \qquad F_2 = \frac{a}{a+b} F_s$$

The resultant displacements between the inner and outer races of the bearings are

$$\delta_1 = \frac{F_1}{k_{b_1}} = \frac{bF_s}{(a+b)k_{b_1}}$$

$$\delta_2 = \frac{F_2}{k_{b_2}} = \frac{aF_s}{(a+b)k_{b_2}}$$

The total deflection of the mass due to bearing compliance is obtained by superposition,

$$\delta = \frac{b}{a+b} \delta_1 + \frac{a}{a+b} \delta_2 = \frac{b^2 F_s}{(a+b)^2 k_{b_1}} + \frac{a^2 F_s}{(a+b)^2 k_{b_2}}$$

$$= \frac{F_s}{(a+b)^2} \left(\frac{b^2}{k_{b_1}} + \frac{a^2}{k_{b_2}} \right)$$

The effective spring constant at the mass from the bearing compliance is

$$k_b = \frac{F_s}{\delta} = \frac{(a+b)^2}{\left(\dfrac{b^2}{k_{b_1}} + \dfrac{a^2}{k_{b_2}} \right)} = \frac{25^2}{\dfrac{20^2}{100{,}000} + \dfrac{5^2}{70{,}000}} = 143{,}443 \text{ lb/in}.$$

In a similar manner, we can determine the effective spring rate at the mass from the compliance of the bearing housings,

$$k_h = \frac{(a+b)^2}{\left(\dfrac{b^2}{k_{h_1}} + \dfrac{a^2}{k_{h_2}} \right)} = \frac{25^2}{\dfrac{20^2}{40{,}000} + \dfrac{5^2}{30{,}000}} = 60{,}345 \text{ lb/in}.$$

The net equivalent spring constant at the mass is obtained from Eq. 11-5,

$$k = \frac{k_r k_b k_h}{k_r k_b + k_r k_h + k_b k_h} = \frac{(22{,}500)(143{,}443)(60{,}345)}{(3227 + 1358 + 8656) \times 10^6}$$

$$= 14{,}708 \text{ lb/in}.$$

The critical speed can now be obtained from Eq. 11-3,

$$\omega_{cr} = \sqrt{\frac{k}{M}} = \sqrt{\frac{14{,}708(386)}{10}} = 753.5 \text{ rad/s}$$

or

$$n_{cr} = \frac{60}{2\pi} \omega_{cr} = 7195 \text{ r/min} \qquad (Ans.)$$

11-4 Other Critical Speed Considerations

If the bearings supporting the shaft cannot be considered self-aligning (i.e., if they allow no angular misalignment of the shaft), the previously developed equations remain valid. The shaft will merely assume a different shape as it whirls (see Fig. 11-6), and the proper spring constant to be used is that corresponding to the true deformation. This situation is not very common, but it would occur if a pair of bearings were used at each end of the shaft.

Example 11-2. Determine the critical speed of the steel rotor illustrated in Fig. 11-7, which has a bearing that allows angular misalignment at one end but not at the other. Assume that the radial compliances of the bearings and the bearing mounts are insignificant

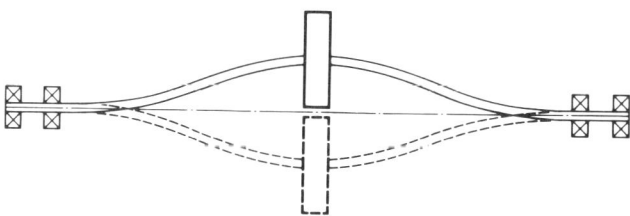

FIG 11-6. Whirl characteristics of a single-mass rotor mounted in bearings that do not allow bearing misalignment.

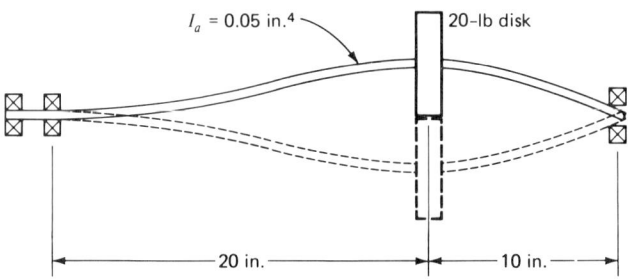

FIG 11-7. Single-mass rotor mounted so that one end of the shaft can have no angular misalignment while the other end is simply supported.

in comparison to that of the rotor shaft, and that the mass of the shaft is negligible.

Solution: We determine the stiffness of the rotor shaft from the beam equation for Case 6 in Appendix A:

$$k = \frac{F_s}{y} = \frac{12EI_a l^3}{a^3 b^2 (3l+b)} = \frac{(12)(30 \times 10^6)(0.05)(30^3)}{(20^3)(10^2)(90+10)} = 6080 \text{ lb/in.}$$

The critical speed is now obtained by direct application of Eq. 11-3,

$$\omega_{cr} = \sqrt{\frac{k}{M}} = \sqrt{\frac{6080(386)}{20}} = \sqrt{117{,}000} = 342 \text{ rad/s}$$

or

$$n_{cr} = \frac{60}{2\pi} \omega_{cr} = 3270 \text{ r/min} \qquad (Ans.)$$

Our analysis becomes somewhat more complicated if we include the effect of damping. With a small amount of damping (the usual situation) the values of the critical speeds are little different from those of the corresponding undamped system. It is therefore seldom worthwhile to include damping in critical speed analyses, since the value of the critical speed is usually the principal thing one is trying to determine, and the radius of the whirl orbit (more sensitive to damping) depends upon factors that are very difficult to predict in most cases.

If there are two masses on a single massless shaft, as illustrated in Fig. 11-8, there are two different modes of whirling that can occur. The lower of the two corresponding critical speeds occurs with the shaft bowed so that the two masses are on the same side of the bearing centerline. At the second critical speed, which is much higher than the first, the shaft takes a *double bend* (Fig. 11-8), so the two masses stay on opposite sides of the centerline during the whirling. Again, as in the case of the single-mass rotor, no continuous flexing of the shaft occurs; the shaft simply assumes

11-4 OTHER CRITICAL SPEED CONSIDERATIONS

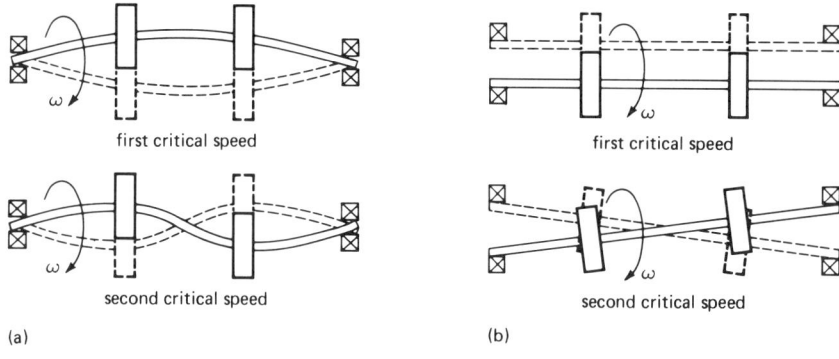

FIG 11-8. Whirl modes of a two-mass rotor.
(a) Very "stiff" bearings.
(b) Very "soft" bearings.

the curvature corresponding to its speed of rotation and maintains the bent position as it rotates. In Fig. 11-8 are also shown the ways whirling will appear at the two critical speeds if most of the compliance is found in the bearings and/or their support structure.

A lumped parameter rotor with additional masses will have the same number of critical speeds as the number of masses. Rayleigh's method (Appendix E) is useful for finding the first critical speed of a lumped parameter rotor with two or more masses. It cannot, however, be used to determine higher critical speeds.

Any real rotor is, of course, a distributed parameter system, and therefore it will have an infinite number of critical speeds. Usually, however, only the first few are of any practical significance. If the rotor has flywheels, or sections with much more mass than the basic shaft, then it may be possible to obtain accurate answers from a simple lumped parameter model. Often, however, the shape of the rotor is such that a lumped parameter model, to be accurate, must consist of many masses connected together by very short massless beam springs. Solutions for such models can be found by methods based partly on trial-and-error techniques, using the digital computer, as is also true for distributed parameter vibrating systems [4, 13, 15]. As a matter of fact, the same techniques apply, since any rotor's critical speeds are the same as its natural frequencies of lateral vibration.[5]

---

[5]The validity of this statement can be readily shown for the lumped parameter rotor with a single mass, since $\omega_{cr} = \omega_n = \sqrt{k/M}$. It can also be shown to be essentially true for any other lumped or distributed parameter system. With a rigorous analysis of a real rotor, however, one will find a slight difference between the critical speed and the natural frequency of lateral vibration because of secondary effects such as gyroscopic moments.

Whirling can occur for rotors mounted other than with a bearing at each end. The rotor may be overhung, as a cantilever, or three or more bearings may be used (see Fig. 11-9).

It is relatively easy to determine the critical speeds of shafts with uniform cross section, since the technique presented in Chapter 10 for lateral vibration of beams can be used (remembering that $\omega_n = \omega_{cr}$). The overhung uniform shaft of Fig. 11-10 will therefore have critical speeds and mode shapes to match the natural frequencies of a rigidly supported cantilever beam of the same size, shape, and material.

In designing a machine that contains a rotor, it is important to check that the rotor will not have to run continuously at or near a critical speed. Generally speaking, it is good practice to keep the speed of operation well below the first critical speed, but this is not always practical. In the case of very high-speed machines such as turbines, staying below the first critical speed might require the rotor to be so large and heavy that other serious problems would result. Operation between the first and second critical speeds is often the most practical solution, provided that the rotor accelerates through the first critical speed rapidly enough that the whirl amplitude is not allowed to build up to its full equilibrium critical speed value. (Deceleration back through the critical speed upon shutdown must also be considered.) Good balance makes it easier to pass through the first critical speed without problems. Operation between the second and third critical speeds or between the third and fourth may also be possible. With extremely precise balancing, some systems have been able to be run continuously at a critical speed without ill effect, but this is not normally good practice.

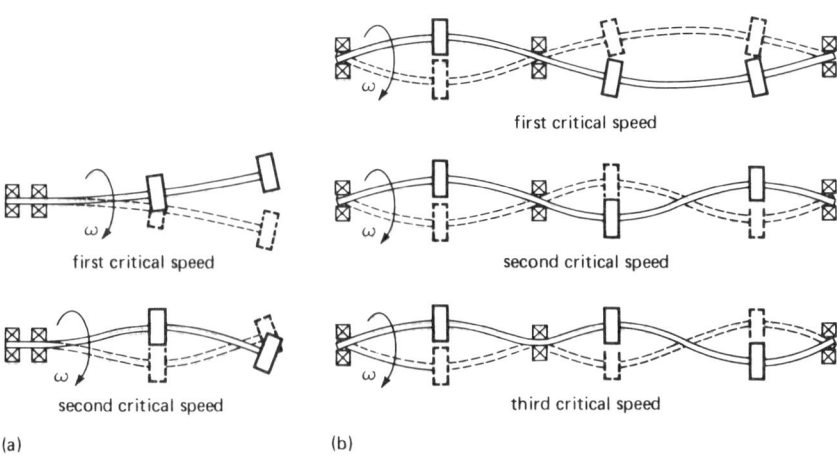

FIG 11-9. Whirl modes of other rotors.
    (a) Cantilevered two-mass rotor.
    (b) Rotor with three bearings and three masses.

11-4 OTHER CRITICAL SPEED CONSIDERATIONS

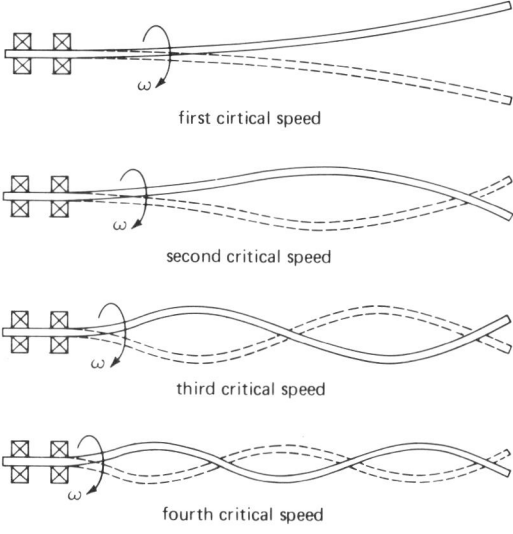

FIG 11-10. Whirl of a distributed parameter cantilevered rotor with uniform cross section.

The frequency band over which whirling is a serious problem is directly determined by the eccentricity ϵ. The whirl characteristics given by Eq. 11-2 and Fig. 11-2, though for an idealized single-mass rotor, are very similar to those of a real rotor. Whirl is a problem when r is appreciable, and at any speed r is seen to be directly proportional to ϵ.

Example 11-3. A highly unbalanced single-mass rotor has an eccentricity of 0.050 inch, and a critical speed of 2000 r/min.

(a) What range of speeds must be avoided in order to keep the whirl amplitude below 0.10 inch?

(b) If the eccentricity is reduced to 0.001 inch by precision balancing, what would the answer be?

Solution:

(a) The r/ϵ ratio below which we would like to operate is, for the original highly unbalanced rotor,

$$\left|\frac{r}{\epsilon}\right| = \frac{0.10}{0.05} = 2.0$$

It can be seen, from Fig. 11-2, that whirl greater than this amount will occur for

$$0.8 < \frac{\omega}{\omega_n} < 1.4$$

We must therefore not operate between 1600 and 2800 r/min if the whirl amplitude is to remain below 0.10 inch.

(b) If ϵ is reduced to 0.001 inch, then a whirl amplitude of 0.10 inch corresponds to the whirl-eccentricity ratios

$$\frac{r}{\epsilon} = \pm \frac{0.10}{0.001} = \pm 100$$

Since Fig. 11-2 does not cover this range, we must use Eq. 11-2, which may be manipulated into the form

$$\frac{\omega}{\omega_{cr}} = \sqrt{\frac{r/\epsilon}{1 + r/\epsilon}}$$

For the rotor under consideration, we find the two corresponding speed ratios

$$\left(\frac{\omega}{\omega_{cr}}\right)_1 = \sqrt{\frac{100}{101}} = \sqrt{0.99} = 0.995$$

$$\left(\frac{\omega}{\omega_{cr}}\right)_2 = \sqrt{\frac{-100}{-99}} = \sqrt{1.01} = 1.005$$

With the well-balanced rotor, whirl amplitude will exceed 0.10 inch only over the speed range between 1990 and 2010 r/min.   (*Ans.*)

Fig. 11-2 and Eq. 11-2 are most accurate for relatively small whirl amplitudes. As the magnitude of the whirl increases, nonlinearities and other factors that were ignored in the derivation become significant. (Note, for example, that r/ϵ can obviously never be infinite for a real rotor.)

11-5 Similar Rotor Instability Phenomena

There are other rotor instability phenomena that, though somewhat similar to critical speed whirling, are caused by different factors and result in a different action of the rotor. Although the theories will not be pursued in this introductory book, it is felt that these phenomena should at least be mentioned, since they are very important to consider in the design of many high-speed rotors.

Journal bearing whirl, or *oil whip* [4, 21], can occur in rotors mounted in hydrodynamic journal bearings, with the action of the oil film of a well-lubricated bearing causing the phenomenon. For the normal type of journal bearing, the *threshold speed* of oil whip is twice the first critical speed of the rotor. Unlike critical speed whirling, it is not possible to reduce this type of whirl by increasing the rotor speed. This will merely increase the problem; hence the term *threshold speed*. Journal bearing whirl is often eliminated by changing to ball or roller bearings or to special types of journal bearings.

Hysteresis whirl [4, 21] is sometimes found in rotors having a large amount of hysteresis. The hysteresis of a solid metal shaft is insufficient to cause this phenomenon, but it has often been observed in rotors that are built up by attaching disks to a shaft with interference fits. Rubbing and friction on the interference fit surfaces during flexing of the rotor give the characteristics of hysteresis that can be shown to cause an instability or whirling at all speeds above the first critical speed—also called the *threshold speed* in this case. One normally expects friction to help damp out instabilities, but this is one case in which friction adds to the problem. The solution to hysteresis whirl normally consists of going to a solid rotor or changing the design so that operation is always below the threshold speed. The phenomenon has also been observed by the authors in a rotating solid rotor driven through a universal joint with a large amount of Coulomb friction.

Half-speed whirl [4, 21] sometimes occurs in *horizontally mounted rotors* that are either noncircular in cross section or else have considerable unbalance. Whirl occurs when the rotor is running at half the first critical speed, hence the name. The phenomenon does disappear, however, if the rotor speed is either lowered or raised somewhat.

Journal bearing whirl, hysteresis whirl, and half-speed whirl all differ from critical-speed whirling in one important respect: Continuous flexing of the rotor shaft occurs with all three phenomena; that is, the whirling speed is not the same as the rotor speed. Actually, the *whirl speed* (i.e., the frequency at which the mass center of gravity goes around its circular orbit) is in each case the same as the first critical speed, whereas the *speed of rotation* of the rotor may have any number of different values. Failure from fatigue is therefore an additional problem to consider with these phenomena, along with high bearing loads, mechanical interference, and vibration excitation.

11-6 Conclusion

Avoidance of critical speed whirling is an important consideration in the design and application of rotating machinery. Whirling is often a violent phenomenon and has been the cause of many mechanical failures.

Although a lumped parameter rotor model will have the number of critical speeds equal to the number of masses, any real (distributed parameter) rotor will have an infinite number of critical speeds. Only the first few of these are normally of any practical significance, however. The values of the critical speeds of a rotor depend upon its mass and the overall flexibility of the rotor-bearing system; compliance in the bearings or bearing mounts is as significant as the flexibility of the rotor shaft itself. Each critical speed of a rotor is equal (except for a slight difference due to secondary effects) to a natural frequency of lateral vibration.

There are several other rotor-bearing instability phenomena that are very similar in their physical manifestations to critical speed whirling. These should also be considered by the designer of rotating machinery.

Problems

11-1. The mass of the rotor shaft shown is negligible compared to that of the single flywheel. When a transverse force $F = 200$ lb is applied to the flywheel, it is found to deflect a total of 0.040 in., 0.015 in. of which is from deflection in the bearings and 0.025 in. from shaft bending. Determine the critical speed in revolutions per minute.

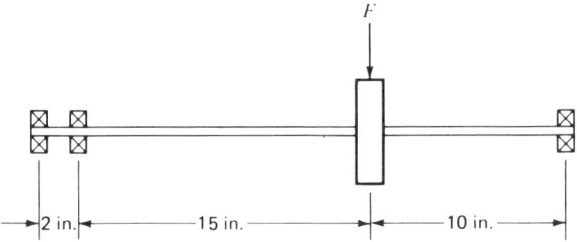

11-2. Is the given shaft rotating *at*, *below*, or *above* its critical speed? Explain your answer briefly.

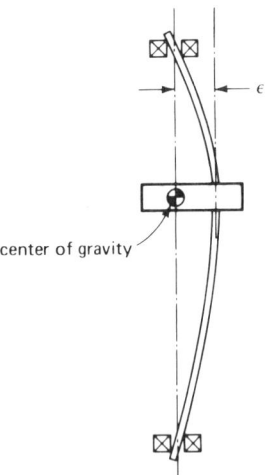

PROBLEMS

11-3. In the illustration shown, the shaft stiffness at point P is k_s and bearing stiffness at each end is k. Determine an expression for the critical speed of the rotor.

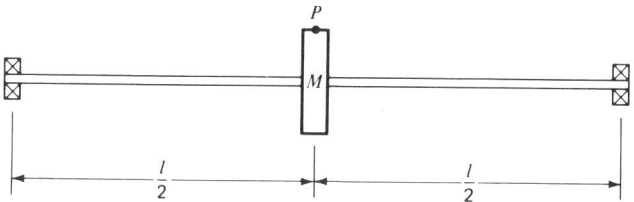

11-4. Determine the critical speed of the given shaft-mass combination. The system is shown in static equilibrium; that is, the weight of the concentrated mass produces a static deflection of 0.0050 meter.

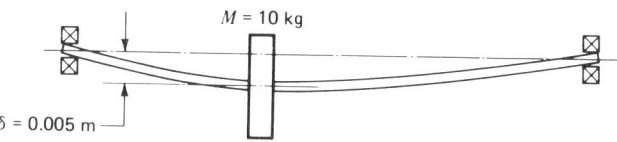

11-5. An automobile drive shaft is made of steel tubing with an outside diameter of 3.0 in. and a wall thickness of 0.050 in. What is the maximum allowable distance between the universal joints if the critical speed is to be no less than 9000 r/min? (The solution to the first part of Prob. 10-6 is required to work this problem.)

11-6. The weight of the flywheel causes it to deflect 1.5 mm from the bearing centerline when the cantilevered rotor is mounted horizontally as shown. The bending of the shaft accounts for 0.4 mm, the rest of the deflection being from bearing compliance. Assuming that the shaft has negligible mass, determine the critical speed.

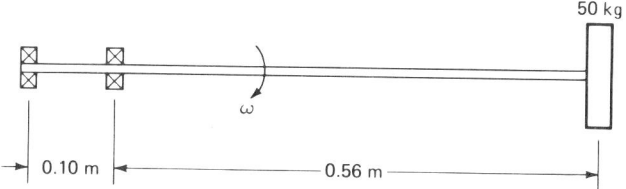

11-7. Each of the two bearings shown has a spring constant of 1500 N/mm. If the shaft is completely rigid and has negligible mass, determine ω_{cr}.

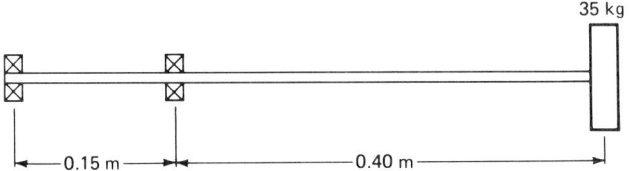

11-8. The two views shown correspond to rotational speeds very close to critical, one being slightly above and the other slightly below ω_{cr}. Identify the two views.

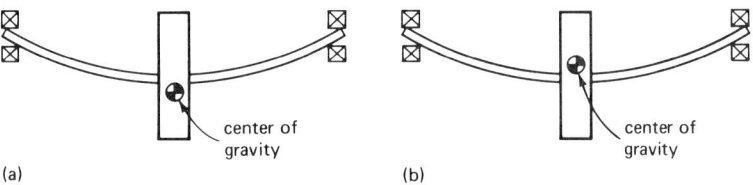

11-9. For the given system determine the degree of rotor balance required if the whirl amplitude is not to exceed 0.030 mm at a speed of 1625 r/min.

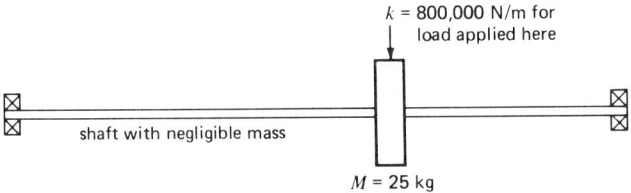

11-10. Determine the whirl amplitude of the given rotor at a speed of 11,500 r/min.

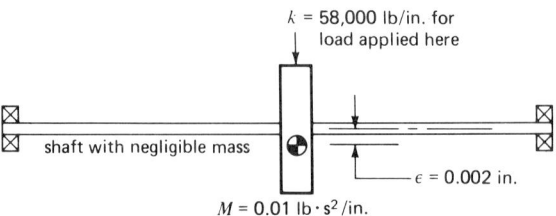

PROBLEMS

11-11. The rotor shown is very stiff with respect to the ball bearing compliance. The stiffness characteristics of the two identical bearings are dependent upon the amount of axial preload, as given by the accompanying curves of deflection versus force. Plot the critical speed as a function of preload.

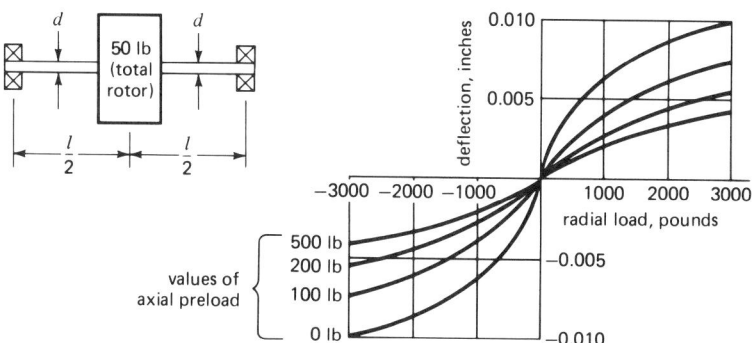

11-12. A steel rotor is mounted in ball bearings and bearing housings having the given spring constants, which are uniform in all radial directions. Determine the critical speed, making the approximation that the shaft is massless.

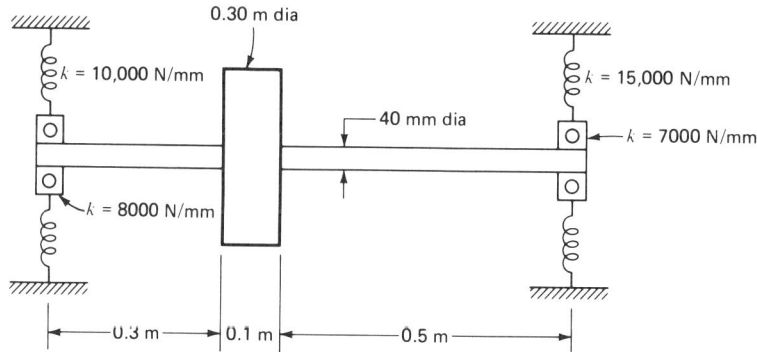

11-13. Assuming bearings and bearing housings are much stiffer than the shaft, determine the critical speed of the given steel rotor. (Assume the mass of the shaft to be negligible.)

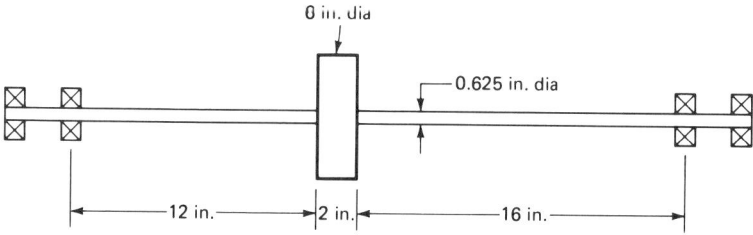

11-14. Determine the critical speed of the steel rotor shown. The given compliances of the ball bearings and bearing housings are uniform in all radial directions. (Assume the mass of the shaft to be negligible.)

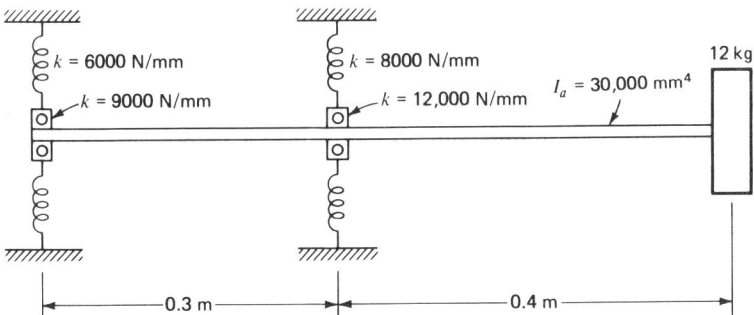

11-15. A 100-lb force was applied radially to each of two disks, with the resulting deflections as shown. Calculate the first critical speed using Rayleigh's energy method (Appendix E). Assume the following values: $M_1 = 0.05$ lb · s²/in., $M_2 = 0.08$ lb · s²/in.

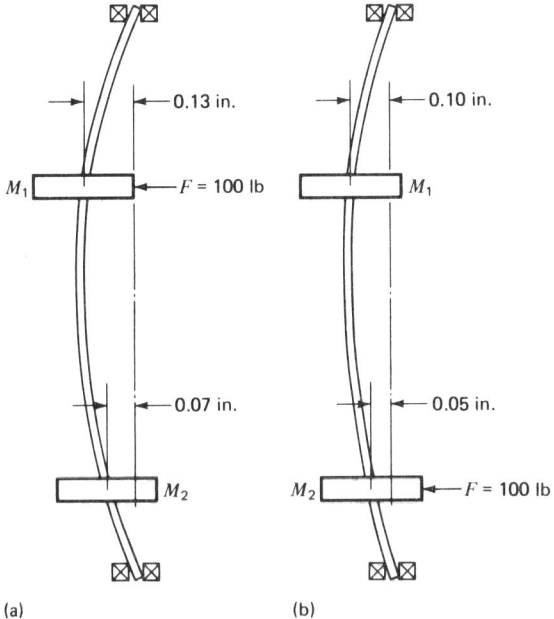

11-16. Using Rayleigh's energy method (Appendix E), determine the first critical speed of the given rotor. Assume that the only significant compliance is in the shaft and that the shaft has negligible mass.

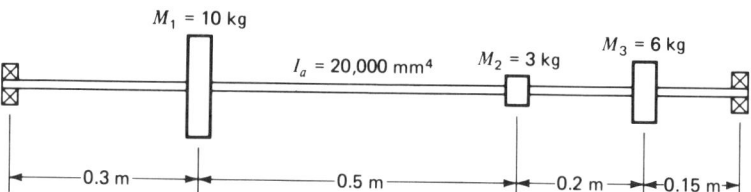

12
Balance of Rotors

12-1 Introduction

Unbalanced rotors are a major cause of vibration problems. A rotor running at high speed will produce a *rotating centrifugal force vector* that has a magnitude proportional to the product of the unbalance and the square of the rotational speed. This force vector can easily excite vibrations of the machinery in a number of different modes: up and down, sideways, and torsionally about various axes. In addition, most unbalanced rotors will also produce a *rotating moment vector* (as will be explained later in this chapter), which further increases the probability of exciting some type of undesirable vibration.

Besides causing vibration problems, a rotor with excessive unbalance may also cause the bearings to be overloaded, and will make critical speed whirling a more severe problem if such a rotor must pass through the critical speed or operate for periods of time near it.

What constitutes excessive unbalance depends entirely upon the parameters and use of any particular rotor. Balance requirements for a shaft in a gearbox that rotates at 100 r/min would have little relationship to the requirements for a gas turbine that operates between its first and second critical speeds at 30,000 r/min.

The material in this chapter is somewhat different from that covered in the remainder of the book, and it could be omitted without adversely affecting the understanding of the basic dynamic system theory. Analysis of rotor unbalance does not involve differential equations and their solution but consists instead of determining the equilibrium of rotating forces and moments. It is analogous in many respects to the static analysis of beams. The topic is important to the field of mechanical vibrations because rotor unbalance often provides the input to a vibrating system and is also one of the principal parameters involved in critical speed whirling problems.

12-2 Static Balance

A rotor has static balance when its center of gravity lies precisely on its axis of rotation. If a balanced rotor is mounted in low-friction bearings with its shaft horizontal, gravity will produce no moment that will tend to rotate it regardless of its angular position. If, however, the rotor is unbalanced (i.e., if its center of gravity is offset from the axis of rotation by an *eccentricity* ϵ), the rotor will tend to rotate except when the heavy side is down.[1]

Static unbalance is equal to the product of rotor mass and eccentricity (with the units of mass · distance). Since there is always a specific direction associated with the static unbalance (the direction from the axis of rotation to the center of gravity), the unbalance can be represented by a vector in this direction having a length equal to the magnitude of the unbalance.

If a perfectly balanced rotor of mass M_0 is caused to become unbalanced by the addition of a mass M_1 set at a distance r_1 from the axis of rotation (Fig. 12-1a), the resultant rotor unbalance will be $M_1 r_1$.

We would expect to get the same answer by finding the position of the new center of gravity of the rotor and calculating the unbalance magnitude as the product of the corresponding eccentricity and the total rotor mass; that is, unbalance = $\epsilon(M_0 + M_1)$. To verify this equivalency, we first of all determine the center of gravity of the two-mass rotor by using the basic principle that the product of the first mass and its distance from the center of gravity must equal the product of the second mass and its distance from the center of gravity and that the center of gravity lies on a straight line connecting the two masses. This principle gives the equation (referring to

---

[1]The heavy side up is also a condition of equilibrium, but it is *unstable equilibrium*. Thus it is very difficult to keep the rotor in this position if the bearings have low friction.

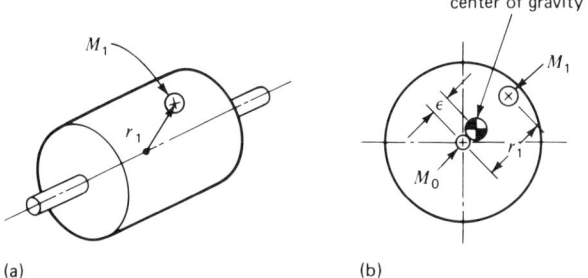

FIG 12-1. (a) Rotor caused to become unbalanced by the addition of M_1.
(b) Figure used to determine the center of gravity of the two-mass system.

Fig. 12-1b)

$$\epsilon M_0 = (r_1 - \epsilon) M_1 \qquad (12\text{-}1)$$

By rewriting Eq. 12-1 in the form

$$\epsilon (M_0 + M_1) = r_1 M_1 \qquad (12\text{-}2)$$

it can be seen that the two alternative expressions for the unbalance of the rotor are indeed equivalent.

It should be noted that the unbalance magnitude $M_1 r_1$ could also be obtained by adding a mass of $\frac{1}{2} M_1$ to the rotor if it were placed at the distance $2r_1$ from the axis of rotation.

If a rotor can be considered to consist of a number of lumped masses situated at various angles and at various distances from the axis of rotation, then each of them will contribute a certain unbalance to the system. The net unbalance of the rotor is the vector sum of these unbalance components.

Example 12-1. Determine the static unbalance of the rotor illustrated in Fig. 12-2a, consisting of four point masses located in the positions shown.

Solution: The unbalance contributed by each point mass is represented by a vector of length equal to the product of its mass and its distance from the axis of rotation. By adding these vectorially, as shown in Fig. 12-2b, the net rotor unbalance is represented by the vector $\overrightarrow{od}$, with the magnitude of 0.025 kg · m, as given in the figure.

Rotor models consisting of several point masses are useful for demonstrating the principles of unbalance, and as a guide for the design of rotors with minimum unbalance. All real rotors are distributed mass objects, however, so that calculation of unbalance with a high degree of accuracy is

12-3 DYNAMIC BALANCE

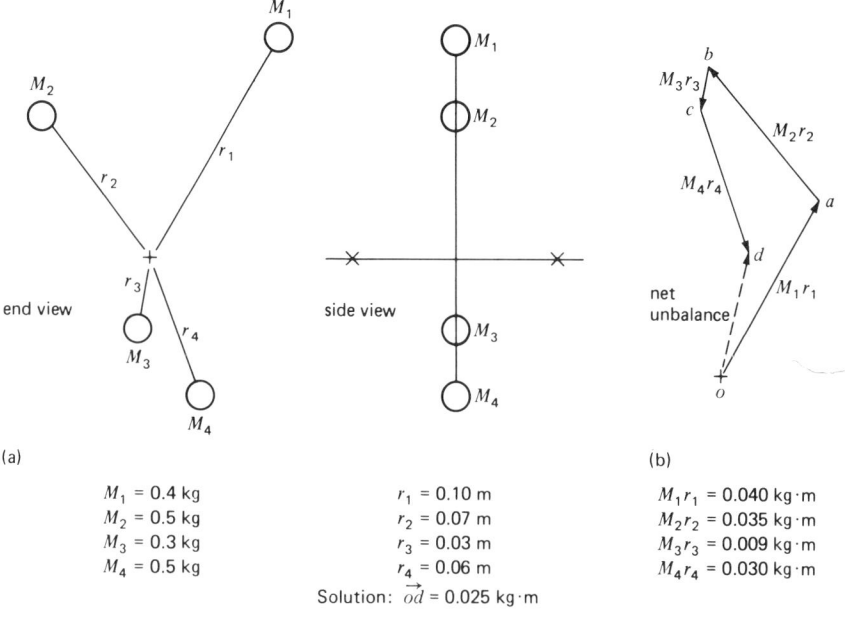

FIG 12-2. Unbalanced rotor having all its mass within a single transverse plane.
(a) Model of rotor (drawn to scale).
(b) Static unbalance vector polygon (drawn to scale).

impractical. Static balancing is normally achieved by experimental techniques, mounting the rotor with its shaft horizontal in low-friction bearings and making corrections until it is found to remain stationary when set at any angular position. Corrections can be made by either adding or removing mass.

12-3 Dynamic Balance

For the analysis of dynamic balance problems, it is convenient to assume an imaginary centrifugal force acting on each mass located at a radius from the shaft centerline. The *principle of D'Alembert* (as introduced in footnote 2 of Chapter 11) can then be used, treating the rotor as an object in static equilibrium under the combined effects of the true forces (bearing reactions) and imaginary inertia forces. A rotor is said to have dynamic balance if, when rotating, there is no net force or net torque due to centrifugal forces that must be overcome by the bearings.

Centrifugal forces always act radially outward from the axis of rotation. In the lumped parameter rotor of Fig. 12-2a, the centrifugal force on mass 1 will have the magnitude

$$CF_1 = M_1 r_1 \omega^2$$

and can be represented by a vector of length equal to this value and having the same direction as the static unbalance vector due to M_1 ($\overrightarrow{oa}$ in Fig. 12-2b). The other three masses will produce similar centrifugal force vectors, each having the same direction as the static unbalance vector caused by that mass.

Since ω^2 is present in each of the centrifugal force vectors, it could be dropped without affecting their relative lengths, and each resultant vector would be identical to the static unbalance vector corresponding to the same mass. The net centrifugal force vector acting on the rotor at any given speed ω can therefore be obtained by simply multiplying the net static unbalance vector ($\overrightarrow{od}$ in Fig. 12-2b) by ω^2.

If the point masses of the rotor all lie within a single transverse plane, as shown in the side view of Fig. 12-2a, then the net centrifugal force vector is the sole dynamic effect of the unbalance. For this limited case, static and dynamic balance are synonymous. As a practical consequence, satisfactory dynamic balance of short rotors (flywheels, fans, propellers, etc.) may often be achieved by static balancing procedures.

It is more common, however, for a rotor *not* to have most of its mass located in a single transverse plane. In this case the problem of dynamic balance becomes somewhat more complex, and static and dynamic balance cannot be considered equivalent. Consider, for example, the rotor illustrated in Fig. 12-3, having two point masses of equal value situated on opposite sides of the axis of rotation and at the same distance from it. The center of gravity of this rotor lies exactly on the axis of rotation, so the rotor has perfect static balance. The net centrifugal force during rotation is also zero, but the locations of the two masses will in this case create a couple that cannot be ignored. Looking at the side view of Fig. 12-3, the two equal centrifugal force vectors can be seen to produce the couple $Mr\omega^2 l$. This couple can be represented by a vector as shown in the end

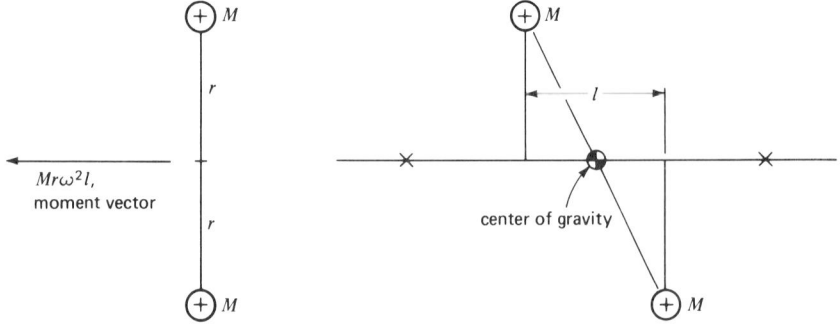

FIG 12-3. Two-mass rotor having static balance but dynamic unbalance.

12-3 DYNAMIC BALANCE

view based on the "right-hand rule."[2] The direction of this vector is fixed with respect to the rotor, and therefore rotates with it. It can cause several types of problems, depending upon the characteristics of the machinery involved. Since it has vertical and horizontal components (with the rotor shaft horizontal) that vary sinusoidally with time, it will tend to excite undesirable rotational vibrations in the structure in which it is mounted. The additional bearing loads necessary to counteract the couple might cause premature bearing failure. The couple might cause significant rotor distortion at high speeds, although this is normally less significant than deformation caused by an unbalance *force* vector.

To have complete dynamic balance of a long rotor, not only must the vector sum of the centrifugal force vectors be zero (which is equivalent to saying the rotor must be statically balanced), but the vector sum of the moment vectors must also be zero. In calculating the moment that any particular centrifugal force vector produces, it is necessary to specify the point about which the moment is measured. In general, the moments can be summed about any point along the axis of rotation, but the proper choice (which depends on the type problem under consideration) will facilitate the solution. The two different types of rotor balance problem—(1) the determination of bearing reactions caused by an unbalance and (2) the determination of masses to be added or subtracted to achieve balance—will now be considered, with an example problem for each.

BEARING REACTIONS CAUSED BY ROTOR UNBALANCE

To determine the bearing reactions caused by an unbalanced rotor, it is best to take moments about one of the bearings so that there will be no moment from the reaction of that bearing. The following example illustrates the proper method of solution.

Example 12-2. Determine the load on each bearing caused by the unbalance of the lumped-mass rotor shown in Fig. 12-4a, which rotates at 1000 rad/s.

Solution: Choosing the left bearing as a convenient point about which to compute moments, the moment due to mass 1 is determined to be $M_1 r_1 \omega^2 l_1$. The corresponding vector $\overrightarrow{o'a'}$ is drawn on an end view of the rotor (Fig. 12-4b) in the direction determined by the right-hand rule. The moment vectors due to the other two masses are determined and added vectorially to complete the moment polygon, with $\overrightarrow{o'c'}$ being the net moment about the bearing on the left from rotor

---

[2]The "right-hand rule" is used to determine the direction of a vector representing a moment. If the fingers of the right hand are curled to point in the direction that the moment tends to pivot the shaft, the extended thumb will point in the direction of the moment vector. An alternative explanation of the rule is that the moment vector should point in the direction that a screw with right-hand threads would advance if caused to rotate by the moment.

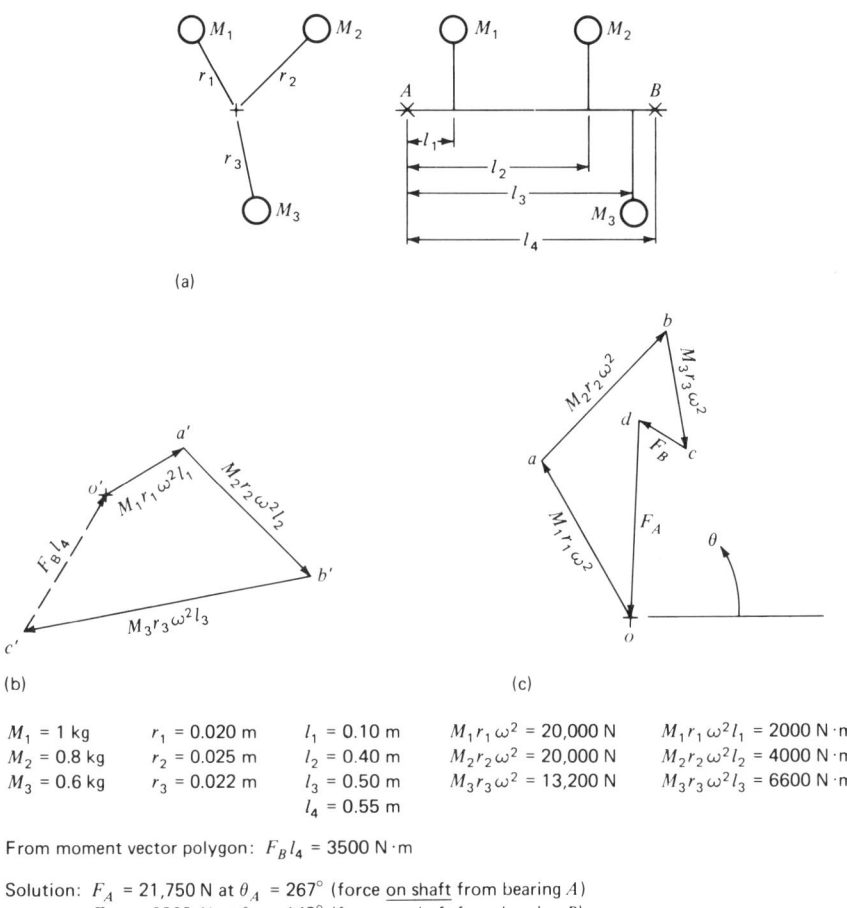

$M_1 = 1$ kg	$r_1 = 0.020$ m	$l_1 = 0.10$ m	$M_1 r_1 \omega^2 = 20{,}000$ N	$M_1 r_1 \omega^2 l_1 = 2000$ N·m
$M_2 = 0.8$ kg	$r_2 = 0.025$ m	$l_2 = 0.40$ m	$M_2 r_2 \omega^2 = 20{,}000$ N	$M_2 r_2 \omega^2 l_2 = 4000$ N·m
$M_3 = 0.6$ kg	$r_3 = 0.022$ m	$l_3 = 0.50$ m	$M_3 r_3 \omega^2 = 13{,}200$ N	$M_3 r_3 \omega^2 l_3 = 6600$ N·m
		$l_4 = 0.55$ m		

From moment vector polygon: $F_B l_4 = 3500$ N·m

Solution: $F_A = 21{,}750$ N at $\theta_A = 267°$ (force on shaft from bearing A)
$\phantom{\text{Solution: }}F_B = 6363$ N at $\theta_B = 148°$ (force on shaft from bearing B)

FIG 12-4. Determining the bearing reactions caused by an unbalanced three-mass rotor (Example 12-2). All drawings are to scale.
(a) Model of rotor.
(b) Vector polygon for moments about bearing A.
(c) Complete force vector polygon.

unbalance.[3] (It should be noted that this moment is *not* in general equal to the product of the net centrifugal force vector and its distance from the left bearing.)

---

[3] It is a property of vectors that they may be moved (in translation) without affecting their validity as long as they retain the same length and still point in the same direction. The moment vectors from dynamic rotor unbalance can therefore all be drawn in a single plane corresponding to an end view of the rotor.

12-3 DYNAMIC BALANCE

The bearing forces must have the proper values to overcome the effect of the rotor unbalance. Equating the sum of all moments about the left bearing to zero, it is determined (from the moment vector polygon of Fig. 12-4b) that the bearing on the right must produce the moment vector $\overrightarrow{c'o'}$. The bearing force F_B must therefore be equal to the magnitude of $\overrightarrow{c'o'}$ divided by the distance l_4. Its direction, determined from the right-hand rule, is as shown in Fig. 12-4c. (Note that F_B is the force acting on the rotor, the force acting on the bearing being equal in magnitude but opposite in direction.)

There are now two alternative ways in which the second bearing reaction, F_A, can be obtained. One is to construct a moment vector polygon for moments about the bearing on the right, determining F_A in the same manner as was used for F_B. The other way is to construct a force polygon and use the requirement that the sum of the *forces* on the rotor must be zero. The second approach, which is slightly easier, will be used here. By vectorially adding the three centrifugal force vectors and the bearing reaction F_B as shown in Fig. 12-4c, the vector F_A representing the force of the left bearing is simply that required to close the polygon. The numerical values of F_A and F_B and their directions are given in the figure.

Note that both vector polygons (Fig. 12-4b and c) are fixed with respect to the rotor and rotate with it.

In comparing Fig. 12-4b with 12-4c it can be seen that the moment vector caused by each mass is rotated 90° clockwise from the centrifugal force vector of that mass when the right-hand rule is used. The moment vector caused by the bearing force F_B on the rotor is also rotated 90° clockwise from that force. Note, however, that a moment vector is rotated 90° *counterclockwise* from the force vector producing it if that force is to the *left* of the arbitrary point about which the moments are calculated. This information can be useful as a check on whether or not the right-hand rule has been properly applied.

ACHIEVING DYNAMIC BALANCE

In order to dynamically balance a rigid rotor, it is necessary to add (or subtract) mass in each of two different transverse planes. Although a single balancing mass could be made to eliminate the moment unbalance about any point, it would be an unlikely coincidence if it also eliminated the net centrifugal force vector to give complete dynamic balance. The two transverse planes to be used for balancing may be at any position along the rotor. The greater the distance between them, however, the smaller the mass changes necessary to achieve dynamic balance. To determine the corrections necessary for balancing a given rotor, it is most convenient to take moments about one of the transverse balancing planes.

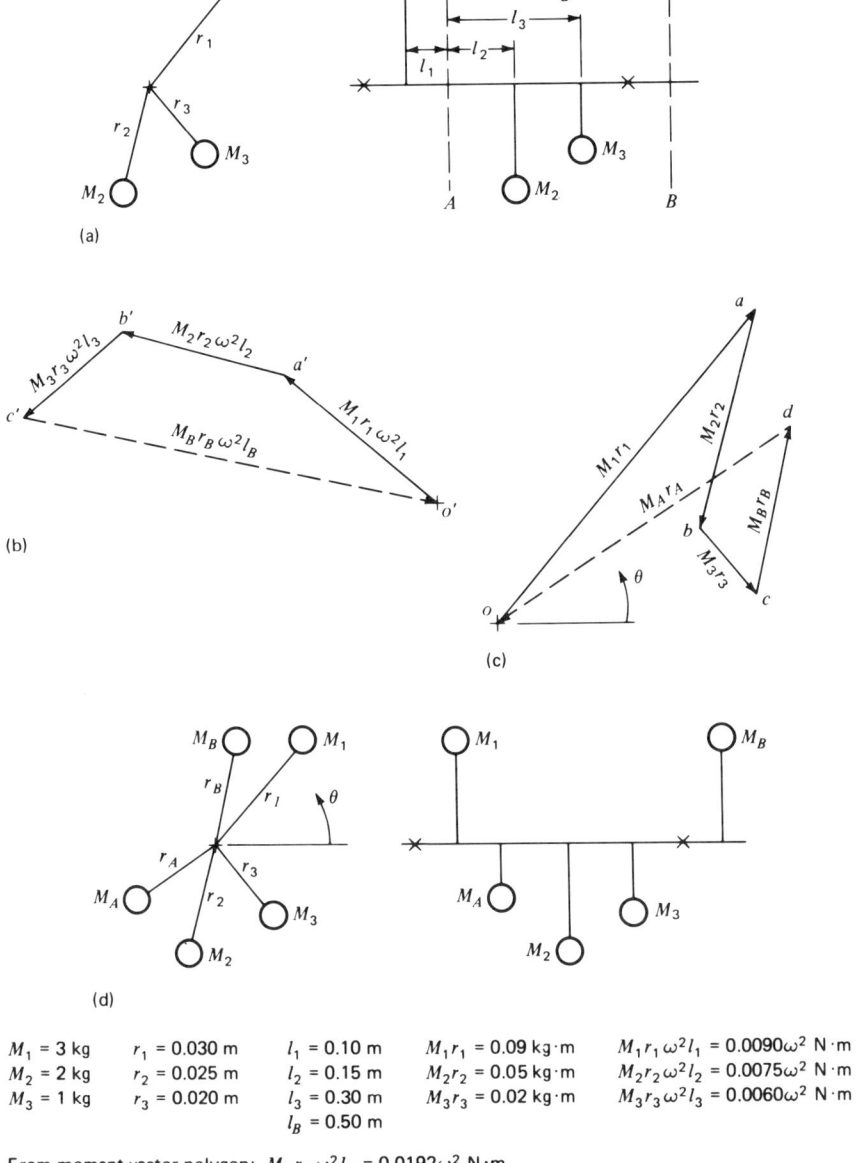

$M_1 = 3$ kg	$r_1 = 0.030$ m	$l_1 = 0.10$ m	$M_1 r_1 = 0.09$ kg·m	$M_1 r_1 \omega^2 l_1 = 0.0090\omega^2$ N·m
$M_2 = 2$ kg	$r_2 = 0.025$ m	$l_2 = 0.15$ m	$M_2 r_2 = 0.05$ kg·m	$M_2 r_2 \omega^2 l_2 = 0.0075\omega^2$ N·m
$M_3 = 1$ kg	$r_3 = 0.020$ m	$l_3 = 0.30$ m	$M_3 r_3 = 0.02$ kg·m	$M_3 r_3 \omega^2 l_3 = 0.0060\omega^2$ N·m
		$l_B = 0.50$ m		

From moment vector polygon: $M_B r_B \omega^2 l_B = 0.0192\omega^2$ N·m

Solution: $M_A r_A = 0.0787$ kg·m at $\theta_A = 213°$
$\quad\quad\quad\;\; M_B r_B = 0.0384$ kg·m at $\theta_B = 78°$

FIG 12-5. Achieving dynamic balance of a rotor by the addition of mass in two arbitrary transverse planes (Example 12-3). All drawings are to scale.
  (a) Rotor to be dynamically balanced.
  (b) Vector polygon for moments about balancing plane A–A.
  (c) Static unbalance vector polygon.
  (d) Rotor balanced by the addition of two masses.

Example 12-3. Determine the magnitudes and locations of the two masses, which, when placed in planes A–A and B–B, will give the rotor of Fig. 12-5a complete dynamic balance.

Solution: The method of solution is similar to that employed in Example 12-2. To find the mass to be added in plane B–B, a vector polygon (corresponding to the end view of the rotor) is drawn for unbalance moments about plane A–A, as illustrated in Fig. 12-5b. Forces from the bearings are not included in the moment vector polygon since they will be zero (ignoring the effect of gravity) when balance has been achieved. Note that the moment vector caused by M_1 is rotated 90° counterclockwise from the corresponding centrifugal force vector, since M_1 is to the left of plane A–A. The mass M_B is placed in plane B–B as shown in Fig. 12-5d so that its centrifugal force produces the moment vector $\overrightarrow{c'o}$ that closes the vector polygon. The required mass of M_B depends upon its radius r_B from the axis of rotation.

Next the static unbalance vector polygon is drawn, including the effect of the new balance mass M_B (Fig. 12-5c). By placing the second balancing mass M_A in plane A–A as shown (Fig. 12-5d), the unbalance vector $\overrightarrow{do}$ is produced to close the polygon and give static balance (and therefore no net centrifugal force). The required mass of M_A depends upon the radius r_A. Note that M_A has no effect upon the previously achieved moment balance. The addition of M_A and M_B gives complete dynamic balance to the rotor.

A real rotor will have distributed mass, so it is not practical to determine how to achieve dynamic balance by calculations alone. Practical dynamic balancing is done on machines specifically designed for that purpose. A balancing machine allows the rotor to be set in flexibly mounted bearings, with a means of driving it at a proper speed. The resultant shaking caused by the rotor unbalance is measured electronically in two planes (the planes to be used for balancing), with both amplitude and rotor angle being determined. These electric signals are processed to give the operator values of mass to add or subtract at specified radii in each of the two balancing planes, and the rotor angles at which the corrections are to be made. For additional information on practical dynamic balancing, the reader should consult one of the references [4, 22].

12-4 Conclusion

The principles of static and dynamic balance have been presented in this chapter. An understanding of the fundamental principles should allow the reader to design rotor systems with a minimum amount of unbalance. The

Problems

12-1. Determine the correction for the given system in which the masses are in a single plane and have the following sizes and positions (with angles measured counterclockwise from the right horizontal axis).

$M_1 = 4$ kg $M_2 = 3$ kg $M_3 = 5$ kg
$r_1 = 0.7$ m $r_2 = 0.9$ m $r_3 = 0.6$ m
$\theta_1 = 90°$ $\theta_2 = 120°$ $\theta_3 = 30°$

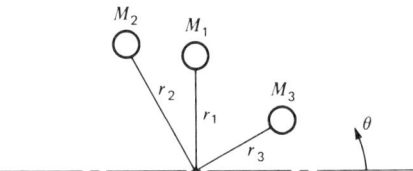

12-2. A 16-kg flywheel is found to have an unbalance force of 100 newtons when rotating at 6000 r/min. Determine the masses to be added (or removed) at points A and B to achieve balance.

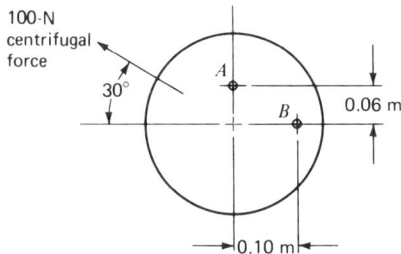

12-3. A 50-lb flywheel has its center of gravity offset 0.005 in. from the axis of rotation. Determine the magnitude of the rotating unbalance force vector at a speed of 1000 r/min.

PROBLEMS

12-4. Determine the directions and magnitudes of the bearing reactions of the rotor in the position shown. System parameters are as follows: $M = 0.002$ lb · s²/in., $r = 0.5$ in., $l = 4$ in., and $\omega = 600$ rad/s.

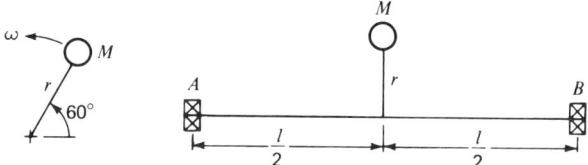

12.5. An automobile tire is to be statically balanced by weights placed on the outside edge of the rim, 7.0 in. from the axle centerline. Although balance could be accomplished by a single weight of the right magnitude, a pair of equal weights is used so that no trimming of the standard sizes is necessary. With a tire unbalance of 11 oz · in., determine the angular locations of two 1-oz weights that will achieve balance.

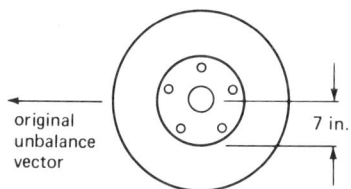

12-6. A shaft rotates at 1500 r/min and carries a single unbalance mass as shown. Determine (a) the corrections in the planes A–A and B–B that will give dynamic balance and (b) the bearing reaction at C before and after balancing.

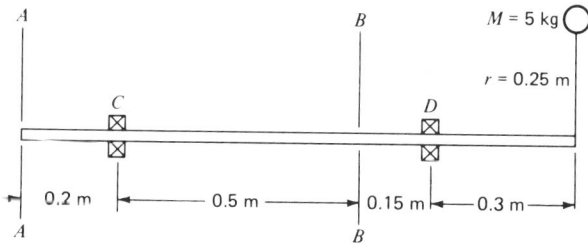

12-7. Find the directions and magnitudes of the two bearing reactions (at A and B) if the rotor shown has a speed of 1000 r/min.

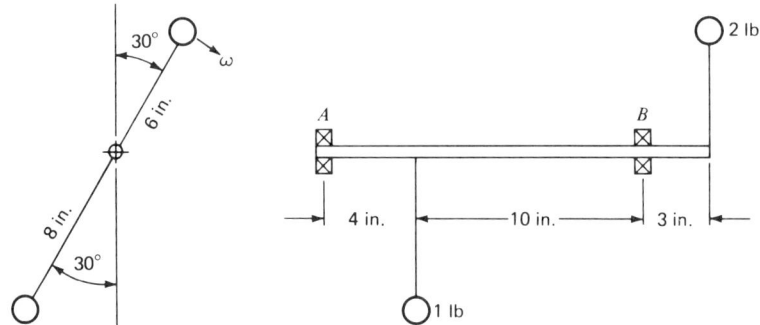

12-8. Determine the magnitudes and angular positions of masses to be added in planes A–A and B–B at 0.4-meter radii in order to dynamically balance the rotor. System parameters are as follows: $r_1 = 0.4$ m, $\theta_1 = 60°$, $M_1 = 0.3$ kg, $r_2 = 0.3$ m, $\theta_2 = 150°$, $M_2 = 0.4$ kg, $r_3 = 0.25$ m, $\theta_3 = 270°$, and $M_3 = 0.6$ kg.

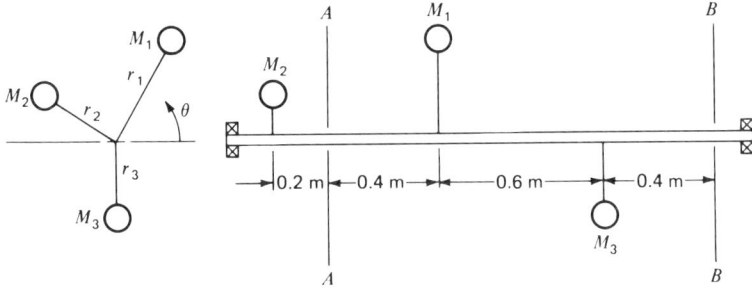

PROBLEMS

12-9. A shaft carrying two disks with holes rotates at 500 r/min. The disks are made of aluminum (0.00277 kg/cm³). Determine the magnitudes and angular positions of the two bearing reactions for the rotor position shown.

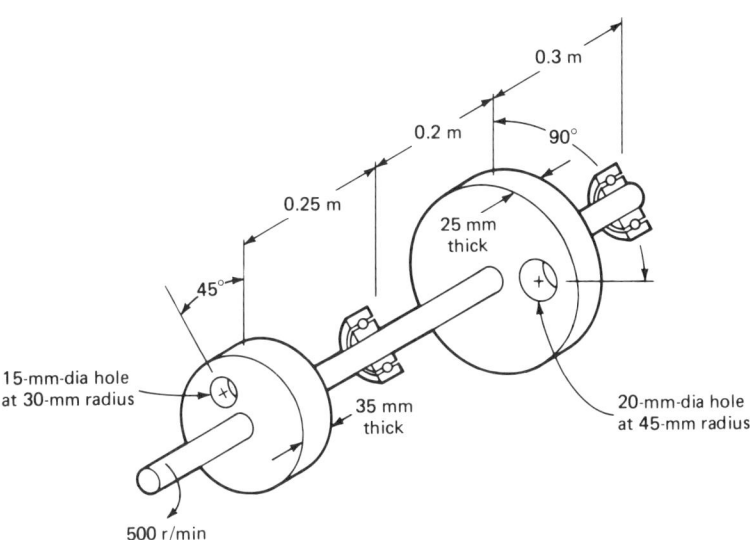

12-10. A crankshaft has four equal cranks at 90° as shown. Each crank is equivalent to 2.5 lb at a radial distance of 3 in. Masses are to be added at 5-in. radii in planes $A–A$ and $B–B$ to achieve dynamic balance. Determine the required masses and show their proper positions on an end view of the crankshaft.

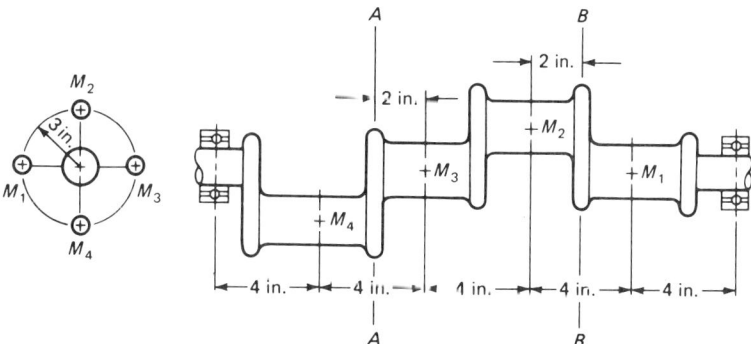

13
The Feedback Control System

13-1 Introduction

The utilization of automatic control systems is widespread in the world today, and there is every indication that the trend toward increased use of automatic equipment will continue in the future. The modern engineer can scarcely avoid being involved in some manner with control systems. For this reason, one unfamiliar with control theory should obtain at least a preliminary knowledge of control system fundamentals. The purpose of Chapters 13 through 16 is to introduce basic control concepts, briefly but with sufficient coverage to make the effort worthwhile. Obviously, such a treatment is by no means exhaustive.

Since the underlying concepts of linear control system theory are essentially the same as those of other linear dynamic systems, we can make use of much of the theory developed in earlier chapters.

Attention is now turned to the development of the nature of the feedback control system. An actual control application will be used as an example.

13-2 Home Heating Application

There are a number of ways in which the concept of feedback control can be introduced. The choice here is to investigate an actual control application—that of home heating—because most readers are probably familiar with it.

Let us assume that we wish to devise a system for controlling room temperature but that we are totally unaware of the nature of similar systems already in existence.[1] If we stop and think for a moment, we realize that the necessity for a heating system arises because the outside air temperature falls significantly below that desirable for the inside of the house. A heat source is required to supply the heat lost through the walls of the structure. Certainly, then, a suitable system for controlling inside temperature must take into account the variations of outside temperature.

One possible control system is illustrated in block form in Fig. 13-1. The controller is designed to sense outside temperature and to transmit a signal to the furnace, calling for variations in heat output to match variations in outside temperature. If the heat input to the house is properly controlled, the inside temperature can be maintained essentially constant. At this point the reader may feel that if the controller were carefully calibrated to match the structure, the control system of Fig. 13-1 would offer satisfactory performance. However, further thought will reveal that this is not likely to be the case.

All control systems are subjected to disturbances of one sort or another. A *disturbance* is exactly what the name implies—something that tends to upset the system, making difficult the control of some variable. As illustrated in Fig. 13-1, the house is subjected to disturbances such as sun and wind. For a constant outside temperature, the proper heat input to the house will depend on whether the day is sunny or cloudy, windy or still. There are other possible disturbances such as the number of people in the house (a party, perhaps), the opening of doors and windows, and so forth. It should now be quite obvious that a heating control system that responds only to outside temperature can be badly fooled, making the control of inside temperature precarious at best. The system of Fig. 13-1 had best be abandoned in favor of some other approach.

The control system illustrated in Fig. 13-2 is the conventional approach to home heating. Here the controller is a thermostat. The thermostat is physically located within the controlled environment and can compare the desired temperature setting (the *input* to the system) with the

---

[1]The authors readily admit that to approach an engineering project without first ascertaining what has already been done is foolhardy. However, the underlying objective here is to introduce the idea of feedback control.

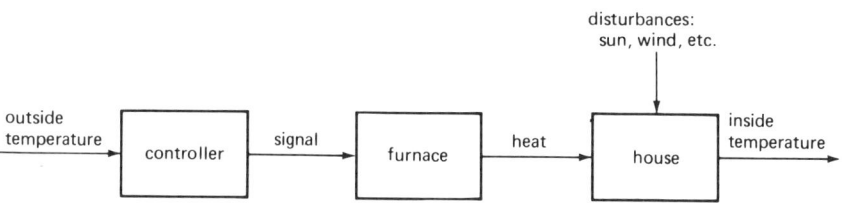

FIG 13-1. Home heating system.

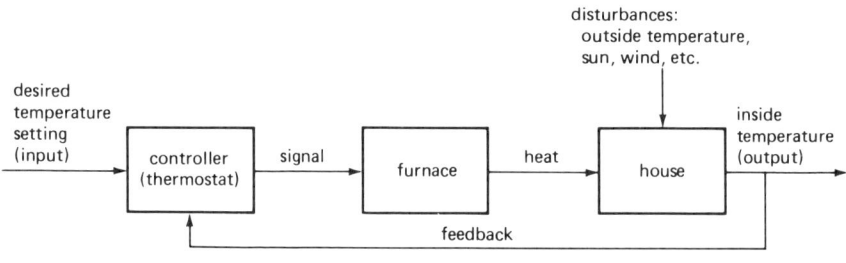

FIG 13-2. Home heating system with feedback.

actual inside temperature (the system *output* or *response*). Should the difference between the input and the output be significant, the thermostat signals the furnace for more or less heat output. The *feedback* line (Fig. 13-2) indicates that the system output is measured and fed back to the controller for comparison with the input.

The house is again subjected to disturbances, now including variations in outside temperature. However, should a disturbance change the inside temperature, the thermostat will sense the change and initiate corrective action.

The control system of Fig. 13-2 is the better of the two considered because, by using feedback, a comparison between the desired temperature setting and the actual inside temperature can be made, and any significant difference between the two will result in corrective action. The system of Fig. 13-1 makes no comparison, and as a result, the inside temperature can change markedly, with no means for modifying the heat input. As might be expected, the system utilizing feedback is called a *feedback control system*.

13-3 Feedback Control Systems

A meaningful definition can now be given; that is, a feedback control system is one in which the output is compared with the input, and any difference between the two (the *error*) is used by the system to reduce that difference. The important distinguishing feature of a feedback control system is that a *comparison* is made. It is this comparison that makes such a system so effective for control purposes.

Feedback control systems are also called *closed-loop systems*. A study of Fig. 13-2 will reveal the reason for the name. The block diagram takes on the form of a closed loop, with the feedback line said to *close the loop*. The physical system also provides a closed loop; that is, there is a continuous path around which signals of one sort or another can flow. For example, in the physical system represented by Fig. 13-2, should the inside temperature change, the thermostat will sense the change because feedback exists. The thermostat will then signal the furnace for a modification of heat output. The resulting change in heat input into the house will affect the inside temperature. The change in inside temperature will influence the thermostat, which will modify the furnace output, with the result that the inside temperature will change, and so forth. The point is that, with a feedback control system, the physical system is arranged so that a closed loop exists for the flow of signals.

There are also *open-loop systems*, that is, systems without feedback. The system illustrated in Fig. 13-1 is an example. Although this particular system has little merit, there are in existence many open-loop systems that perform satisfactorily. However, with open-loop systems, calibration is all-important because no comparison between output and input is made.

Feedback control is desirable because comparison between the system output and input (and the resulting correction) is made possible. However, feedback can cause difficulties since closing the loop provides a continuous path for the flow of signals. Suppose that the signals grow larger as they circulate around the loop. (Feedback control systems typically have an external source of power, so the situation described is entirely possible.) If the signals do indeed increase in magnitude, the system will not be able to achieve the desired steady response. The system is then said to be *unstable*. Further consideration of system stability will be deferred until Chapter 14, at which point a quantitative definition can be advanced. The idea to be gained here is that, while beneficial, feedback can also be a source of difficulty.

13-4 Conclusion

Two key ideas were introduced in this chapter:

1. A feedback control system involves a comparison between the system output and the system input, with corrective action being taken if the difference between the two (the system error) is significant. The comparison feature offers effective control, especially when disturbances are present.

2. Feedback is a source of difficulty, since feedback control systems can become unstable.

From these two facts the reader can correctly deduce that feedback control theory is largely concerned with capitalizing on the virtues of feedback while minimizing its vices.

Control system behavior is defined by differential equations. The methods presented in Chapter 2 for obtaining differential equations representing various dynamic systems can be applied directly to feedback control systems, as can the methods of solution given in Chapter 3.

System response and stability are the subjects of Chapter 14, the discussion centering around how dynamic systems respond to various inputs. Part of the topic of response has been previously covered in Chapters 6 and 7, but additional information is necessary for a study of feedback control systems.

Chapter 15 is concerned specifically with control systems. The subject matter deals with the ways in which system error can be utilized to produce appropriate corrective action. Chapter 16 introduces the use of block diagrams, an alternative method for developing the differential equations describing control systems.

14
System Response and Stability

14-1 Introduction

The response of dynamic systems to various inputs is of extreme importance in control work. A control system constantly receives inputs of one sort or another, and its response to them must be known in order to achieve a satisfactory system. The purpose of this chapter, then, is to study certain aspects of dynamic-system response.

The inputs to actual control systems are, of course, quite varied in nature. However, for both analytical and experimental purposes, two inputs are commonly used. They involve forcing functions which are either constants (K) or sinusoidals $(K \sin \omega t)$. These inputs are shown plotted against time in Fig. 14-1. In both cases the input is assumed to begin at $t = 0$. In Fig. 14-1a the input is zero until $t = 0$, at which time it takes on a value K, which remains constant for $t > 0$. Control engineers call this input a *step input* because of the appearance of the plot. The sinusoidal input, shown in Fig. 14-1b, is generally called by that name—a *sinusoidal*

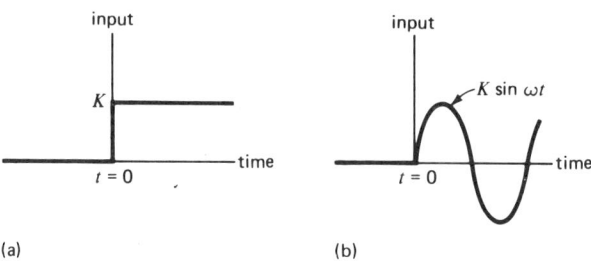

FIG 14-1. System inputs.
(a) Step.
(b) Sinusoidal.

input. These inputs are popular because they facilitate both analytical and experimental work.

In Chapters 6 and 7 we studied the response of mechanical vibrating systems to both step inputs and sinusoidal inputs. Single-degree-of-freedom mechanical vibrating systems are, because of the presence of inertia elements, all second-order systems. The theory developed for them can be directly applied to all other second-order dynamic systems that can be modeled by differential equations of the same general form. The frequency-response curves developed for mechanical vibrating systems therefore apply to a wide variety of second-order dynamic systems.

Not all dynamic systems are of second order, however. A very important category is, in fact, the *first-order system*. The response of first-order dynamic systems will be illustrated with the liquid-level system first introduced in Chapter 2.

14-2 Response of a Liquid-Level System

A linear liquid-level system is shown in Fig. 14-2a. The flow rate q_i will be considered as the input, and the head h as the response. The system differential equation (Eq. 2-29) is

$$A \frac{dh}{dt} + G_v h = q_i \qquad (14\text{-}1)$$

Let us assume that the tank is initially empty and that a step input is then applied. The input is shown graphically in Fig. 14-2b; $q_i = Q_i$ (a constant) for $t \geqslant 0$. Thus, the equation to be solved for the response h is

$$A \frac{dh}{dt} + G_v h = q_i = Q_i \qquad (14\text{-}2)$$

with the initial condition that $h = 0$ at $t = 0$.

14-2 RESPONSE OF A LIQUID-LEVEL SYSTEM

The complementary function h_c is obtained by solving the associated homogeneous equation

$$A \frac{dh}{dt} + G_v h = 0 = h(Ap + G_v)$$

The characteristic equation and its root are

$$Ap + G_v = 0 \quad \text{and} \quad p = \frac{-G_v}{A}$$

Therefore,

$$h_c = C_1 e^{pt} = C_1 e^{-G_v t/A} \quad (14\text{-}3)$$

The forcing function is a constant, so the particular integral is assumed to be a constant; that is, $h_p = K$. The derivative is $dh_p/dt = 0$. Substituting in Eq. 14-2,

$$A(0) + G_v K = Q_i$$

$$h_p = K = \frac{Q_i}{G_v} \quad (14\text{-}4)$$

The complete solution is

$$h = h_c + h_p = C_1 e^{-G_v t/A} + \frac{Q_i}{G_v}$$

From the initial condition $h = 0$ when $t = 0$,

$$0 = C_1 + \frac{Q_i}{G_v} \quad \text{and} \quad C_1 = \frac{-Q_i}{G_v}$$

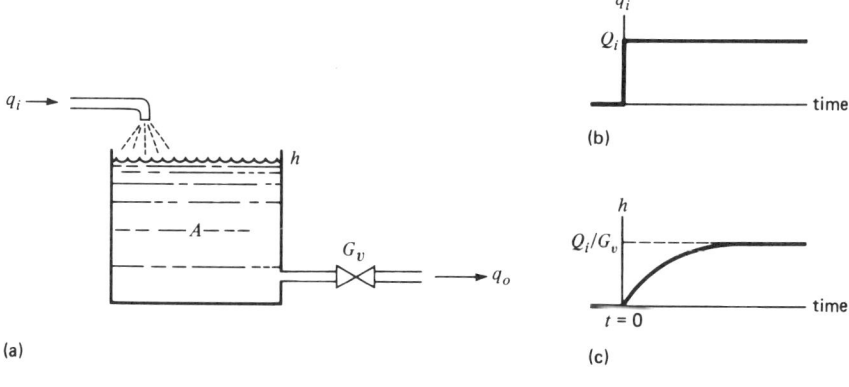

FIG 14-2. (a) Liquid-level system.
(b) Step input.
(c) Its response to the step input.

Therefore,

$$h = \frac{Q_i}{G_v}(1 - e^{-G_v t/A}) \tag{14-5}$$

The response is illustrated in Fig. 14-2c; h starts at zero with $t = 0$ and approaches the value Q_i/G_v with increasing time.

The response obtained is entirely plausible from an intuitive viewpoint. For small values of head, the outflow is small, so the level rises rapidly. The limiting value of head is that which will produce an outflow equal to the inflow.

Since the approach used to determine the response of the liquid-level system is entirely general, it can be used with more complex systems and for inputs other than step inputs.

14-3 First-Order System Response to a Step Input

A first-order system is, by definition, one whose behavior is defined by a first-order differential equation. The liquid-level system of Section 14-2 is an example. The purpose of this section is to discuss the response of first-order systems, in general, to step inputs. In particular, the idea of a *time constant* will be developed, because it is common to all first-order systems. A mechanical system will be used for illustrative purposes.

A spring-dashpot system is shown in Fig. 14-3. With the system initially at rest, a step input is applied; that is, $x = X$ for $t \geq 0$. The system differential equation is

$$c\frac{dy}{dt} + ky = kx = kX$$

or

$$\frac{c}{k}\frac{dy}{dt} + y = X = \tau\frac{dy}{dt} + y \tag{14-6}$$

where $\tau = c/k$ is the *time constant* for the system. The name comes from the fact that the units of τ are time:

$$\tau = \frac{c}{k} = \frac{\text{lb} \cdot \text{s/in.}}{\text{lb/in.}} = \text{s}$$

The time constant in Eq. 14-6 was obtained by making the coefficient of y unity. With first-order dynamic system equations in general, if the coefficient of the zeroth-derivative term is made unity, the coefficient of the first-derivative term will be the system time constant. For example, the time constant for the liquid-level system of Section 14-2 can be obtained from Eq. 14-2 and is $\tau = A/G_t$. The reader should check the units to show that they are indeed time.

14-3 FIRST-ORDER SYSTEM RESPONSE TO A STEP INPUT

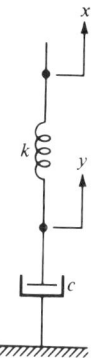

FIG 14-3. First-order mechanical system.

The next step is to solve Eq. 14-6 so that the usefulness of the time constant, in predicting system response, can be illustrated. To obtain the transient solution y_t,

$$\tau p + 1 = 0 \quad \text{and} \quad p = \frac{-1}{\tau}$$

Therefore,

$$y_t = C_1 e^{-t/\tau}$$

The steady-state solution is $y_{ss} = X$, and the total solution then is

$$y = C_1 e^{-t/\tau} + X$$

The system was assumed to be initially at rest, so the response equation is found to be

$$y = X(1 - e^{-t/\tau}) \tag{14-7}$$

Equation 14-7 can be put in dimensionless form by dividing through by X. The result is

$$\frac{y}{X} = 1 - e^{-t/\tau} \tag{14-8}$$

In Eq. 14-8, the ratios y/X and t/τ are, of course, dimensionless. A plot of Eq. 14-8 is given in Fig. 14-4. Since the plot is in dimensionless form, it can be regarded as providing the mathematical response for *all* first-order systems subjected to a step input; that is, the ratio y/X can be replaced by the ratio of any response function to its steady-state value.

The use of the time constant for predicting response times can now be discussed. In Eq. 14-8, if $t = \tau$, $t/\tau = 1$, and

$$\frac{y}{X} = 1 - e^{-1} = 1 - 0.368 = 0.632$$

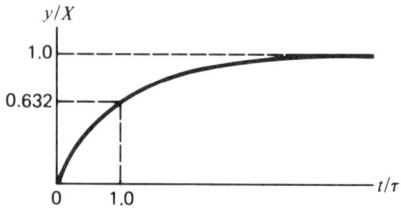

FIG 14-4. Response of first-order mechanical system to a step input X.

or

$$y = 0.632X \qquad (14\text{-}9)$$

The important idea to be gained from Eq. 14-9 is that, in response to a step input, the output of a first-order system proceeds 63.2 percent of the way to its steady-state value in a period of time equal to the numerical value of the time constant.

Furthermore, if $t = 4\tau$, $t/\tau = 4$, and

$$y = X(1 - e^{-4}) = 0.982X \qquad (14\text{-}10)$$

Speaking mathematically, first-order systems do not respond completely in finite time; however, physical systems do. A commonly used estimate of the time required for complete response is $t = 4\tau$ (98.2 percent of the way mathematically).

Note, in particular, that the magnitude of the step input in no way influences the response time for a first-order system. Thus, a knowledge of the time constant for a physical system is quite valuable in predicting response characteristics.

The reader should realize that the system time constant plays an important role in the response to inputs other than a step input. This fact becomes apparent in the following section, in which sinusoidal forcing functions are considered.

14-4 First-Order System Response to a Sinusoidal Input

If we apply the periodic input $x = X \sin \omega t$ for $t \geq 0$ to the system of Fig. 14-3, the differential equation of motion will be

$$\frac{c}{k}\frac{dy}{dt} + y = x = X \sin \omega t \qquad (14\text{-}11)$$

This equation can be put into the form

$$\tau \frac{dy}{dt} + y = K_2 \sin \omega t \qquad (14\text{-}12)$$

14-4 FIRST-ORDER SYSTEM RESPONSE TO A SINUSOIDAL INPUT

The general form of Eq. 14-12 will also describe a great many other first-order dynamic systems with a sinusoidal input, so the solution will have wide application.

The solution to Eq. 14-12 contains two parts: the transient (complementary) solution and the steady-state solution (particular integral). But since the transient solution disappears (quickly if the damping is heavy, but more slowly with light damping), it is normally ignored, the response being considered simply the steady-state solution.

To obtain the steady-state solution to Eq. 14-12, we follow the method of Chapter 3 and assume that

$$y_{ss} = A \sin \omega t + B \cos \omega t$$

Substituting the above into Eq. 14-12 and solving for the constants A and B, we obtain

$$A = \frac{K_2}{1 + \tau^2 \omega^2}$$

$$B = -\frac{\tau \omega K_2}{1 + \tau^2 \omega^2}$$

The solution to Eq. 14-12 is therefore

$$y_{ss} = \frac{K_2}{1 + \tau^2 \omega^2} \sin \omega t - \frac{\tau \omega K_2}{1 + \tau^2 \omega^2} \cos \omega t \qquad (14\text{-}13)$$

For plotting frequency response curves, it is advantageous to manipulate this equation to the alternative form (see Appendix B),

$$y_{ss} = Y \sin(\omega t + \phi) \qquad (14\text{-}14)$$

where

$$Y = \frac{K_2}{\sqrt{1 + \tau^2 \omega^2}} \qquad (14\text{-}15)$$

and

$$\phi = -\tan^{-1}(\tau \omega). \qquad (14\text{-}16)$$

In Fig. 14-5 are given the two frequency-response curves, based on Eqs. 14-15 and 14-16, which show the amplitude ratio and phase angle of a first-order system when excited by a harmonic forcing function $K_2 \sin \omega t$ over a wide range of frequencies. These are plotted in dimensionless form, so they are convenient to use for any such system. With proper application of these curves, one can avoid having to make calculations for each particular first-order system. For problems in which the response to harmonic inputs at a number of different frequencies is needed, the curves can save a great deal of labor.

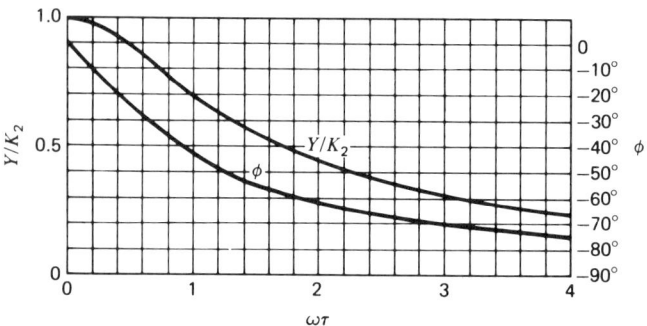

FIG 14-5. Normalized frequency-response curves (with linear coordinates) for first-order dynamic systems described by Eq. 14-12.

As in the case of second-order systems, the first-order frequency-response cures are often plotted with logarithmic coordinates. This allows a greater range of frequencies and amplitude ratios to be accurately covered and also allows the use of a straight-line approximation, which is often valuable.

For the analysis of control systems, however, as well as certain other types of dynamic systems, it has become customary to plot *amplitude ratio on a linear ordinate scale* in units of *decibels* (dB). (A decibel is defined as $20 \log_{10}$ of the amplitude ratio.) The accompanying abscissa scale (frequency) is normally logarithmic. The phase angle curve is plotted on the same graph, with degrees on the linear ordinate scale versus frequency on the logarithmic abscissa scale.

Such a set of frequency response curves is referred to as a *Bode diagram*. Note that plotting the amplitude ratio in decibels on a linear scale is equivalent to plotting the basic amplitude ratio on a logarithmic scale, so the shape of a Bode plot amplitude ratio curve is the same as the equivalent log-log plot. One advantage of using decibels as an alternative to a logarithmic scale is that the phase angle ϕ can conveniently be plotted on a linear scale using the same set of grid lines as the amplitude ratio curves, allowing a more compact set of curves.

Figure 14-6 is a Bode diagram for first-order dynamic systems represented by Eq. 14-12. Although it contains exactly the same information as Fig. 14-5, the logarithmic scales cause the shape of the curves to be quite different. Note that the significance of the first-order time constant is readily apparent on the Bode diagram but not on the curve of Fig. 14-5. The straight-line approximation for the amplitude ratio curve of the Bode diagram that follows is often used as an aid in quick plotting.

14-4 FIRST-ORDER SYSTEM RESPONSE TO A SINUSOIDAL INPUT

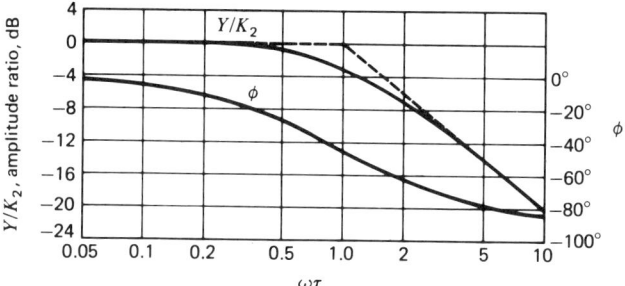

FIG 14-6. Bode diagram (frequency-response curves) for first-order dynamic systems described by Eq. 14-12.

STRAIGHT-LINE APPROXIMATION

The solution for the amplitude ratio of the system under consideration, Eq. 14-15, can at frequencies appreciably below $1/\tau$ be approximated by the form

$$\frac{Y}{K_2} \approx 1.0 \qquad (14\text{-}17)$$

which is the equation of a straight line of zero slope whether plotted with a linear scale or a logarithmic scale. At frequencies appreciably above $1/\tau$, the approximation

$$\frac{Y}{K_2} \approx \frac{1}{\tau\omega} \qquad (14\text{-}18)$$

is valid. Eq. 14-18 may also be written in the logarithmic form

$$\log \frac{Y}{K_2} = -\log \tau - \log \omega$$

$$= \text{constant} - \log \omega \qquad (14\text{-}19)$$

Equation 14-19, when plotted on log-log paper (or with decibels on a linear ordinate scale and a logarithmic abscissa), is a straight line. The slope of this line with Bode diagram coordinates is -20 dB/decade; that is, the amplitude ratio drops 20 dB for a factor of 10 increase in frequency. (On the equivalent log-log plot the slope would be -1.)

The two straight lines used in this approximate construction intersect at $\omega = 1/\tau$. (Note the dashed lines showing this approximation in Fig. 14-6.) At this frequency the error between the true curve and the straight line approximation is a maximum. A comparison of Eqs. 14-15 and 14-17 shows that the true value here is $1/\sqrt{2}$ times the approximation, or, on the Bode diagram, 3 dB below the approximate curve. The reciprocal of the time constant τ is called the *break frequency*, or *corner frequency*, since it is the frequency at which the straight-line approximate curve has a sharp corner.

It should be noted that the phase angle curve is symmetrical, with a logarithmic abscissa scale, about the frequency $\omega = 1/\tau$. The phase angle ϕ is approximately $0°$ at very low frequencies, exactly $-45°$ at the frequency $\omega = 1/\tau$, and asymptotically approaches $-90°$ as the frequency becomes very large.

It should be reemphasized at this point that the curves, either Fig. 14-5 or 14-6, can be used to evaluate the response of any first-order system, with a sinusoidal forcing function, described by the general form of Eq. 14-12.

Appendix F has been prepared to simplify the process of converting from a pure ratio to decibels, and vice versa.

14-5 Second-Order System Response

The preceding sections have brought out the fact that all first-order systems have a time constant, and that the time constant is a factor that unifies the response of all first-order systems, despite the diverse nature of the physical systems.

For second-order systems there are two such factors that unify system response. These are the *damping ratio*[1] and the *undamped natural frequency*. In Chapters 6 and 7, solutions were obtained for second-order mechanical vibrating systems, and the concepts of natural frequency and damping ratio were developed for such vibrating systems. At this point, however, it should be reemphasized that *all* second-order dynamic systems will have a natural frequency ω_n even though there may be no corresponding motion of a mechanical part. Similarly, all second-order systems will have a damping ratio ζ even though there may be no mechanical dissipation of energy. In fact, some second-order systems may, unlike the mechanical vibrating systems previously studied, have a *negative* value of ζ, indicating that energy is added to the system rather than being dissipated.

As previously explained in Chapters 6 and 7, values of ω_n and ζ can be obtained for any second-order dynamic system by putting its differential

---

[1]The *time constant* τ is sometimes used as an alternative to the damping ratio for system characterization. With a step input, the *envelope* of the second-order system response has a shape identical to a first-order step response. The significance of τ as discussed in Section 14-3 can therefore be applied to the second-order step response envelope. In this book, ω_n and ζ are normally used for system characterization because it is felt that this is the better approach.

14-5 SECOND-ORDER SYSTEM RESPONSE

equation in the following general form (by making the coefficient of the zeroth derivative term unity):

$$\frac{1}{\omega_n^2} \frac{d^2 y}{dt^2} + \frac{2\zeta}{\omega_n} \frac{dy}{dt} + y = \text{forcing function} \quad (14\text{-}20)$$

By equating the coefficients of any particular system to the corresponding terms in Eq. 14-20, the natural frequency and damping ratio can be readily obtained.

Second-order dynamic systems having the same values for ω_n and ζ will have identical response characteristics even if they have no physical similarity. The family of free-vibration curves presented in Chapter 6 (Fig. 6-6) can be used to determine the step response characteristics of any second-order dynamic system even though the curves were drawn for free vibration caused by an initial condition of displacement. Figure 14-7 is

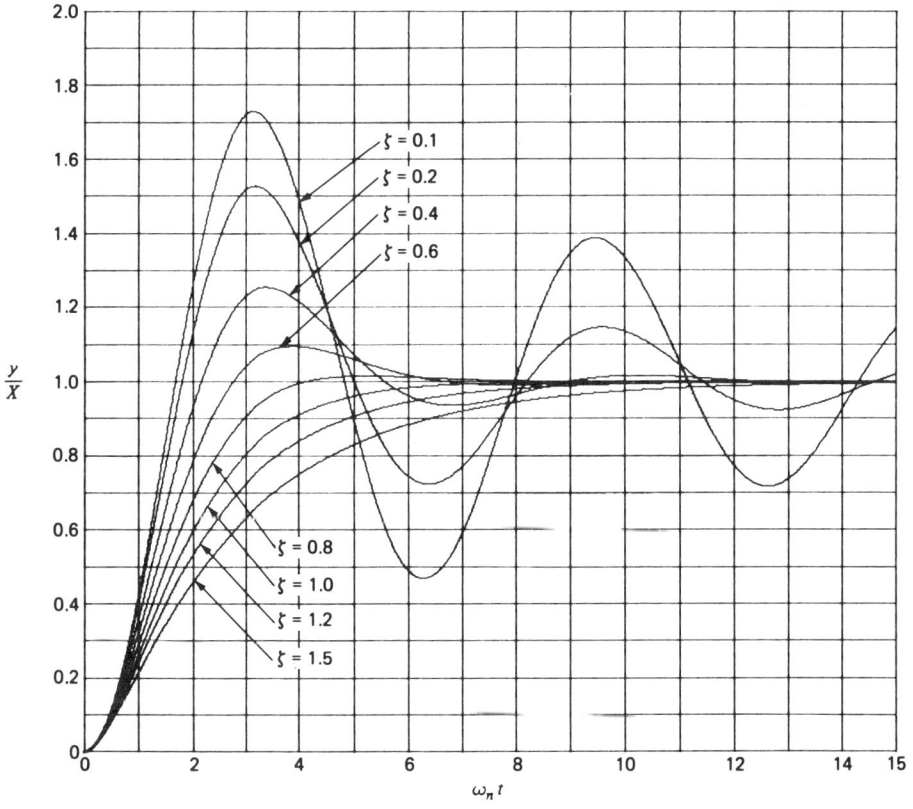

FIG 14-7. Response of a second-order dynamic system to a step input X.

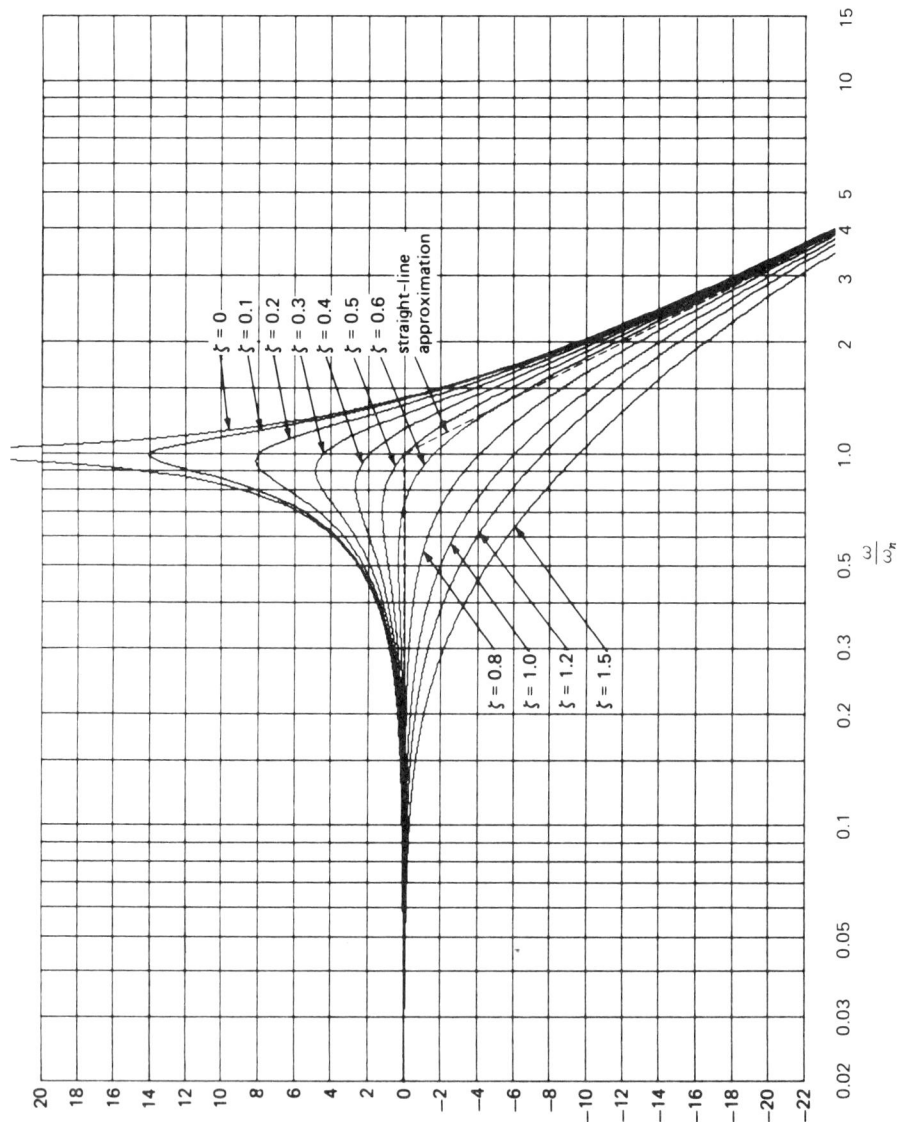

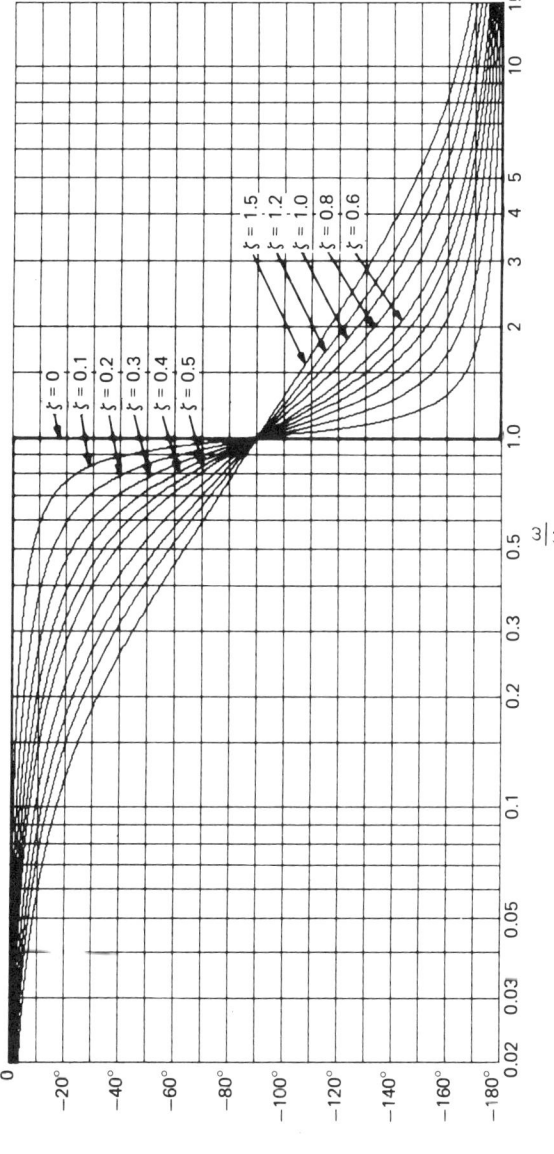

FIG 14-8. Bode diagram giving frequency-response characteristics of second-order dynamic systems described by Eq. 14-20 and excited by the forcing function $K_1 \sin \omega t$.

basically the same set of curves, but it is drawn with an ordinate scale that makes it apply more specifically to step inputs.

Note that the definition of damped natural frequency (see Section 6-3),

$$\omega_d = \omega_n(1 - \zeta^2)^{1/2}$$

is applicable not only to mechanical vibrating systems but to any second-order dynamic system. Also, the *logarithmic decrement* method for determining the damping ratio ζ (see Section 6-4) can be applied to the experimental step response characteristics of any second-order dynamic system.

When studying the response of second-order dynamic systems to sinusoidal inputs, the use of nondimensional frequency-response curves can be of considerable help and save a great amount of labor in many cases. The frequency-response curves given for vibrating systems (Fig. 7-5) can be applied to other types of second-order dynamic systems.

Since amplitude ratio is normally measured in decibels for control systems work, the second-order frequency-response curves are more useful for that field when presented in the form of Bode diagrams such as that of Fig. 14-8. Note that this Bode diagram has the same shape as the corresponding log-log curve of Fig. 7-5. In terms of decibels, the straight-line approximation of the family of curves has a 0-dB slope below ω_n and a slope of -40 dB/decade at frequencies appreciably above ω_n. The phase angle curve for each damping ratio is symmetrical about ω_n when given in Bode diagram form.

14-6 Response of Higher-Order Systems to Simple Sinusoidal Inputs

Frequency-response curves are also useful tools for the design and analysis of third and higher-order dynamic systems, but it is not practical to draw normalized curves or families of curves for such systems that can be used for all possible parameter variations because the number of variables is too great. For such higher-order systems the Bode diagram is especially convenient to use because it can be plotted fairly quickly by means of approximate straight-line techniques that are an extension of those described for first- and second-order systems and by means of other simple rules.

The development and use of such higher-order Bode diagrams are beyond the scope of this book but are covered in most control system books [7].

14-7 The Concept of System Stability

The fact that feedback control systems can respond in undesirable ways (that is, become unstable) was mentioned in Chapter 13. The purpose of this section is to develop more fully the concept of system stability.

From a practical viewpoint, a stable system is one whose response to an input reaches and maintains some useful value within a reasonable period of time. For this to be the case all transients must die out, or, looked at from a mathematical viewpoint, all roots (values of p) of the system characteristic equation must be either negative real numbers or complex numbers with negative real parts. The latter statement constitutes the concept of stability for linear systems. Several characteristic equations will now be considered to illustrate the idea of stability.

Suppose that the characteristic equation for a first-order system is

$$p + C = 0$$

where C is a positive real number. Then, $p = -C$, and the system should be stable because the root is a negative real number. The response equation is

$$y = C_1 e^{-Ct} + y_{ss}$$

A sketch of the response to a step input is shown in Fig. 14-9a. The transient dies out, and the system is well behaved (stable).

Suppose that the characteristic equation for a first-order system is

$$p - C = 0$$

Then $p = C$, and the response equation is

$$y = C_1 e^{Ct} + y_{ss}$$

Figure 14-9b illustrates the unstable response of the system to a step input. In this case the transient increases with time.

The characteristic equation for a second-order system might be

$$\frac{1}{\omega_n^2} p^2 + \frac{2\zeta}{\omega_n} p + 1 = 0 \tag{14-21}$$

(a)
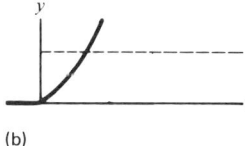
(b)

FIG 14-9. Response of (a) stable and (b) unstable first-order systems to a step input.

With the assumption that the damping ratio is less than 1, the roots are

$$p = -\zeta\omega_n \pm j\omega_n(1 - \zeta^2)^{1/2} = -\zeta\omega_n \pm j\omega_d$$

where ω_d is the *damped natural frequency*. The response equation, then, is

$$y = e^{-\zeta\omega_n t}(C_1 \sin \omega_d t + C_2 \cos \omega_d t) + y_{ss} \qquad (14\text{-}22)$$

The response to a step input is illustrated in Fig. 14-10a. The system is clearly stable.

Suppose that in the characteristic equation for a second-order system (Eq. 14-21) we find ζ to be a negative number. The response equation (Eq. 14-22) will, with ζ negative, contain an exponential term that increases the amplitude of oscillation with time. The resulting unstable response to a step input is illustrated in Fig. 14-10b.

At this point the reader should readily understand why the roots of the system characteristic equation must have negative real parts if the system is to be stable. However, there exists the possibility of having zero real parts. As an example, consider the characteristic equation

$$\frac{1}{\omega_n^2} p^2 + 1 = 0$$

Here the roots are

$$p = \pm j\omega_n$$

Figure 14-11 illustrates the response of the system to a step input, the response equation being

$$y = C_1 \sin \omega_n t + C_2 \cos \omega_n t + y_{ss}$$

Persistent oscillation is the result, with the amplitude neither increasing nor decreasing with time. Such a system is said to possess *limited stability*.

An interesting point can be appreciated from a review of Figs. 14-9b, 14-10b, and 14-11. No steady-state response exists for any of these systems. The reason is that the transients do not die out. Thus, for an unstable system, the particular integral (which can be determined) is not the steady-state solution, because a steady-state response does not exist. Only stable systems have steady-state responses.

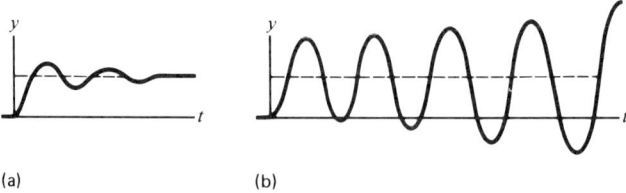

(a)   (b)

FIG 14-10. Response of (a) stable and (b) unstable second-order systems to a step input.

FIG 14-11. Response of second-order system of limited stability.

A physical system can be stable or unstable, or it can have limited stability. From a practical viewpoint, limited stability is instability because the desired response cannot be attained. For a stable system *all* roots of the characteristic equation must have negative real parts, that is, be either negative real numbers or complex numbers with negative real parts.

Since solving a higher degree characteristic equation for its roots is time-consuming, the *Routh criterion* can be used to determine whether or not there are positive roots without actually solving for the roots. Routh's criterion for stability is presented in Appendix *G*.

14-8 Conclusion

A knowledge of system response is fundamental to the study of control systems. However, it should be reemphasized that the concepts presented in this chapter are general in nature, being equally applicable to other dynamic systems for which the defining equations are similar. Because of this, some of the material in this chapter (on second-order system response, to be specific) should be recognizable by the reader as very similar to that presented in earlier chapters on mechanical vibrating systems.

Problems

14-1. For the system shown in the accompanying figure, determine the response y if the initial conditions are (a) $+1$ in. and (b) -0.5 in. Sketch the response curves.

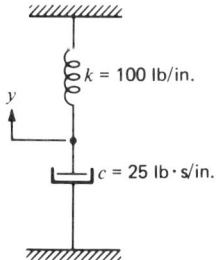

14-2. For the system shown, determine the response y for a step input ($x = 0.5$ in.) at x. Assume that the system is initially at rest. Sketch the input and response curves.

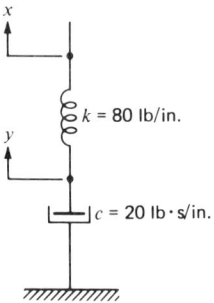

14-3. For the system shown, determine the response y for a step input ($x = 1$ in.) at x. Assume that the system is initially at rest. Sketch the input and response curves.

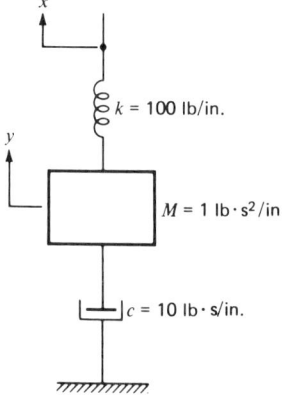

14-4. For the system shown, determine the steady-state response y_{ss} for a sinusoidal input ($x = 2 \sin 10t$) at x.

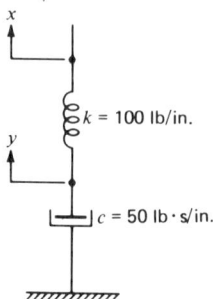

PROBLEMS

14-5. A hydraulic actuator is shown in the accompanying illustration.
   (a) Obtain the system differential equation which relates the output displacement y with the input displacement x.
   (b) What is the time constant for the actuator? If a faster response is desired, should the numerical value of the valve constant K_v be increased or decreased?
   (c) Determine the response y for a step input X.

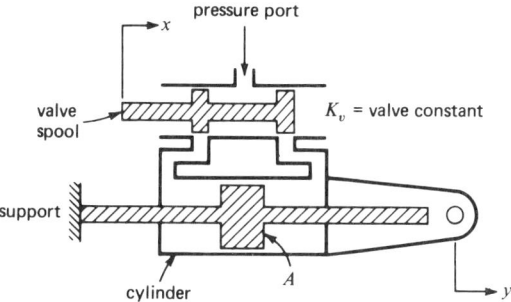

14-6. For the tank shown, obtain the differential equation which relates the output flow rate q_o with the input flow rate q_i. What is the system time constant? Assume linear valve characteristics.

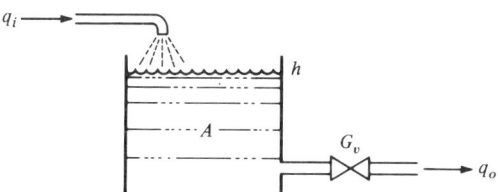

14-7. For the rotating system shown, obtain the differential equation which relates the response speed ω with the input torque T. What is the system time constant?

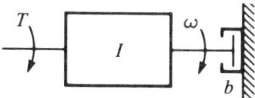

14-8. For the system shown in the accompanying illustration, a) obtain the differential equation in which θ_o is the response and θ_i is the input displacement. b) What is the damping ratio? c) Determine the steady-state response $\theta_{o_{ss}}$ for a step input Θ. d) Sketch θ_o versus t for $\zeta < 1$. Assume the step input of part c and θ_o initially at zero.

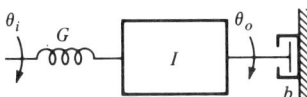

14-9. For the tank system shown in the accompanying illustration,
  (a) obtain the differential equation which relates the head h_2 with the input flow rate q_i. Assume linear valve characteristics.
  (b) What is the damping ratio?
  (c) Determine the steady-state response $h_{2_{ss}}$ for a step input Q_i.
  (d) Sketch h_2 versus t. Assume the step input of (c) and that h_2 is initially at zero.

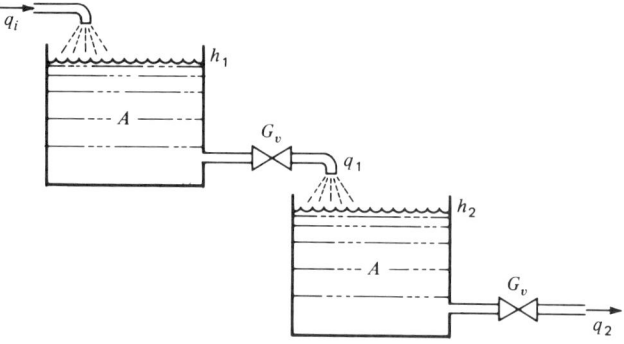

14-10. The equations given below describe certain control systems. Ascertain the nature of stability and sketch the response of each system to a step input. The systems are initially at rest.

  (a) $\dfrac{d^2y}{dt^2} + Ky = Kx$

  (b) $\dfrac{dy}{dt} - Ky = Kx$

  (c) $-\dfrac{dy}{dt} - Ky = Kx$

  (d) $\dfrac{d^2y}{dt^2} + D\dfrac{dy}{dt} + Ky = Kx$

  (e) $(p-3)(p+4)(p+5)y = Kx$

14-11. Prove that a straight line tangent at the origin to the normalized first-order step response curve (Fig. 14-4) will intersect the point $y/X = 1.0$, $t/\tau = 1.0$.

14-12. Develop a mathematical expression for the relationship between the time constant of a second-order system and its damping ratio and undamped natural frequency.

PROBLEMS

14-13. Determine the time constant of a second-order system with the given response to a step input.

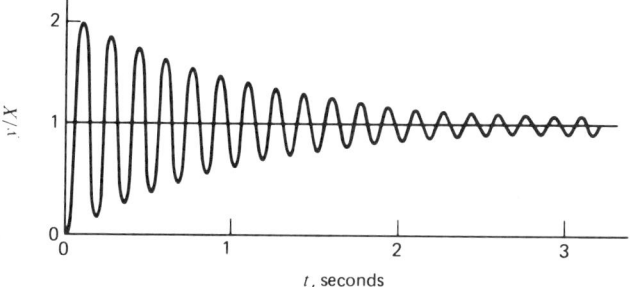

14-14. Sketch the response of the spring-dashpot system to the given square-wave input. (*Hint*: The response is simply a series of step responses.)

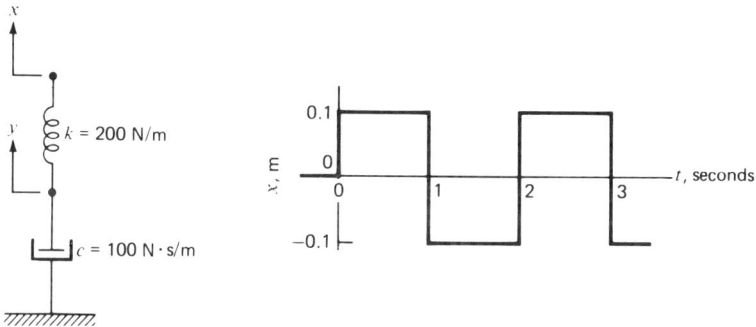

14-15. Using Routh's criterion (see Appendix G), determine the stability of systems represented by the following characteristic equations.
(a) $p^3 + 3p^2 - 2p + 3 = 0$
(b) $p^3 + 2p^2 + 5p + 1 = 0$
(c) $p^4 + 2p^3 + 4p^2 + 12p + 1 = 0$
(d) $p^5 + 7p^4 + 5p^3 + p^2 + 6p + 4 = 0$
(e) $p^5 + 2p^4 + 3p^3 + 9p^2 + 10p + 2 = 0$

14-16. (a) Determine the time constant of the given electric system.
(b) Sketch i versus t when the switch is closed.

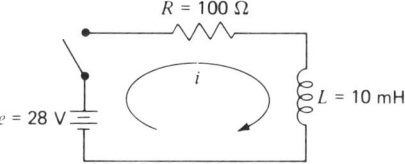

14-17. A building and its contents are initially at the same temperature as the outside ambient air, 270 K. The contents have an overall "heat capacitance" of 12×10^6 J/K, and an overall coefficient of heat transfer through the walls of 1500 W/K. A 30,000 W furnace is turned on.
  (a) Determine the steady-state temperature that the inside of the building will eventually reach.
  (b) What is the time constant of the system?
  (c) Sketch temperature versus time.

14-18. Assume that the linear valve shown gives flow proportional to head difference. System parameters are as follows: $A_1 = 10$ m², $A_2 = 4$ m², $G_v = 0.10$ (m³/s)/m, $h_1 = 4$ m at $t = 0$, and $h_2 = 1.5$ m at $t = 0$.
  (a) Develop the differential equation that describes the variation of head h_1 with time.
  (b) What is the system time constant?
  (c) Sketch h_1 and h_2 versus t for the given initial conditions.

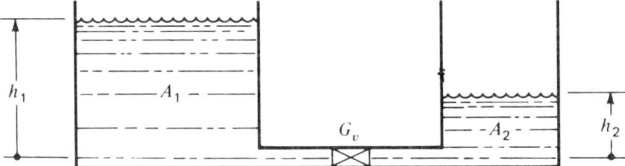

PROBLEMS

14-19. The effective gross weight of an automobile is 3000 lb. The given curves show the engine torque (in high gear) required to maintain different constant speeds, and the engine torque as a function of the accelerator position and the vehicle speed. Each ft-lb of torque at the engine produces 3.2 lb thrust on the road in high gear.

(a) Write the linearized differential equation for the change in velocity δv caused by small changes in accelerator position δx if the average velocity is 50 mi/h.

(b) Determine the system time constant at 50 mi/h—that is the time required to achieve 63.2 percent of the steady-state change in car velocity caused by a small step change in accelerator position.

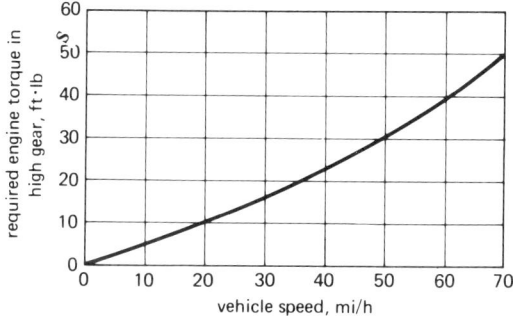

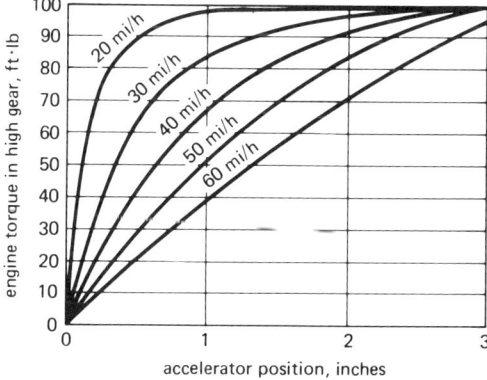

15
Control Actions

15-1 Introduction

The contents of the preceding chapter are general in nature and therefore applicable to a wide variety of dynamic systems. The material to be developed in this chapter, in addition to discussing control actions, will illustrate how the concepts of dynamic-system behavior are applied to feedback control systems.

The term *control action* refers to the manner in which the system error is utilized in making a correction, or, conversely, how the correction is related to the error. An understanding of control actions is beneficial to the control engineer because certain general inferences about system response can be drawn if it is known what control action is being used. Several commonly encountered control actions will be treated in detail in this chapter. A liquid-level control application will be used for illustrative purposes; this will be a comparatively simple system and also one for which the response is easy to visualize.

15-2 Proportional Control

The nature and characteristics of proportional control action will be developed with the use of the liquid-level system shown in Fig. 15-1. A float senses the liquid level h and, through the linkage shown, controls the input flow rate q_i. A rise in level results in a decreased flow rate, whereas a fall in level produces an increased flow rate. With the setting of the outlet valve fixed, the system linkage can be adjusted to give the desired equilibrium value of the head. However, a disturbance flow rate q_d will upset the system. The analysis to be presented will be concerned with the response of the system to a disturbance flow rate.

The equation for the tank is based on Eq. 2-27 and can be written as

$$A \frac{dh}{dt} = q_{\text{net}} = q_i + q_d - q_o \tag{15-1}$$

in which q_o is the output flow rate and A is the cross-sectional area of the tank. The disturbance flow rate q_d is assumed to begin at $t = 0$, with the system previously in equilibrium. It is convenient to define the variables in Eq. 15-1 as

$$\begin{aligned} h &= H + \delta h \\ q_i &= Q_i + \delta q_i \\ q_o &= Q_o + \delta q_o \\ q_d &= 0 \quad \text{for } t < 0 \\ q_d &= q_d \quad \text{for } t \geqslant 0 \end{aligned} \tag{15-2}$$

where H, Q_i, and Q_o are the initial constant equilibrium values, and δh, δq_i, and δq_o are deviations resulting from the disturbance.[1]

Substituting Eqs. 15-2 into Eq. 15-1 yields

$$A \frac{dH}{dt} + A \frac{d\delta h}{dt} = Q_i + \delta q_i + q_d - Q_o - \delta q_o \tag{15-3}$$

However, at the initial equilibrium condition (before the disturbance),

$$A \frac{dH}{dt} = Q_i - Q_o = 0$$

Therefore, Eq. 15-3 becomes

$$A \frac{d\delta h}{dt} = \delta q_i + q_d - \delta q_o \tag{15-4}$$

---

[1]Considering deviations away from equilibrium values is convenient because (1) the control system is expected to minimize the head *deviation*, and (2) the differential equation written in terms of deviations is readily linearized if one or more system elements (e.g., inlet and outlet valves) is nonlinear (see Example 4-2 in Section 4-3).

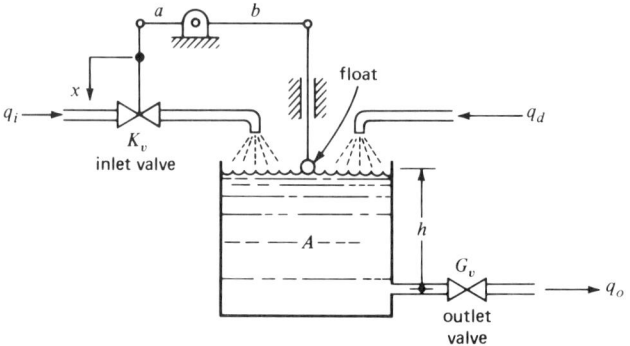

FIG 15-1. Liquid-level control system utilizing proportional control.

The outflow deviation can be expressed as

$$\delta q_o = G_v\, \delta h$$

with the result that Eq. 15-4 becomes

$$A\frac{d\delta h}{dt} = \delta q_i + q_d - G_v\, \delta h$$

or

$$A\frac{d\delta h}{dt} + G_v\, \delta h = \delta q_i + q_d \tag{15-5}$$

where G_v is the outlet valve constant.

The inlet valve stem displacement deviation δx, for small movements, is

$$\delta x = \frac{a}{b}\, \delta h \tag{15-6}$$

The valve stem deviation results in an inflow deviation

$$\delta q_i = -K_v\, \delta x \tag{15-7}$$

where K_v is the inlet valve constant. Note, in particular, the negative sign. A positive level deviation produces a positive stem deviation; however, such a condition must produce a decreased inflow. Thus the negative sign in Eq. 15-7 is required to define the relationship properly. Combining Eq. 15-6 and 15-7 yields

$$\delta q_i = -\frac{aK_v}{b}\, \delta h \tag{15-8}$$

With Eqs. 15-5 and 15-8, the overall system equation can be obtained as

$$A\frac{d\delta h}{dt} + G_v\delta h = -\frac{aK_v}{b}\,\delta h + q_d$$

15-2 PROPORTIONAL CONTROL

or

$$A \frac{d\delta h}{dt} + \left(\frac{bG_v + aK_v}{b}\right)\delta h = q_d \tag{15-9}$$

Equation 15-9 can be solved to determine the liquid-level deviation for any disturbance input.

As an illustration, let us assume that a step-disturbance input occurs; that is, $q_d = Q_d$ (a constant). Equation 15-9 can be solved for the steady-state deviation, which is

$$\delta h_{ss} = \left(\frac{b}{bG_v + aK_v}\right)Q_d \tag{15-10}$$

Note that a steady-state deviation proportional to the magnitude of the step disturbance results. This response is a deviation from the desired value of level, so it is also a system error. Plots illustrating the response of the system are given in Fig. 15-2.

The results of the analysis are easily checked intuitively. The additional flow into the tank, in the form of a disturbance, requires that the flow through the inlet valve be reduced. This condition requires that the inlet valve stem be moved toward the closed position, which can be accomplished only by an increase of liquid level.

The liquid-level control system of this section illustrates the use of proportional control action. *With proportional control, a correction is made which is proportional to the error.* In the system considered, a change of liquid level (an error) resulted in a proportional change of inflow (a correction), as given by Eq. 15-8.

One important characteristic of proportional control is that if a sustained correction (owing to a sustained disturbance) is necessary, a steady-state error will result. Figure 15-3 illustrates this point; the steady-state value of the head is plotted against the disturbance flow rate expressed as a percentage. The desired value H for the head can be maintained only when no disturbance is present.

Another important feature of proportional control is that it is the most stable control action.

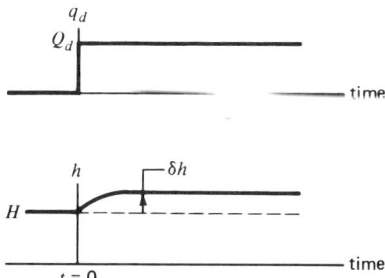

FIG 15-2. Response of liquid-level system to a step disturbance input.

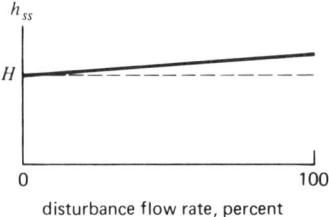

FIG 15-3. Error characteristics of liquid-level control system.

15-3 Integral Control

Integral control is another widely used control action. Its nature and characteristics will be developed with the use of the system shown schematically in Fig. 15-4. The system is much the same as that of Fig. 15-1, except that the float controls the displacement y of the spool in a hydraulic valve. This valve, in turn, controls the displacement x of a piston that is directly connected to the inlet valve stem. A constant valve stem velocity results from any fixed position of the tank float that acts to open the hydraulic valve. This behavior is in contrast to that obtained with proportional control, wherein a fixed float position yields a fixed inlet valve stem displacement. With integral control, correction will continue to take place until the float maintains the fixed position required to shut off the hydraulic valve. Thus, integral control tends to eliminate steady-state errors.

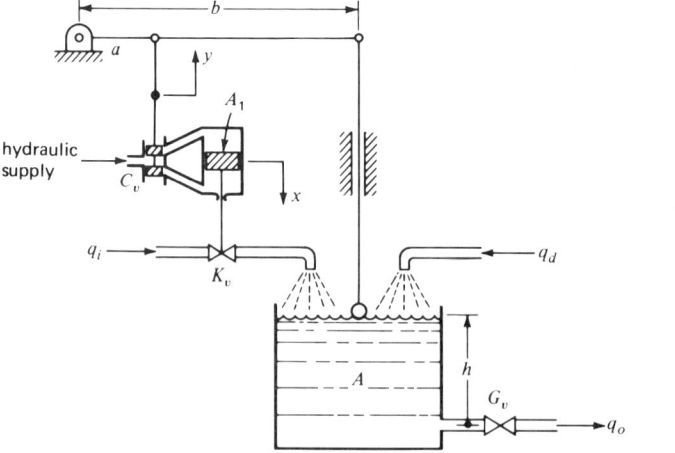

FIG 15-4. Liquid-level control system utilizing integral control.

15-3 INTEGRAL CONTROL

The approach to obtaining the overall system equation is the same as in the preceding section. The system is assumed to be in equilibrium at the time a disturbance occurs. The system variables are defined, in exactly the same manner as for the proportional control system, by Eqs. 15-2.

The basic tank equation developed for the proportional control system (Eq. 15-5, repeated below for convenience) will also apply in this case.

$$A \frac{d\delta h}{dt} + G_v \, \delta h = \delta q_i + q_d$$

The expression for spool displacement y is

$$y = \frac{a}{b} \delta h \tag{15-11}$$

The flow rate q through the hydraulic valve can be expressed as

$$q = C_v y \tag{15-12}$$

where C_v is the valve constant. The cylinder equation is

$$q = A_1 \frac{dx}{dt} \tag{15-13}$$

in which A_1 is the piston area. Combining Eqs. 15-12 and 15-13 yields

$$q = C_v y = A_1 \frac{dx}{dt} \tag{15-14}$$

Equation 15-14 can be put in operator form, with the following result:

$$C_v y = A_1 p x$$

or

$$x = \frac{C_v}{A_1} \left(\frac{y}{p} \right) \tag{15-15}$$

By defining $x = 0$ for the valve stem position corresponding to an inlet flow rate $q_i = Q_i$ (i.e., corresponding to $\delta q_i = 0$), the input flow rate is related to the stem displacement x by

$$\delta q_i = -K_v x \tag{15-16}$$

Combining Eqs. 15-11, 15-15, and 15-16 yields

$$\delta q_i = -\frac{K_v C_v a}{A_1 b} \left(\frac{\delta h}{p} \right) = -K_0 \left(\frac{\delta h}{p} \right) \tag{15-17}$$

in which K_0 is an overall system constant. However,

$$\frac{\delta h}{p} = \int_0^t \delta h \, dt$$

so Eq. 15-17 can be written as

$$\delta q_i = -K_0 \int_0^t \delta h \, dt \tag{15-18}$$

Equation 15-18 reveals the reason for the term *integral control*. The correction made is proportional to the time integral of the head deviation (which is equal in magnitude to the system error).

Substituting Eq. 15-18 into Eq. 15-5 yields

$$A \frac{d\delta h}{dt} + G_v \, \delta h + K_0 \int_0^t \delta h \, dt = q_d \tag{15-19}$$

Differentiation provides the overall system differential equation

$$\frac{A}{K_0} \frac{d^2 \delta h}{dt^2} + \frac{G_v}{K_0} \frac{d\delta h}{dt} + \delta h = \frac{1}{K_0} \frac{dq_d}{dt} \tag{15-20}$$

Equation 15-20 can be solved for a step disturbance input. The steady-state response is

$$\delta h_{ss} = 0 \tag{15-21}$$

Thus, integral control action is ultimately able to eliminate the deviation (and system error) resulting from a step disturbance. We should recall that this was not the case with proportional control. However, with proportional control the system equation was of first order, whereas with integral control a second-order equation (and the possibility of an oscillatory response) results.

The transient solution of Eq. 15-20 for the case of a step disturbance input requires some explanation. The forcing function, $(1/K_0) \, dq_d/dt$, is zero except at the very instant $t = 0$, at which time it is infinite. Such a forcing function, having an infinite value for an infinitesimal period of time, is known as an *impulse*. Response to an impulse is quite important in advanced control system theory as well as in other disciplines. It will not, however, be covered in any detail in this introductory book. Here it will merely be pointed out how to handle an impulse forcing function for an equation having the form of Eq. 15-20.

Since the forcing function is zero except for the instant of time $t = 0$, the best way to handle the solution is to find the condition of the system immediately after the impulse has been applied ($t = 0+$) and then to solve the homogeneous form of the equation with those initial conditions. If Eq. 15-19 is applied to the system at the instant immediately following the application of the step input, the second and third terms on the left-hand side of the equation will be zero since a finite period of time must pass before δh can become greater than zero. The equivalent initial conditions are therefore

$$\delta h \big|_{t=0+} = 0$$
$$\frac{d\delta h}{dt} \bigg|_{t=0+} = \frac{Q_d}{A} \tag{15-22}$$

That is, δh is zero, whereas its first derivative is equal to the value of the step disturbance (which produced the impulse forcing function in Eq. 15-20) divided by the tank area.

15-4 PROPORTIONAL-PLUS-INTEGRAL CONTROL

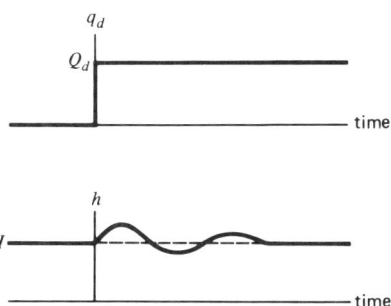

FIG 15-5. Response of a liquid-level system to a step disturbance input.

Figure 15-5 illustrates the typical system response, determined from the solution of the homogeneous equation

$$\frac{A}{K_0}\frac{d^2\delta h}{dt^2} + \frac{G_v}{K_0}\frac{d\delta h}{dt} + \delta h = 0$$

with the initial conditions given by Eqs. 15-22.

There is an alternative way to look at the action of an integral control system. If we take the derivative of Eq. 15-18,

$$\frac{d\delta q_i}{dt} = -K_0\,\delta h \tag{15-23}$$

we see that *the rate of change of the correction is directly proportional to the error* (while in a system with proportional control action, the change itself is directly proportional to the error).

With integral control a correction is made that is proportional to the time integral of the error. The desirable feature of integral control is that it tends to eliminate steady-state error. Its undesirable feature is that it tends to produce an oscillatory response and, with some systems, instability.

15-4 Proportional-plus-Integral Control

It is possible to combine the two control actions discussed in the previous two sections to produce what is known as *proportional-plus-integral* (PI) control. The idea behind such a system is to combine the advantageous features of both proportional and integral action at the expense of additional hardware complexity and cost. A PI system will eliminate steady-state error from a step disturbance because of the integral action. The proportional action, on the other hand, can be used to reduce the undesirable oscillatory action (and sometimes actual instability) that often occurs with a pure integral control system.

For easy comparison with the systems already discussed in this chapter, a liquid-level control system will be used to illustrate proportional-plus-integral-control action. The system illustrated in Fig. 15-6 has two parallel lines through which the liquid may enter the tank, and each of these is controlled by a separate valve. A single float is used to measure the liquid level, but the signal is used in two different ways. First of all, the float level will change the opening of one inlet valve in a proportional manner, exactly as in the proportional control system of Fig. 15-1. The other valve, controlled by the hydraulic valve and piston, gives an integral control action. The net result of the two-valve system is proportional plus integral control. (It should be pointed out, however, that Fig. 15-6 is not meant to illustrate the way a practical PI liquid-level system would be designed. It has been drawn only for the purpose of clearly illustrating the *principle* of PI control. A practical system would most likely incorporate a single input line with a single valve specifically designed to accommodate both the proportional and integral control functions.)

To develop the system equation, we begin with the basic tank equation (Eq. 15-5),

$$A \frac{d\delta h}{dt} + G_v \, \delta h = \delta q_i + q_d$$

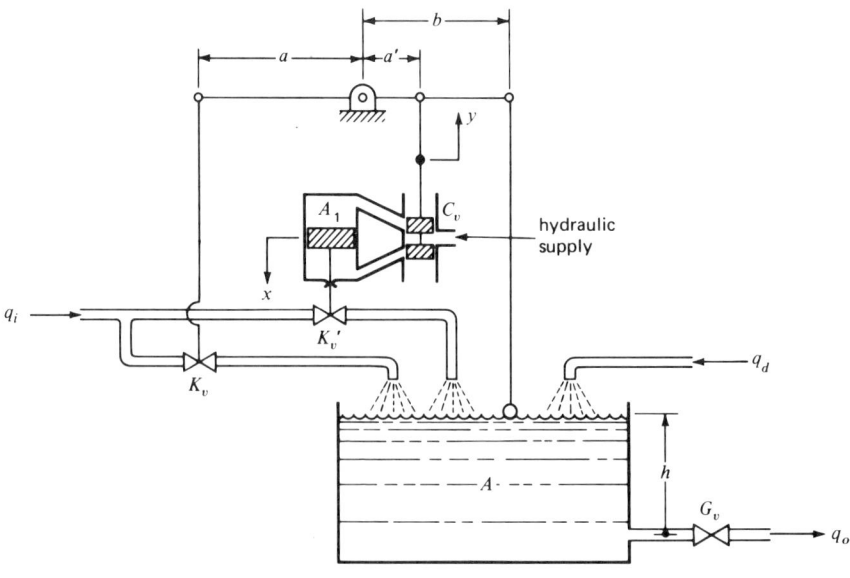

FIG 15-6. Liquid-level control system utilizing proportional plus integral (PI) control.

15-5 "BANG-BANG" CONTROL

The change in inlet flow, δq_i, has two parts, one given by Eq. 15-8 and the other by Eq. 15-18, so that

$$\delta q_i = -\frac{aK_v}{b}\delta h - K_0 \int_0^t \delta h \, dt \tag{15-24}$$

Substitution into the basic tank equation yields

$$A\frac{d\delta h}{dt} + \left(G_v + \frac{a}{b}K_v\right)\delta h + K_0 \int_0^t \delta h \, dt = q_d \tag{15-25}$$

With a single differentiation, this becomes

$$\frac{A}{K_0}\frac{d^2\delta h}{dt^2} + \frac{G_v + (a/b)K_v}{K_0}\frac{d\delta h}{dt} + \delta h = \frac{1}{K_0}\frac{dq_d}{dt} \tag{15-26}$$

Comparison of Eq. 15-26 with Eq. 15-20 shows them to be identical except for the coefficient of $d\delta h/dt$. Eq. 15-26 can also be written (see Section 14-5) as

$$\frac{1}{\omega_n^2}\frac{d^2\delta h}{dt^2} + \frac{2\zeta}{\omega_n}\frac{d\delta h}{dt} + \delta h = \frac{1}{K_0}\frac{dq_d}{dt} \tag{15-27}$$

By comparing Eqs. 15-20, 15-26, and 15-27 it is apparent that the damping ratio of the PI system can be readily adjusted to a higher value than that of the integral control system. The exact damping characteristics will depend upon the relative values of all system parameters. The steady-state error to a step disturbance input is zero for the PI system, based on the same analysis as used for Eq. 15-21.

The PI system has been shown to have the potential of eliminating any steady-state error to a step input while allowing the damping ratio to be adjusted to suit system requirements. For the liquid-level example, this has been accomplished only with greater mechanical complexity. In some cases PI control action can be partially implemented by electronic circuitry, so the additional expense and complexity are small.

15-5 "Bang-Bang" Control

One relatively simple concept often used for control systems is to apply corrective action only when the system has an error which is greater than some predetermined value. When the system output is within the allowable error band, no correction is applied. When the allowable error limit is exceeded, however, the corrective action (e.g., force or torque) is applied *at a fixed level of intensity* until the error returns to an allowable value. Such a system is called "bang-bang" because of the intermittent feature.

Assuming that a certain error can be tolerated by a control system with no ill effects, such control action has certain advantages. First of all, it can often be implemented with relatively simple and inexpensive equipment. In addition, a considerable saving of energy is often possible since all components that consume appreciable amounts of energy can be turned off except when the allowable error is exceeded.

A good example of an application where "bang-bang" control may be advantageous is a satellite or spacecraft whose attitude must be maintained within fairly close limits rather than being allowed to tumble freely [14]. This may be necessary to keep solar panels properly aligned with the sun, cameras aimed at a specific path on the earth, and so on. One way commonly used to obtain the required corrective torque on a satellite is to use the reaction from small gas jets (Fig. 15-7). Since only a limited amount of compressed gas for the jets can be carried aboard the satellite, the energy-saving feature of "bang-bang" control is quite important in this application. Cost of equipment is a secondary consideration, but weight must be minimized.

A separate system is required for control about each of the three orthogonal satellite axes. Since these work individually and independently of each other (with only a small amount of dynamic cross coupling), consideration of a single axis will be adequate for the explanation of the control action.

Referring to Fig. 15-8a, and assuming that the spacecraft has an initial small angular velocity about the axis under consideration but a negligible value of external torque (which would come from such low-level effects as solar radiation and the gravity gradient), the angular position of the

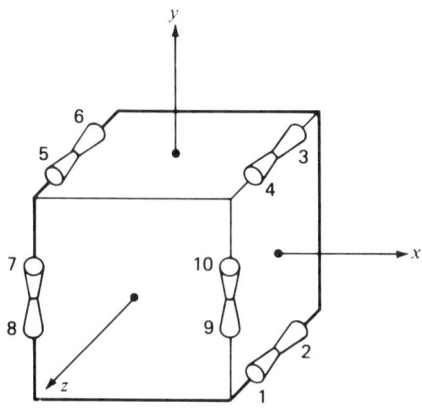

FIG 15-7. A gas-jet attitude control system for a satellite. (Rotation about x-axis is controlled by jets $1 + 3$ and $2 + 4$; rotation about y-axis is controlled by jets $3 + 5$ and $4 + 6$; rotation about z-axis is controlled by jets $7 + 9$ and $8 + 10$).

15-5 "BANG-BANG" CONTROL

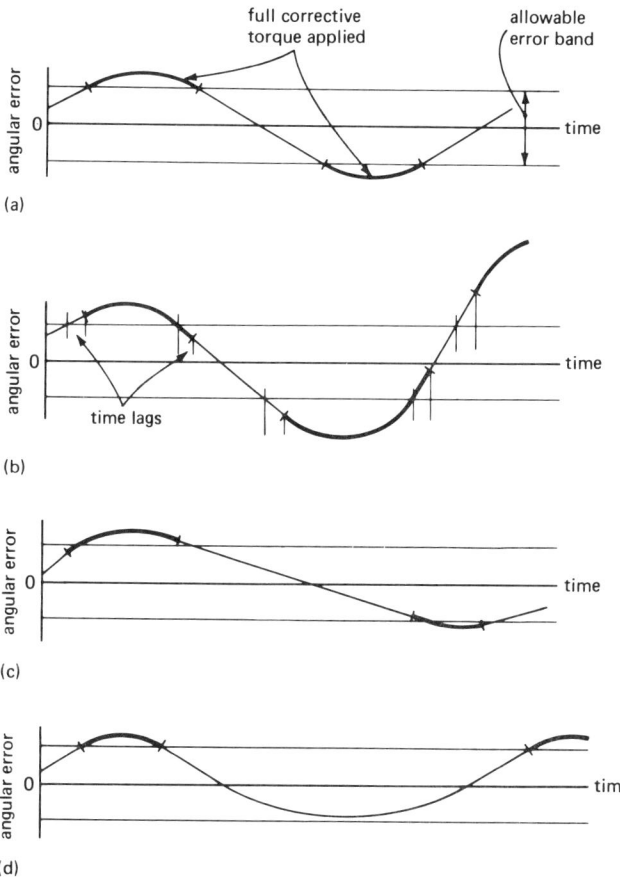

FIG 15-8. Control action about a single axis for the "bang-bang" gas jet attitude control system of Fig. 15-7.
(a) Basic system.
(b) A real system with time lags between error sensing and gas jet actuation.
(c) A system with "anticipation" circuitry.
(d) A system acted upon by a steady external torque of relatively high magnitude.

spacecraft will slowly drift to the limit of allowable error. At that time, *full corrective torque* is applied (a relatively low value of torque for a typical spacecraft of this type) to cause an angular acceleration that will return the spacecraft to the allowable error band. The corrective torque then stops, and the spacecraft angular position slowly drifts across the band until it reaches the other limit. At that time full corrective torque in the opposite direction is applied until the spacecraft again returns to the allowable error band. This process is continuously repeated as illustrated in Fig. 15-8a, with the spacecraft executing what is known as a *limit cycle*, oscillating back and forth within relatively firm limits. Even though the oscillation does not die out, it is not considered an instability in this case.

The basic "bang-bang" control system for a spacecraft as described above, however, must have some slight modifications in order for it to work properly in practice. If there were damping present, the basic approach just described would probably work satisfactorily, but an object in space has virtually no damping. In a real control system there will be a slight delay between the time of the torque signal and the instant of time at which full corrective torque is actually applied (because of mechanical inertia, electronic time delays, etc.). The resultant spacecraft action with this delay is illustrated in Fig. 15-8b. The corrective torque is not applied until after the spacecraft has rotated to a position somewhat beyond the allowable error. Since the factors causing a delay in turning on the jet will similarly affect the turning-off operation, the jet will remain on until the spacecraft angular position is a small distance within the allowable error band. The net result of these delays is that the velocity across the allowable error band becomes slightly larger for each successive half-cycle, so that the actual error and the amount of energy required for each correction become increasingly greater. The inherent time delays in combination with the lack of damping therefore cause the system to become unstable.

In order to eliminate the instability, a practical satellite attitude control system of this type must include some type of compensation. By the use of circuitry that takes the angular velocity into account as well as the angular position, the system can be made to "anticipate" when the allowable error will be exceeded so that it can apply the corrective torque somewhat earlier. Similarly, it can also anticipate when the error will return to the allowable band and turn off the torque a little before this occurs. The net result of such anticipation circuitry is to reduce the velocity within the allowable band, and to cause the stable action illustrated in Fig. 15-8c. Although its period can be reduced with this approach, the limit cycle itself can never be completely eliminated with the "bang-bang" control action.

If an appreciable external torque (in comparison with the corrective torque available from the gas jets) is continually present, the action of the spacecraft may be as illustrated in Fig. 15-8d. In this case the external torque will prevent the limit at one side of the allowable error band from ever being reached, and a corrective torque in only one direction will ever be needed.

A system using "bang-bang" control action is not linear, so the analysis techniques presented earlier in this book are not directly applicable. The system is "piecewise linear," however, and therefore fundamental linear equations based on Newton's laws of motion can be applied to each distinct part of the operation. Analysis of the limit cycle of a particular system using "bang-bang" control is relatively easy to obtain by means of step-by-step calculations. This approach lends itself readily to a digital computer program (see Chapter 17).

15-6 Other Control Actions

The detailed treatment of other control actions is beyond the scope of this book. However, the reader should at least be aware of the fact that other possibilities exist.

Another control action is *derivative control*, in which a correction is made that is proportional to the time derivative of error. Derivative control cannot be used alone because it will not respond to an error that is constant with time. Rather it is used in combination with other control actions in order to obtain its stabilizing influence on the system.

Control actions can be used in combination; that is, a correction can be made that is the sum of the corrections produced by individual control actions. The features and advantages of proportional plus integral control have already been explained. Another possibility is the use of *proportional plus derivative control*. The stabilizing influence of derivative control permits the proportional control action to be utilized so that its steady-state error characteristic is minimized. There are systems for which *proportional plus integral plus derivative control* is used because very serious control problems exist. Although combined control actions provide better control, the cost of the equipment involved is increased.

15-7 Conclusion

Feedback control systems can be characterized by the manner in which the system error is used in making a correction. A number of alternative control actions have been discussed in this chapter. The type of control action that is best for a given application depends upon a number of factors, such as the degree of accuracy required, whether or not oscillatory action is undesirable, the significance of cost, and the characteristics of the basic system components that will be used.

Problems

15-1. For the tank system shown in the accompanying illustration:
(a) Obtain the system differential equation that relates the change in head δh_2 with the disturbance flow rate q_d.
(b) Determine $\delta h_{2_{ss}}$ for a step input $q_d = Q_d$.
(c) Sketch δh_2 versus t. Assume the step input of (b) and that h_2 is initially at H_2.

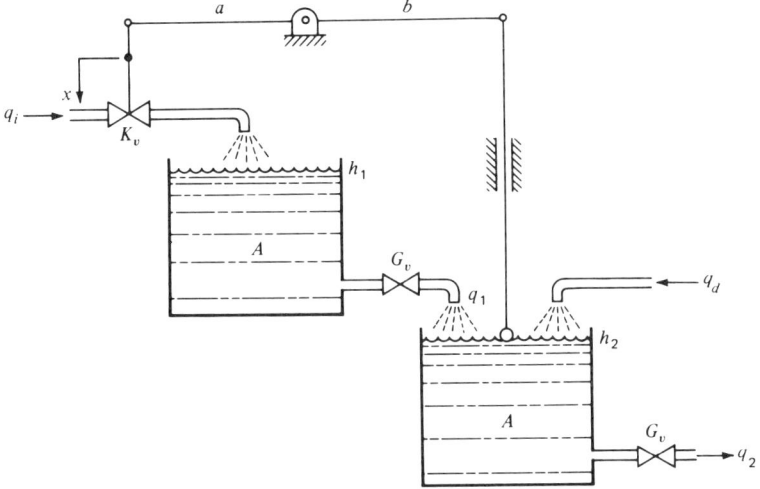

15-2. For the tank system shown in the accompanying illustration:
(a) Obtain the system differential equation that relates the change in head δh with a change in the downstream head δh_d. (*Hint*: $\delta q_o = G_v \, \delta(h - h_d)$.)
(b) Determine δh_{ss} for a step disturbance $h_d = H_d$.

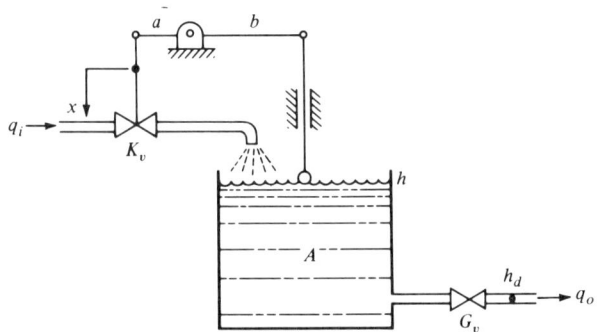

PROBLEMS

15-3. A schematic block diagram for a *speed control* system is given. For the centrifugal governor,

$$\frac{1}{\omega_n^2}\frac{d^2x}{dt^2} + \frac{2\zeta}{\omega_n}\frac{dx}{dt} + x = K_a(\omega_i - \omega_o)$$

For the fuel valve,

$$q = K_b x$$

For the turbine,

$$T = K_c q$$

For the load,

$$T + T_d = I\frac{d\omega_o}{dt} + b\omega_o$$

(a) Obtain the differential equation relating the output speed ω_o with the two possible inputs: speed setting ω_i and torque disturbance T_d.
(b) Determine the steady-state change in response $\delta\omega_{o_{ss}}$ for a step change of speed setting $\delta\omega_i$. (Assume that there is no torque disturbance during the change.)
(c) Determine the steady-state change in response $\delta\omega_{o_{ss}}$ for a step *increase* in load torque of magnitude δT_d. (Assume that there is no change of speed setting during the change.)

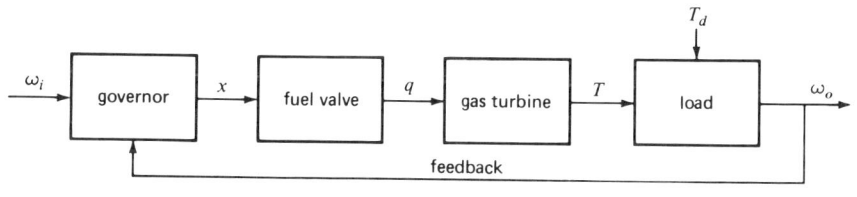

ω_i = speed setting
ω_o = output speed
T = turbine output torque

T_d = disturbance torque
= change in load torque
x = governor displacement
q = fuel flow rate

15-4. A positioning control system, shown schematically in the accompanying illustration, utilizes a bevel-gear differential to make the comparison between the input displacement θ_i and the output displacement θ_o. The output displacement ϕ of the differential positions the wiper of a potentiometer which, through an amplifier, controls the output torque T of a dc motor. The motor torque is used to position the load inertia I. System damping is idealized as viscous friction with damping coefficient b. Assume that the potentiometer-amplifier-motor combination can be adequately represented as $T = K\phi$.

(a) Obtain the overall system differential equation in which θ_o is the response and θ_i is the input. (*Hint*: $\phi = \frac{1}{2}(\theta_i - \theta_o)$.)

(b) Is the system stable?

(c) Determine the steady-state response $\theta_{o_{ss}}$ for a step input $\theta_i = \Theta_i$.

(d) Determine the steady-state error e_{ss} for a ramp input $\theta_i = Ct$. Error is defined as $e = \theta_i - \theta_o$.

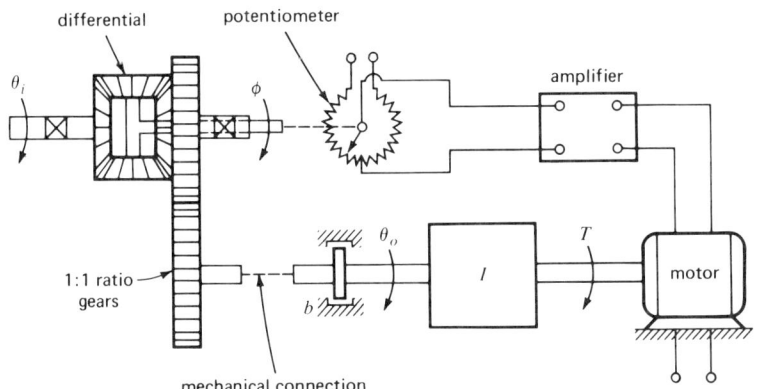

PROBLEMS

15-5. A hydraulic positioning system, shown schematically in the accompanying illustration, utilizes a pair of potentiometers to make the comparison between the input displacement x and the output displacement y. The voltage v is a measure of system error and can be represented as $v = K_a(x - y)$. The amplifier drives an electrohydraulic servovalve that supplies fluid to the cylinder. Assume that the amplifier-valve relationship is $\tau \, dq/dt + q = K_b v$.
  (a) Obtain the overall system differential equation in which y is the response and x is the input.
  (b) Is the system stable?
  (c) Determine the steady-state response y_{ss} for a step input $x = X$.
  (d) Most control systems provide some means for adjustment. Assume that K_b can be varied. Obtain the expression for K_b such that the system will be critically damped.

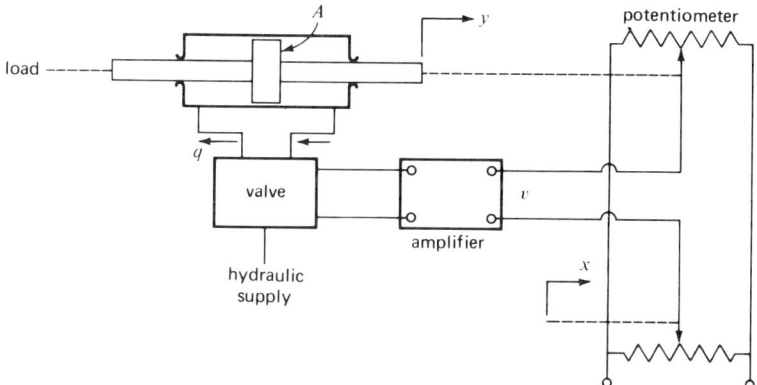

15-6. A hydraulic positioning system, shown schematically in the accompanying illustration, utilizes a pair of potentiometers to make the comparison between input displacement x and output displacement y. The voltage v is a measure of system error and can be represented as $v = K_a(x - y)$. The amplifier drives an electrohydraulic *pressure control* valve so that the amplifier-valve-piston combination can be adequately represented as $F = K_b v$, where F is the hydraulic force exerted on the piston.

(a) Obtain the overall system differential equation in which y is the response and x is the input.
(b) Is the system stable?
(c) Determine the steady-state response y_{ss} for a step input $x = X$.

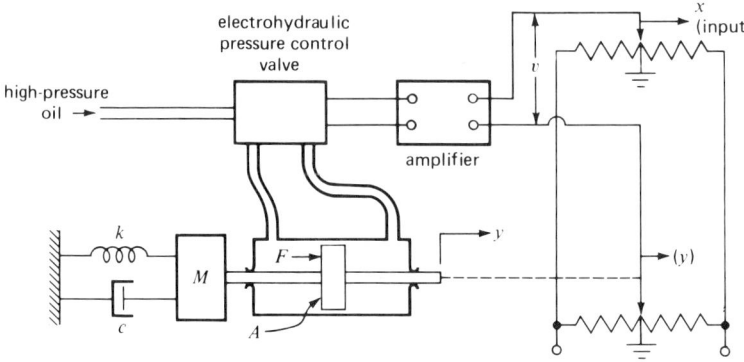

15-7. The temperature control system diagrammed in the figure is designed to control T_o, the temperature of a room, based on an input temperature setting T_i. The system error is

$$e = T_i - T_o$$

The characteristic of the control equipment is

$$\frac{dq}{dt} + q = Ke$$

where q = flow rate of thermal energy. The thermal characteristics of the room are

$$10\frac{dT_o}{dt} + T_o = 0.1q$$

(a) Obtain the defining equation relating T_o to T_i.
(b) Determine a value for K to yield a system damping ratio of 1.

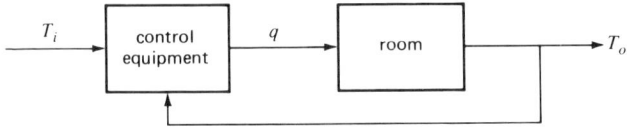

16
Block Diagrams

16-1 Introduction

The analysis of a dynamic system is frequently furthered by representing the system in some diagrammatic, or schematic, form. For control systems, a useful form is the *block diagram*. With block diagrams each system component, or element, is represented by a rectangle (a *block*). Input and output variables are shown as lines, or arrows, directed into and out of the block. The completed block diagram is obtained by arranging the individual blocks to correspond with the way in which the actual system elements are connected. The figure for Prob. 15-3 provides an example of an elementary type of block diagram.

The usefulness of a block diagram is extended if information about the dynamic behavior of each system element is included in the diagram in a suitable mathematical form. This objective is met through the use of a mathematical relationship called a *transfer function* (to be considered in detail in the next section). The completed block diagram (including transfer functions) readily yields any desired overall system differential equation, from which questions regarding matters such as stability and system response to various inputs can be considered.

Attention is first turned to the nature of transfer functions and their relationship to individual blocks. Then the formulation and utilization of block diagrams is discussed and illustrated.

16-2 The Transfer Function

The concept of the *transfer function* is very important for block diagram work. For a given process, mechanism, or system, the transfer function is defined as *the ratio of the output variable to the input variable*. It may be a simple constant or a more complex mathematical function. In a block diagram representation of a system, transfer functions are placed in blocks, or boxes, to show schematically the relation between the input and output variables, as illustrated in Fig. 16-1.

A transfer function is easily obtained if the equation relating the input and output variables is available. Although other approaches could be used, the transfer function is most useful if written in operator form when the equation contains derivatives or integrals. Several examples will now be given to illustrate how transfer functions are obtained.

The transfer function for a pair of gears (Fig. 16-2) is readily obtained. If we consider the speed ω_1 as the input and ω_2 as the output, the transfer function is simply the ratio of gear teeth,

$$\frac{\omega_2}{\omega_1} = \frac{n_1}{n_2}$$

We may alternatively write a transfer function relating the input and output torques,

$$\frac{T_2}{T_1} = \frac{n_2}{n_1}$$

Which of these two alternative transfer functions (Fig. 16-3) is the right one to use in any particular case depends upon the type of analysis being done—that is, upon whether speed or torque is the variable of primary interest.

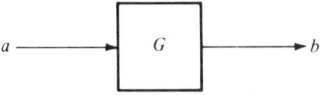

FIG 16-1. Block representation of the transfer function $G = b/a$.

16-2 THE TRANSFER FUNCTION

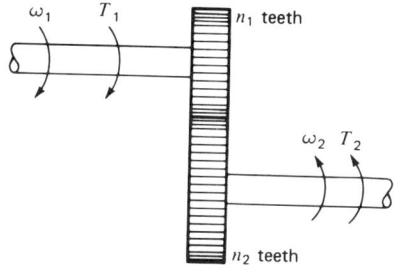

FIG 16-2. A pair of gears.

Consider next the tank of Fig. 16-4. Considering the flow rate q_i as the input and the head h as the output, the transfer function of this system is the ratio h/q_i. The differential equation relating h to q_i is (as obtained in Section 2-3 for a system with a linear outlet valve)

$$A\frac{dh}{dt} + G_v h = q_i \tag{16-1}$$

Rewriting in operator form allows the desired transfer function to be obtained:

$$G = \frac{h}{q_i} = \frac{1}{Ap + G_v} \tag{16-2}$$

If we are working with *deviations from average values of operation*, the required transfer function (obtained in the same manner as Eq. 16-2) will be

$$G = \frac{\delta h}{\delta q_i} = \frac{1}{Ap + G_v} \tag{16-3}$$

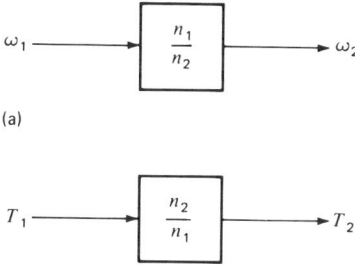

FIG 16-3. Two alternative transfer functions for the gear pair of Fig. 16-2.

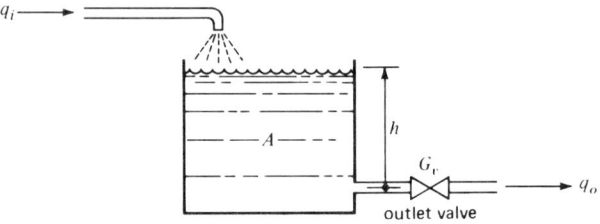

FIG 16-4. Liquid-level system.

It should be carefully noted, however, that the valve coefficients G_v in Eqs. 16-2 and 16-3 are defined differently

$$G_v \equiv \frac{q_0}{h} \qquad \text{for Eq. 16-2}$$

$$G_v \equiv \frac{\delta q_0}{\delta h} \qquad \text{for Eq. 16-3}$$

(This difference has been covered in detail in Section 4-3.)

The valve-controlled piston, Fig. 16-5, is described by the differential equation (as developed in Section 2-3)

$$A \frac{dy}{dt} = K_v x \qquad (16\text{-}4)$$

Putting Eq. 16-4 into operator form and solving for the ratio of output to input yields the transfer function

$$G = \frac{y}{x} = \frac{K_v}{Ap} \qquad (16\text{-}5)$$

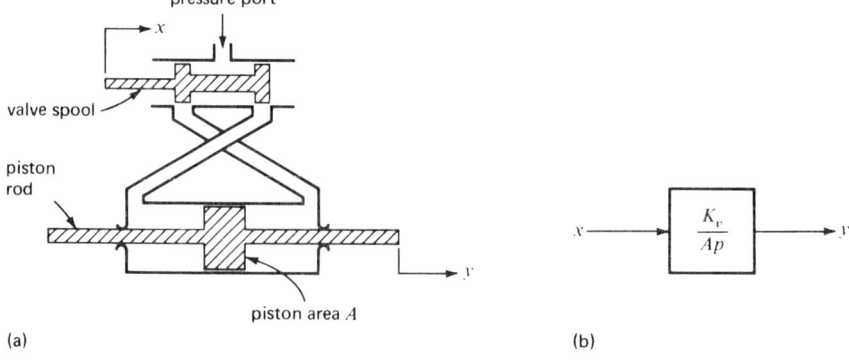

FIG 16-5. Valve-controlled hydraulic piston and its block diagram representation.

16-3 The Block Diagram

Section 16-2 has described the technique required to obtain transfer function blocks to represent various components (or subsystems) of a given system. Before a complete block diagram can be drawn, however, it is necessary to also understand the concept of the *summing point*, or *summing junction*. The summing point, drawn as in Fig. 16-6, is used to represent the part of the system in which system variables are added or subtracted. Figure 16-6a represents the situation in which a single variable is subtracted from another variable (common in feedback control systems where the value of the output is subtracted from the input to obtain the error or actuating signal). It is the block diagram representation of the equation

$$a_3 = a_1 - a_2$$

Figure 16-6b represents a situation in which three system variables are added. It is the block diagram equivalent of the equation

$$a_4 = a_1 + a_2 + a_3$$

Another principle that should be understood is that the taking of *information* about a block diagram variable does not directly affect that variable at the *takeoff point*. For example, as illustrated in Fig. 16-7, if the value of the system variable a_3 is needed for a summing point ahead and/or behind it, drawing lines between a_3 and the summing points does not affect the equations for the blocks G_1 and G_2, and we have

$$a_3 = G_1 a_2$$
$$a_4 = G_2 a_3$$

regardless of whether or not the feedback and feedforward lines are used. Taking information *from* a takeoff point is entirely different from feeding information *into* a summing point.

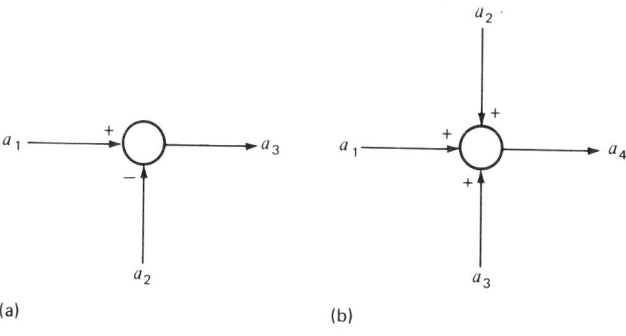

(a)   (b)

FIG 16-6. Typical summing points.

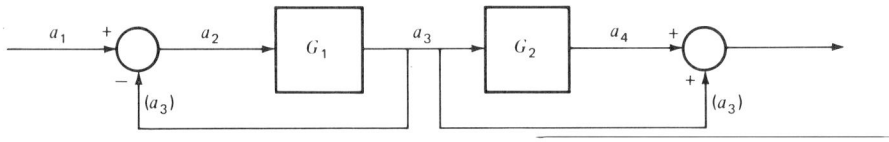

FIG 16-7. A block diagram that illustrates the significance of the *takeoff point*.

With an understanding of the two basic block diagram tools, the transfer function and the summing point, the complete block diagram of a system can readily be developed. First the blocks and summing points are developed for all system elements. These are then combined by lining them up in the proper sequence and joining all lines (arrows) representing each system variable. The details of this process will be illustrated in Section 16-5.

16-4 Block Diagram Algebra

One value of the block diagram is that it serves as an aid in understanding the interrelationships among the various elements and functions of the system it represents. It is also valuable for obtaining objective data on system response, stability, and other dynamic characteristics. Before it can be used for obtaining this type of information, however, it must usually be reduced to a simpler form. Such reduction is accomplished by using what is known as *block diagram algebra*. The basic rules of block diagram algebra will be presented in this section.

BLOCKS IN CASCADE

Blocks in a block diagram often occur in "cascade," or series, with the output variable of one serving as the input to the next, as illustrated in Fig. 16-8. A set of blocks in cascade can be reduced to a single block whose transfer function is the product of the transfer functions of the original cascaded blocks. Mathematically, the equation representations of the three blocks,

$$a_2 = G_1 a_1$$
$$a_3 = G_2 a_2$$
$$a_4 = G_3 a_3$$

can be combined into the single equation

$$a_4 = G_1 G_2 G_3 a_1$$

which is represented by Fig. 16-8b.

16-4 BLOCK DIAGRAM ALGEBRA

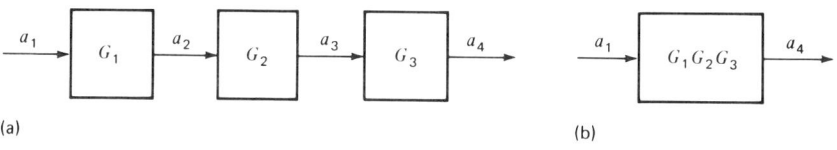

FIG 16-8. (a) Blocks in cascade.
(b) Single-block representation.

FEEDBACK LOOPS

One basic form of the block diagram that is quite important because of its common occurrence is that illustrated in Fig. 16-9a. Since the symbols chosen for the system variables and transfer functions are those commonly used in feedback control systems work [7], the information presented in this chapter will coordinate well with more advanced control systems textbooks and papers. The variable r refers to the *reference input*, c to the *controlled variable*, and e to the *actuating signal*. G is the *forward loop transfer function*, and H is the *feedback loop transfer function*.

Figure 16-9a can be reduced by block diagram algebra to the single block (i.e., to the equivalent transfer function) of Fig. 16-9b. To obtain the required algebraic relationship for the reduction, we note that

$$c = Ge \qquad (16\text{-}6)$$

$$e = r - Hc \qquad (16\text{-}7)$$

Combining to eliminate e yields

$$(r - Hc)G = c$$

which is readily manipulated to the desired expression for the overall transfer function:

$$\frac{c}{r} = \frac{G}{1 + GH} \qquad (16\text{-}8)$$

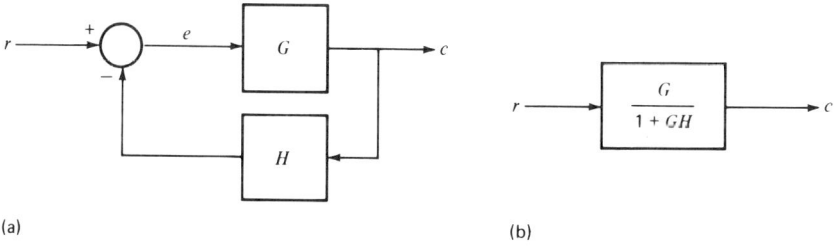

FIG 16-9. (a) A block diagram with a single negative feedback loop.
(b) Its single-block representation.

It should be noted that the transfer function H is often equal to 1, as is the case in a simple feedback control system where the output c has the same units as the input r and is subtracted from it directly to obtain the error, which is used as the actuating signal e. It is also important to remember that Eq. 16-8 applies only to a loop with negative feedback. If positive feedback is used, a different equation is required.

Equation 16-8 is useful not only for reducing a single block diagram of the form of Fig. 16-9a but also for the step-by-step reduction of block

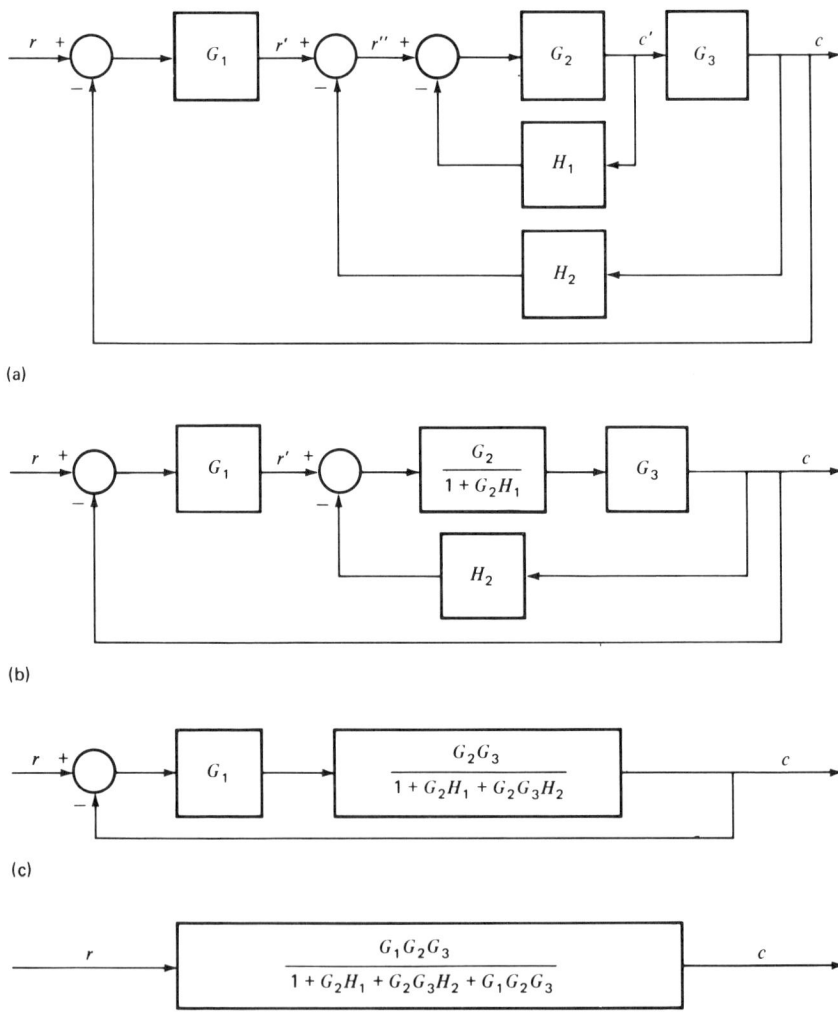

FIG 16-10. (a) A block diagram consisting of nested feedback loops. (b)–(d) Its reduction to a single transfer function.

16-4 BLOCK DIAGRAM ALGEBRA

diagrams of greater complexity, such as that illustrated in Fig. 16-10. To reduce the block diagram of Fig. 16-10a we start by applying Eq. 16-8 to the innermost loop, replacing that loop by a single block as in Fig. 16-10b. Next, Eq. 16-8 is applied to the *new* innermost loop, reducing the block diagram to the form of Fig. 16-10c. A third application of Eq. 16-8 produces the final result, the single block of Fig. 16-10d, which contains the overall system transfer function. (It would be a useful exercise for the student to perform the algebra required in the above reduction.)

In many cases, a block diagram does not consist of a convenient set of nested loops as in Fig. 16-10a. It may consist of a more complex arrangement of feedback loops as in Fig. 16-11. The best way to handle a system of this type is to make a series of adjustments to produce an equivalent diagram of nested loops, at which point Eq. 16-8 can be applied as explained before. To perform the required manipulations, there are certain rules that must be followed. In general, *the adjustments allowed are those that do not result in any change in the mathematical relationship between any two system variables*. In adhering to this basic rule it is, however, permissible to eliminate system variables from the block diagram.

MOVING SUMMING POINTS

If there is no block or takeoff point between them, it is permissible to simply interchange two adjacent summing points as illustrated in Fig. 16-12. This change can be shown to be valid by noting that it does not change any of the three inputs (a_1, a_2, and a_4) and that the output a_5 is equal to $a_1 - a_2 - a_4$ in either case. Note, however, that variable a_3 disappears when the interchange is made. This is usually of no concern, since block diagram manipulation is normally done with the goal of reducing the entire diagram to a single block or transfer function.

Sometimes it is necessary to move a summing point from one side of a block to the other. This is accomplished by adding a block in the feedback

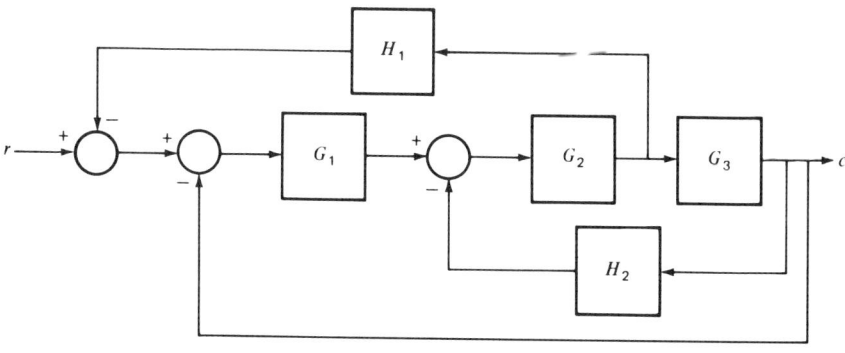

FIG 16-11. Example of a complex block diagram to which Eq. 16-8 cannot be applied without preliminary adjustments.

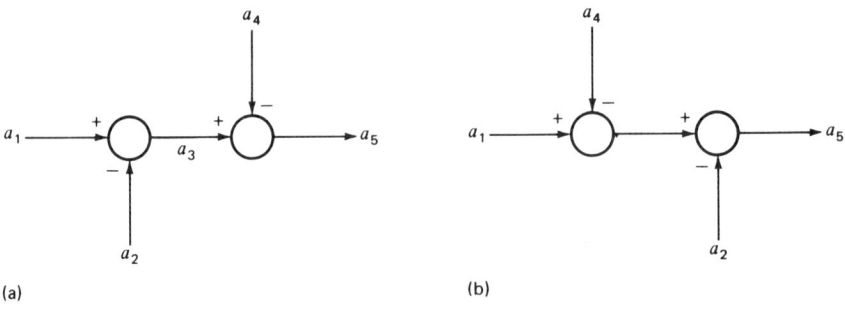

(a)　　　　　　　　　　　　　　　(b)

FIG 16-12. The interchange of two adjacent summing points.

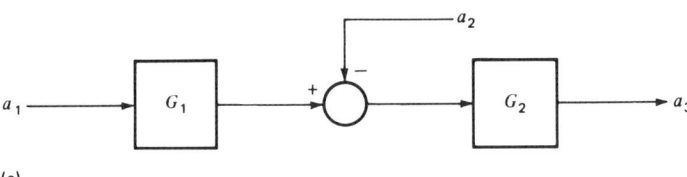

(a)

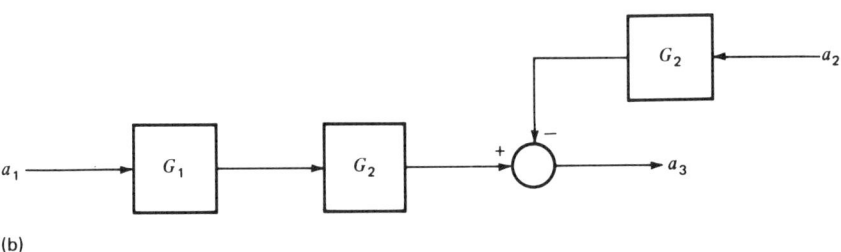

(b)

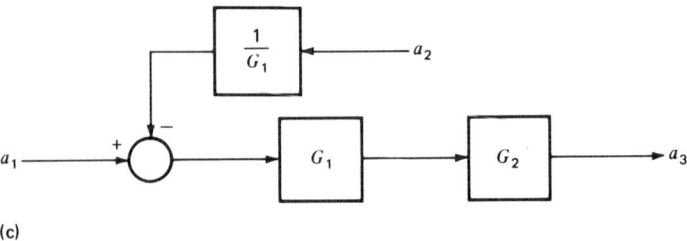

(c)

FIG 16-13. The moving of a summing point across a block.

16-4 BLOCK DIAGRAM ALGEBRA

(or feedforward) line that will cause the variable brought to the summing point to be unchanged. To move the summing point of Fig. 16-13a to the right of the block G_2 requires the addition of the new block with transfer function G_2 in the feedback line as illustrated in Fig. 16-13b. Similarly, to move the summing point to the left of the block G_1 requires the addition of the block with transfer function $1/G_1$ in the feedback line (Fig. 16-13c).

MOVING A TAKEOFF POINT

It is often necessary to move the point at which information is taken from a block diagram to be fed back (or forward) to a summing point. Such changes are very similar to the summing point changes just described. When a takeoff point is moved across a block, a new block must be added so that the variable brought to the summing point is not changed. Figures 16-14b and 16-14c illustrate two permissible changes that can be made to the original diagram of Fig. 16-14a.

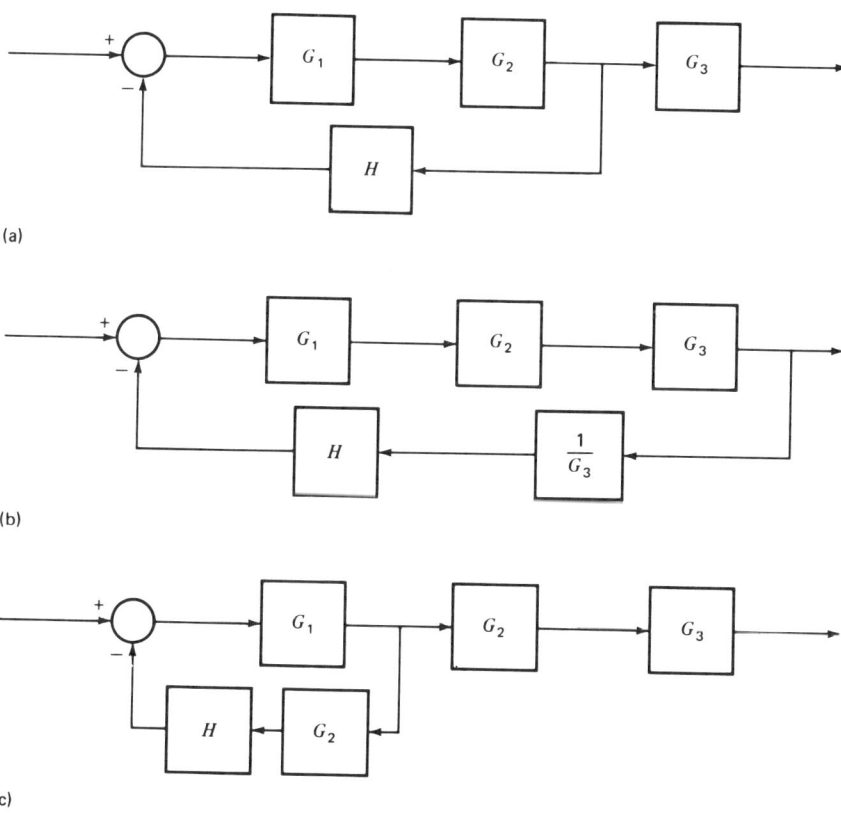

FIG 16-14. The moving of a takeoff point across a block.

INITIAL CONDITIONS

The development of system block diagrams and their reduction to system equations has been presented so far without consideration of initial conditions. With the p-operator approach used in this book, it is best to postpone the consideration of the initial conditions until the solution of the system equation is to be accomplished for a specific situation (which defines the initial conditions). The solution can then be obtained by using the methods described and illustrated in the earlier parts of the book.

16-5 Liquid-Level Integral Control System Example

The liquid-level system utilizing integral control (Fig. 16-15, which is identical to Fig. 15-4), with which the reader should already be familiar, will be used here as an example to illustrate the details involved in the development of a typical block diagram and its reduction to a final overall system transfer function.

The first step is to develop the parts of the block diagram that correspond to each element and process of the system. In essence, block diagram representation is developed for each of the basic equations (the

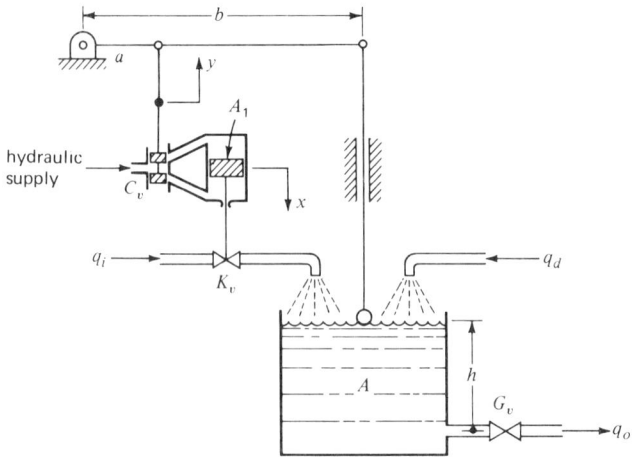

FIG 16-15. Liquid-level system utilizing integral control.

16-5 LIQUID-LEVEL INTEGRAL CONTROL SYSTEM EXAMPLE

same equations used in the derivation of Section 15-3) that apply to the different parts of the system.

The net flow into the tank is the algebraic sum of three components,

$$q_{net} = \delta q_i + q_d - \delta q_0 \qquad (16\text{-}9)$$

Equation 16-9 is represented by the summing point of Fig. 16-16a. The conservation relationship for the liquid level in the tank

$$q_{net} = A \frac{d\delta h}{dt}$$

is manipulated to the desired operator form:

$$\delta h = \frac{q_{net}}{Ap} \qquad (16\text{-}10)$$

with the block diagram representation shown in Fig. 16-16b. The valve in the outlet line has the equation

$$\delta q_0 = G_v \delta h \qquad (16\text{-}11)$$

and is represented by Fig. 16-16c. The similar inlet valve relationship

$$\delta q_i = -K_v x \qquad (16\text{-}12)$$

is represented by Fig. 16-16d. For the float, the kinematic linkage, and the

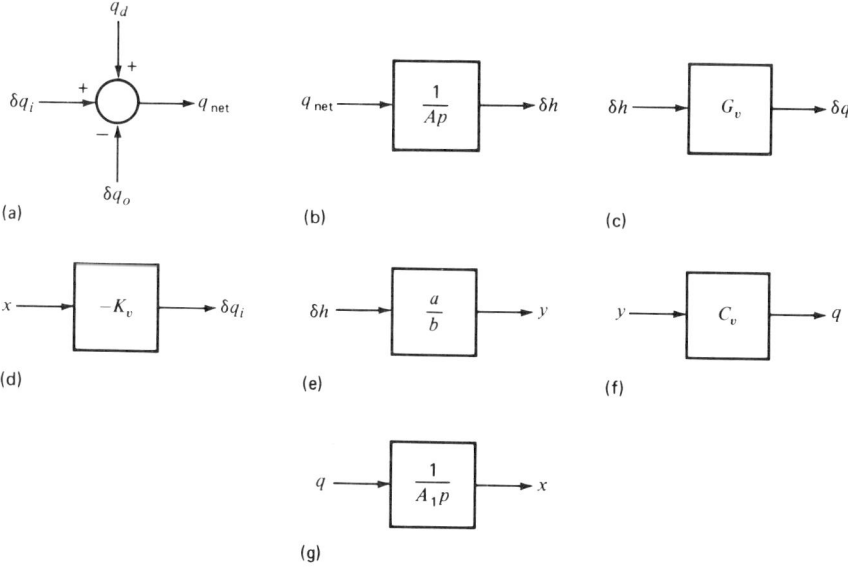

FIG 16-16. Block diagram representations of the components and processes of the liquid-level system of Fig. 16-15.

hydraulic valve-piston unit, the following equations as developed in Section 15-3 apply:

$$y = \frac{a}{b} \delta h \tag{16-13}$$

$$q = C_v y \tag{16-14}$$

$$q = A_1 \frac{dx}{dt} \qquad x = \frac{q}{A_1 p} \tag{16-15}$$

The block diagram representations of Eqs. 16-13 through 16-15 are given by Fig. 16-16e, f, and g, respectively.

By lining up the block diagram components in the proper sequence, and joining all lines representing each system variable, the overall block diagram of the system shown in Fig. 16-17 is obtained. Study and comparison of Figs. 16-16 and 16-17 will help clarify this final step in obtaining a system block diagram. Note the slight change in the representation of the inlet valve (Fig. 16-16d): a change of the $-K_v$ block to $+K_v$ so that the upper loop of Fig. 16-17 will give negative feedback at the summing point.

The completed block diagram can now be reduced to a single block that contains the overall system transfer function. First of all, the blocks in cascade are combined into a single block (Fig. 16-18a). Next, noting that the two feedback loops have the same takeoff point and the same summing junction, they are combined into a single block (Fig. 16-18b). For the final step, Eq. 16-8 is applied, giving the equation

$$\frac{\delta h}{q_d} = \frac{1/Ap}{1 + (1/Ap)(G_v + K_v C_v a / A_1 bp)} \tag{16-16}$$

This reduces to

$$\frac{\delta h}{q_d} = \frac{p}{Ap^2 + G_v p + K_v C_v a / A_1 b} \tag{16-17}$$

which is the transfer function for the single block of Fig. 16-18c. By

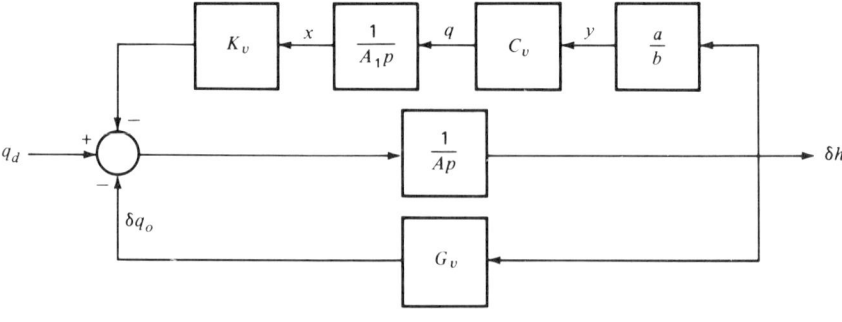

FIG 16-17. Block diagram of the liquid-level control system of Fig. 16-15.

16-5 LIQUID-LEVEL INTEGRAL CONTROL SYSTEM EXAMPLE

defining

$$K_0 = \frac{K_v C_v a}{A_1 b} \qquad (16\text{-}18)$$

as in Section 15-3, Eq. 16-17 can be written as

$$\left(\frac{A}{K_0} p^2 + \frac{G_0}{K_0} p + 1\right)\delta h = \frac{1}{K_0} p q_d \qquad (16\text{-}19)$$

By changing the operator terms to derivatives, the system differential equation is obtained:

$$\frac{A}{K_0}\frac{d^2\delta h}{dt^2} + \frac{G_v}{K_0}\frac{d\delta h}{dt} + \delta h = \frac{1}{K_0}\frac{dq_d}{dt} \qquad (16\text{-}20)$$

Equation 16-20 is identical to Eq. 15-20, which was obtained for the same system by an ordinary mathematical derivation.

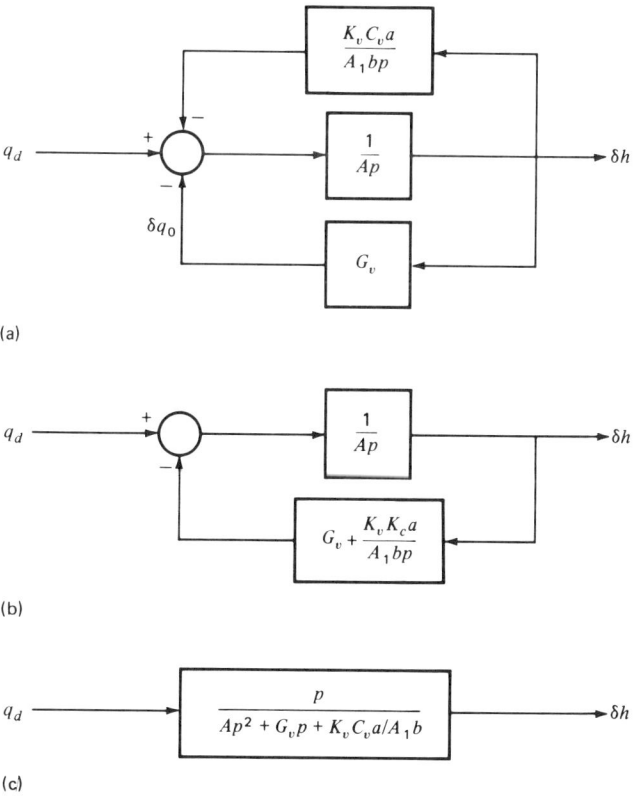

FIG 16-18. Reduction of the block diagram of Fig. 16-17 to a single block.

It should be mentioned at this point that there are usually a number of variations in the way a block diagram can be drawn for any specific system. Figure 16-17 was drawn on the basis of having a separate block or summing point for each individual basic component of the system. This approach has the advantage of giving a good pictorial representation of all details of the system operation. On the other hand, if such fine detail is not required, a less detailed block diagram may be developed by using equations applying to subsystems containing two or more of the basic system components. For example, instead of considering separately the hydraulic control valve and the piston it controls, they might be considered together as a subsystem, described by Eq. 15-17,

$$\delta q_i = - \frac{K_v C_v a}{A_1 bp} \delta h$$

and represented as a single block. If this is done, the system block diagram will appear as in Fig. 16-18a; that is, it will be the same as the first reduction of the more detailed block diagram.

The best approach to use in developing a block diagram depends upon how it is to be used. The choice is often strictly a matter of personal preference.

16-6 Systems with Two or More Inputs

It is fairly common for a dynamic system to have two or more inputs. For example, the liquid-level system of Fig. 16-19 is designed to allow an input from a human operator, as well as having a disturbance flow input that may be uncontrollable. By turning the crank, the operator can raise or lower the pivot point at the left through a screw-and-nut arrangement, thereby commanding a new equilibrium reference height for the liquid. If the screw has N threads per unit length, the change in height of the left pivot is given by the equation

$$z = \frac{\theta}{2\pi N} \qquad (16\text{-}21)$$

The valve position y is obtained by superposition,

$$y = \frac{a}{b} \delta h - \frac{b-a}{b} z \qquad (16\text{-}22)$$

A block diagram of the system, showing the two inputs, is given by Fig. 16-20. Since the equations used in developing the block diagram are linear, the combined effects of the two inputs can be handled by superposition (as explained in Section 3-1). To accomplish this with block diagram algebra and obtain the system equations with the two inputs, the block diagram is reduced to a single transfer function twice, with one of the two

16-6 SYSTEMS WITH TWO OR MORE INPUTS

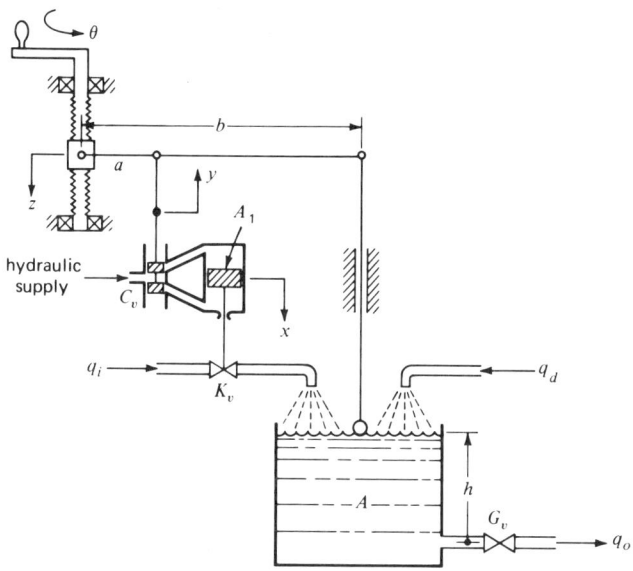

FIG 16-19. Liquid-level system with operator control of reference level.

inputs set to zero each time. With the operator input θ set to zero, the block diagram reduces to that of Fig. 16-17, which has already been reduced to a single transfer function in Fig. 16-18.

The next step is to set q_d to zero and consider θ as the sole input. The manipulation and reduction of the resulting block diagram is illustrated in Fig. 16-21.

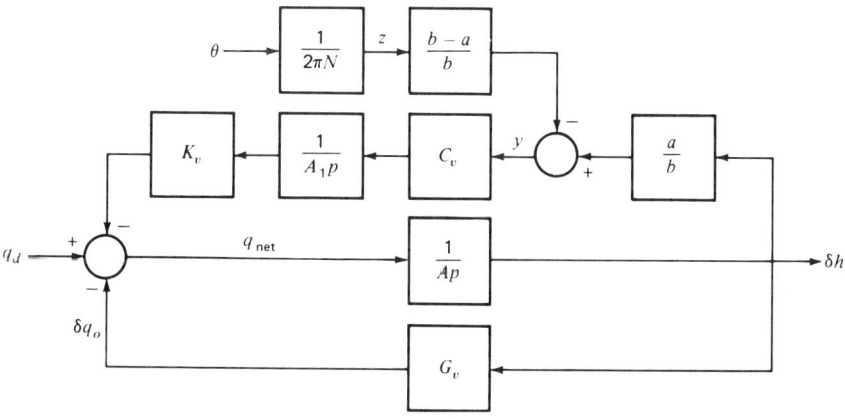

FIG 16-20. Block diagram of the liquid-level system of Fig. 16-19.

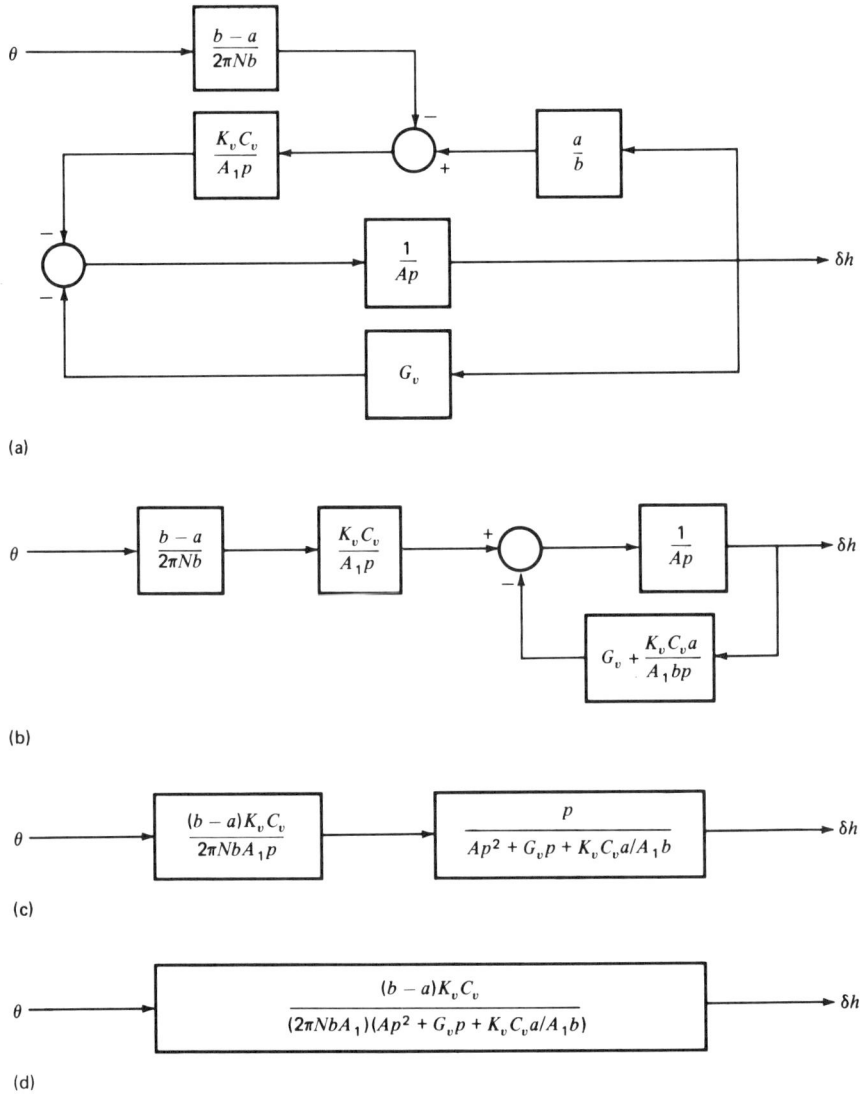

FIG 16-21. The manipulation and reduction to a single transfer function of the block diagram of Fig. 16-20 with q_d set to zero.

With a separate transfer function for each of the two inputs, the equation describing the total system can now be developed. Using Figs. 16-18c and 16-21d, and applying the principle of superposition,

$$\delta h = \left[\frac{1}{Ap^2 + G_v p + K_v C_v a/A_1 b} \right] \left[pq_d + \frac{(b-a)K_v C_v}{2\pi NbA_1} \theta \right] \quad (16\text{-}23)$$

16-7 FEEDBACK COMPENSATION

Multiplying through by the quadratic term and changing the operator terms to derivatives, the system equation is obtained:

$$A\frac{d^2\delta h}{dt^2} + G_v\frac{d\delta h}{dt} + \frac{K_v C_v a}{A_1 b}\delta h = \frac{dq_d}{dt} + \frac{(b-a)K_v C_v}{2\pi NbA_1}\theta \quad (16\text{-}24)$$

Equation 16-24 can be solved with the fundamental differential equation techniques presented earlier. With two forcing functions, the principle of superposition must be used. It is important to include the proper initial conditions for the specific situation under investigation.

16-7 Feedback Compensation

Often a basic feedback control system will have undesirable operating characteristics. It may be oscillatory due to insufficient damping, for example, or may even be unstable. Some type of compensation may be necessary to produce a usable system. A change in the type of *control action* used may improve system performance. This approach has already been discussed in Chapter 15. An alternative approach, the use of feedback compensation, will be discussed now, since the use of the block diagram simplifies its analysis (although it can be analyzed by use of differential equations alone).

Consider the system of Fig. 16-22, which is a simple type of dc electric motor feedback control system. The system is designed to produce an angular position output θ proportional to a dc voltage input e_i. If the motor produces a torque directly proportional to its input voltage, irrespective of its speed of rotation (within its "usable" speed range), the system block diagram is that given by Fig. 16-23. Application of Eq. 16-8 to the block diagram results in the overall transfer function

$$\frac{\theta}{e_i} = \frac{AK_m}{Ip^2 + bp + AK_m K_p} \quad (16\text{-}25)$$

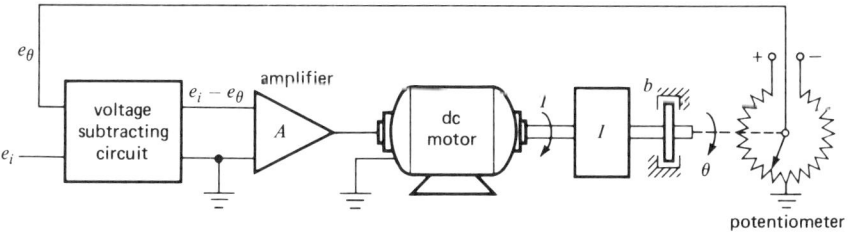

FIG 16-22. A dc motor feedback control system.

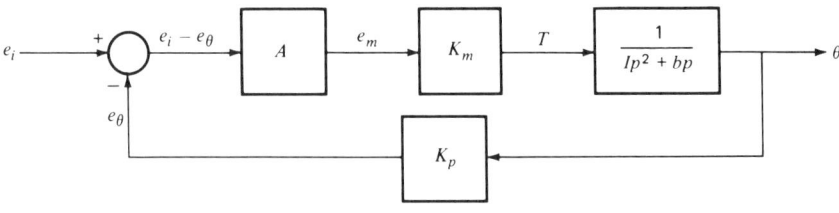

FIG 16-23. Block diagram of the dc motor feedback control system of Fig. 16-22.

which can be converted to the system differential equation

$$\frac{I}{AK_mK_p}\frac{d^2\theta}{dt^2} + \frac{b}{AK_mK_p}\frac{d\theta}{dt} + \theta = \frac{1}{K_p}e_i \qquad (16\text{-}26)$$

Comparison of Eq. 16-26 to the general form of the second-order equation (Eq. 7-5) shows that the damping ratio ζ is equal to $b/(2I\omega_n)$; that is, the damping ratio is dependent entirely upon the damping that is present at the shaft of the motor and load (from bearing friction, windage, etc.). If the friction is small, the system will be underdamped and have an undesirable oscillatory response. To increase the damping ratio by designing the motor and/or load to have more friction is seldom an attractive solution because it results in power loss and the possibility of heat and wear problems. The use of *feedback compensation* by means of a tachometer is a technique often used to increase the effective damping (i.e., the damping ratio) in systems of this type.

The system with tachometer feedback added is illustrated in Fig. 16-24a. The tachometer (generator) produces a voltage proportional to its angular velocity; that is,[1]

$$e_t = K_t\frac{d\theta}{dt} \qquad (16\text{-}27)$$

The transfer function of the tachometer is therefore

$$\frac{e_t}{\theta} = K_tp \qquad (16\text{-}28)$$

The block diagram of the system, Fig. 16-24b, shows how the tachometer operates upon the system output θ to produce the desired feedback signal.

One should be able to see intuitively that the feedback acts as synthetic damping, since reducing the voltage input to the motor (and thereby reducing its torque) in proportion to its speed has virtually the same effect as increasing the viscous friction of the motor shaft. To show

---

[1]The tachometer in Fig. 16-24a is shown attached directly to the load shaft. It is often connected through gearing so that it will operate over a more favorable speed range, but Eq. 16-27 will still apply as long as K_t is defined to take the gearing into account.

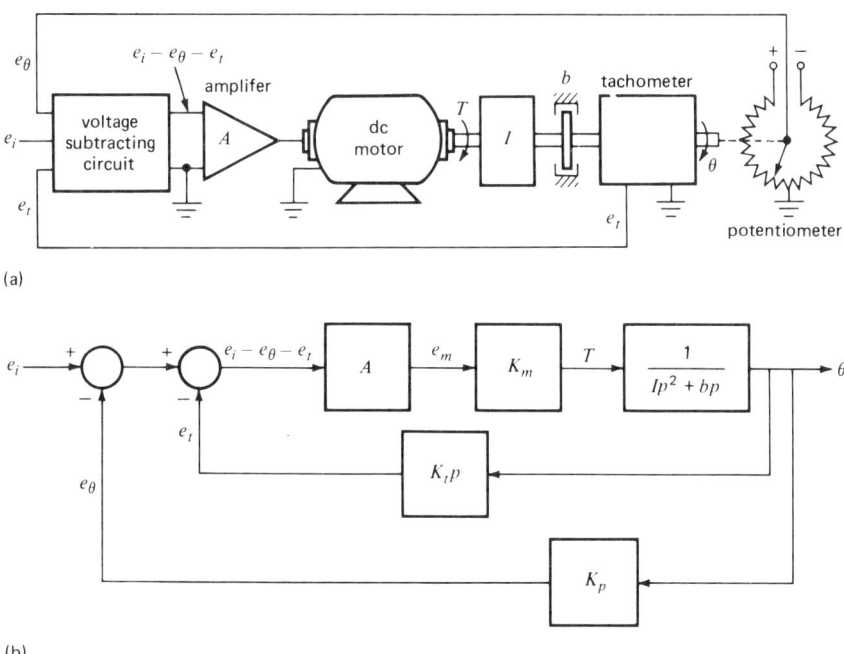

FIG 16-24. (a) A dc motor feedback control system with tachometer feedback. (b) Its block diagram.

mathematically the effect of tachometer feedback, the overall system transfer function must be obtained. Application of block diagram algebra to Fig. 16-24b produces the result

$$\frac{\theta}{e_i} = \frac{AK_m}{Ip^2 + (b + AK_mK_t)p + AK_mK_p} \qquad (16\text{-}29)$$

The damping coefficient b now has the term AK_mK_t added to it. By designing for adequate control over the tachometer gain, it should be possible to set the damping ratio at any desired value. If the true damping in the system should happen to give a damping ratio higher than desired (but note that this is not too likely), it could be reduced by using positive rather than negative tachometer feedback.

16-8 Conclusion

Block diagram representation of feedback control systems has been found to be a very useful tool. The block diagram provides insight into the functional interrelationships among the various system components. The reduction of a block diagram to a single block (transfer function) by means of block diagram algebra is often simpler than developing the transfer function by working with the differential equations alone. Two or more inputs, applied at different points in the system, are readily handled with block diagrams by applying the principle of superposition.

Classical feedback control system theory is based on the use of the types of block diagrams described in this chapter. Although this introductory book will not go any further into control system theory, the block diagram information presented here will serve as an excellent background for further study. In more advanced feedback control systems work, block diagrams are normally constructed on the basis of the Laplace transform (usually represented by the operator s) rather than the simple p operator used here [7]. The Laplace operator adds a certain amount of flexibility, particularly in terms of handling initial conditions, but the resultant block diagrams are virtually the same. The use of Laplace transforms results in no real difference in the block diagram algebra and other techniques and principles presented here.

Problems

16-1. If the system of Fig. 16-9a is changed to have positive feedback, what will the transfer function be?

16-2. Perform the algebra associated with the block diagram reduction illustrated in Fig. 16-10.

16-3. Determine the required transfer function for each of the empty blocks in order for all the block diagrams shown to be equivalent.

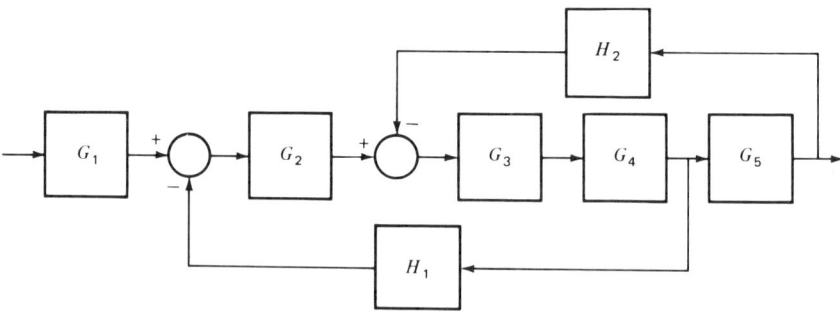

(a)

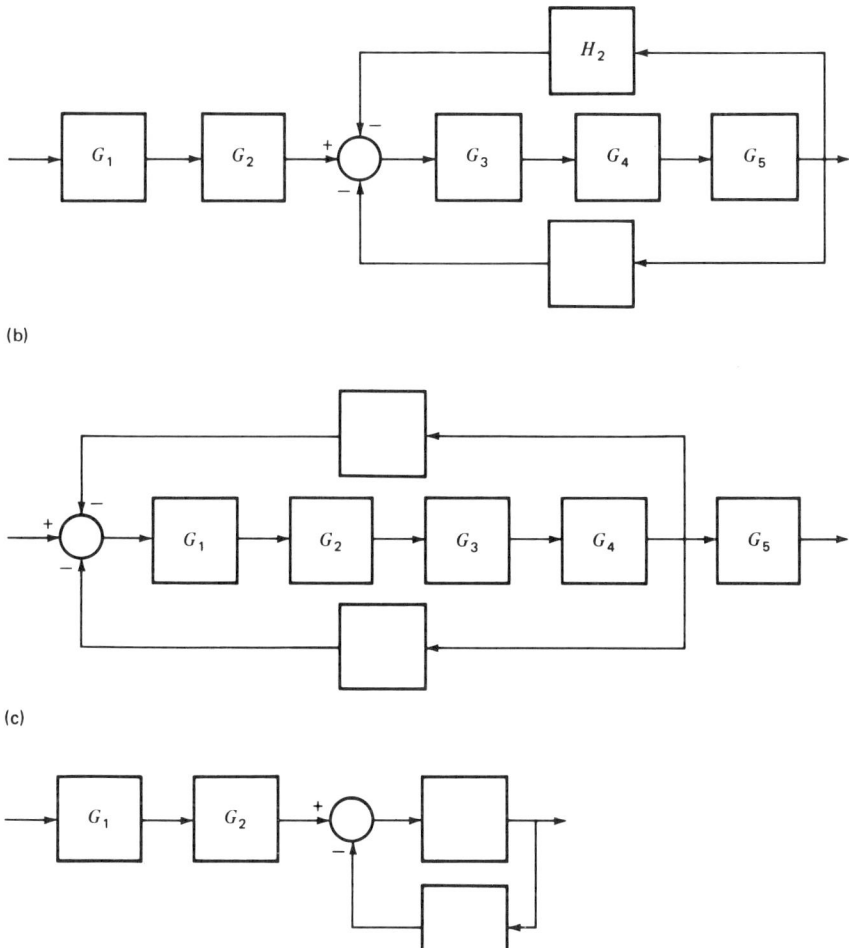

(b)

(c)

(d)

16-4. Reduce each of the given block diagrams to a single block.

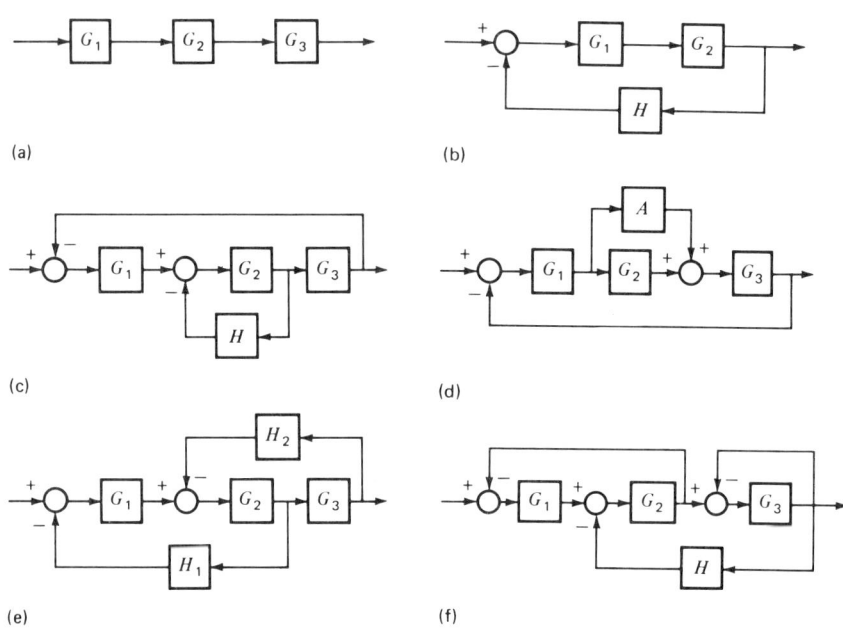

16-5. (a) Draw a block diagram for the system of Prob. 15-1. Reduce it, by means of block diagram algebra, to a single transfer function $\delta h_2 / q_d$.
(b) Do the same for the system of Prob. 15-2, obtaining the transfer function $\delta h / \delta h_d$.
(c) Do the same for the system of Prob. 15-3, obtaining the two transfer functions ω_0 / ω_i and ω_0 / T_d.
(d) Do the same for the system of Prob. 15-4, obtaining the transfer function θ_0 / θ_i.
(e) Do the same for the system of Prob. 15-5, obtaining the transfer function y/x.
(f) Do the same for the system of Prob. 15-6, obtaining the transfer function y/x.
(g) Do the same for the system of Prob. 15-7, obtaining the transfer function T_0 / T_i.

17
State-Variable Formulation and Computer Solutions

17-1 Introduction

Up to this point in the book we have defined each system in terms of an nth-order differential equation in the variable of primary interest. However, it is also possible to define a system by a set of n first-order differential equations. When this approach is followed, we have what is known as the *state-variable* formulation. The n variables involved are called *state variables* because they can be used to define the *state*, or condition, of the system at any instant of time.

The state-variable formulation of system equations is strongly related to both analog and digital computer solutions. The basic relationships will be presented in this chapter. The state-variable formulation is also important in topics beyond the scope of this book.

We will begin by exploring two basic ways by which state-variable equations can be obtained. Then analog and digital computer solutions will be considered.

17-2 State-Variable Formulation

The state-variable formulation of system equations defines the system by a set of first-order differential equations. By way of illustration, Eq. 17-1 shows the general form appropriate for a system with two state variables, x_1 and x_2, and a single input, $f(t)$. The equation set is

$$\frac{dx_1}{dt} = a_{11}x_1 + a_{12}x_2 + b_1 f(t)$$
$$\frac{dx_2}{dt} = a_{21}x_1 + a_{22}x_2 + b_2 f(t) \qquad (17\text{-}1)$$

where the a's and b's are constants.

State-variable equations (also called simply *state equations*) can be obtained in two basic ways. The first way is to manipulate an nth-order equation into the desired set of first-order equations. The second way is to formulate the state-variable equations directly from the physical system. Both approaches are now illustrated, starting with the first.

Example 17-1. A mechanical system with an applied force F is shown in Fig. 17-1. Determine the state-variable equations that describe it.

Solution: The defining differential equation is readily determined to be

$$M\frac{d^2y}{dt^2} + c\frac{dy}{dt} + ky = F(t) \qquad (17\text{-}2)$$

By definition, we can write

$$\frac{dy}{dt} = v \qquad (17\text{-}3)$$

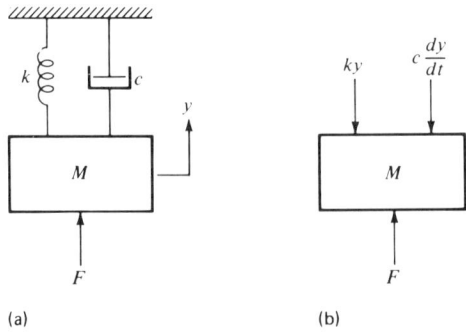

FIG 17-1. (a) Mechanical system with applied force F.
(b) Free-body diagram of the mass.

17-2 STATE-VARIABLE FORMULATION

where v is velocity. Recognizing that acceleration is the first derivative of velocity, Eq. 17-2 can be expressed as

$$M \frac{dv}{dt} + cv + ky = F(t)$$

Rearrangement yields

$$\frac{dv}{dt} = -\frac{k}{M} y - \frac{c}{M} v + \frac{1}{M} F(t) \qquad (17\text{-}4)$$

Equations 17-3 and 17-4 are the two required first-order differential equations. Thus, the set of state-variable equations defining the mechanical system is

$$\begin{aligned} \frac{dy}{dt} &= v \\ \frac{dv}{dt} &= -\frac{k}{M} y - \frac{c}{M} v + \frac{1}{M} F(t) \end{aligned} \qquad (17\text{-}5)$$

where displacement y and velocity v are the state variables. Note that Eq. 17-5 is of the form illustrated in Eq. 17-1. Of course, some of the coefficients indicated in Eq. 17-1 are zero for the system considered.

In general, an nth-order differential equation can be modified to produce a set of n first-order differential equations. In the specific case of Example 17-1, a second-order differential equation produced two first-order differential equations.

The second way to obtain state-variable equations is to formulate them directly from the physical system being considered. This approach will be illustrated through the use of a liquid-level system.

Example 17-2. Determine the set of state-variable equations that describes the liquid-level system illustrated in Fig. 17-2.

Solution: The general relationship for a tank, namely,

$$A \frac{dh}{dt} = \Sigma q$$

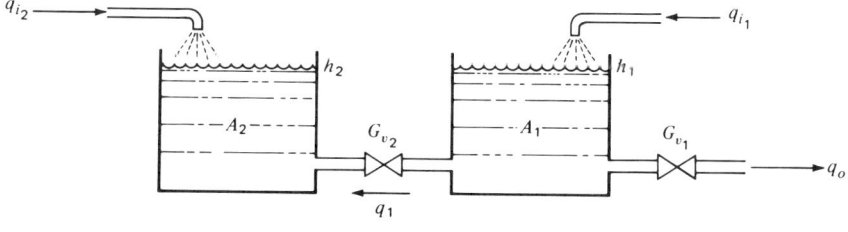

FIG 17-2. Liquid-level system.

can be used to obtain

$$A_1 \frac{dh_1}{dt} = q_{i_1} - q_o - q_1 \tag{17-6}$$

and

$$A_2 \frac{dh_2}{dt} = q_{i_2} + q_1 \tag{17-7}$$

For defining flow rates, we assume that the flow restrictions produce linear flow-pressure relationships and use

$$q = G_v h$$

where h is understood to be the head *difference* across a restriction. Thus[1]

$$q_o = G_{v_1} h_1 \tag{17-8}$$

and

$$q_1 = G_{v_2}(h_1 - h_2) \tag{17-9}$$

Combination of Eqs. 17-6, 17-7, 17-8, and 17-9, and rearrangement yields

$$\begin{aligned}\frac{dh_1}{dt} &= -\frac{G_{v_1} + G_{v_2}}{A_1} h_1 + \frac{G_{v_2}}{A_1} h_2 + \frac{1}{A_1} q_{i_1} \\ \frac{dh_2}{dt} &= \frac{G_{v_2}}{A_2} h_1 - \frac{G_{v_2}}{A_2} h_2 + \frac{1}{A_2} q_{i_2}\end{aligned} \tag{17-10}$$

Equation 17-10 provides a set of two first-order differential equations that defines the liquid-level system. The equations represent a state-variable formulation with the levels h_1 and h_2 being the state variables.

Example 17-3. Determine the set of state-variable equations that describes the two-mass system illustrated in Fig. 17-3.

Solution: There are several different ways in which this problem might be attacked. Two of them, which produce different sets of state variables, will be illustrated here.

Assuming that we are interested primarily in the motion of mass M_1, we can start with the fourth-order differential equation of the system that was previously developed as Eq. 9-5. The equation is

$$M_1 M_2 \frac{d^4 y_1}{dt^4} + \left[(k_1 + k_2) M_2 + k_2 M_1 \right] \frac{d^2 y_1}{dt^2} + k_1 k_2 y_1 = 0 \tag{17-11}$$

---

[1]Note that the flow coefficients G_{v_1} and G_{v_2} as used here are defined in the same manner as the G_v in Eq. 4-10, as opposed to the definition of Eq. 4-16.

17-2 STATE-VARIABLE FORMULATION

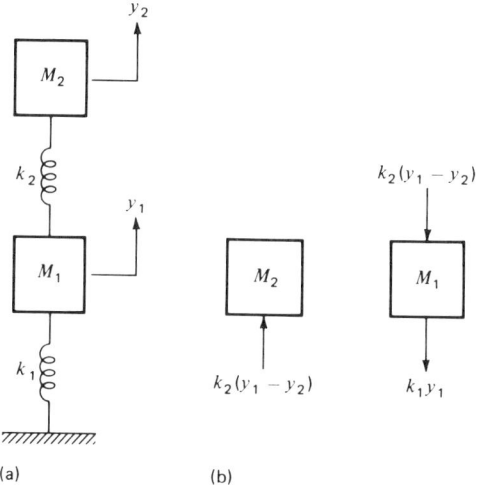

FIG 17-3. (a) Two-mass system.
(b) Free-body diagrams of the two masses.

By defining the four state variables as

$$x_1 = y_1 \qquad x_2 = \frac{dy_1}{dt} \qquad x_3 = \frac{d^2y_1}{dt^2} \qquad x_4 = \frac{d^3y_1}{dt^3}$$

it follows directly that

$$\frac{dx_1}{dt} = x_2 \qquad \frac{dx_2}{dt} = x_3 \qquad \frac{dx_3}{dt} = x_4$$

and Eq. 17-11 becomes

$$M_1 M_2 \frac{dx_4}{dt} + [(k_1 + k_2)M_2 + k_2 M_1]x_3 + k_1 k_2 x_1 = 0$$

The set of state-variable equations can therefore be written as

$$\frac{dx_1}{dt} = x_2$$

$$\frac{dx_2}{dt} = x_3 \qquad\qquad (17\text{-}12)$$

$$\frac{dx_3}{dt} = x_4$$

$$\frac{dx_4}{dt} = -\frac{k_1 k_2}{M_1 M_2} x_1 - \left[\frac{(k_1 + k_2)}{M_1} + \frac{k_2}{M_2}\right] x_3$$

For an alternative solution to this problem, we start from basic principles. By the definition of velocity we can write

$$\frac{dy_1}{dt} = v_1$$

$$\frac{dy_2}{dt} = v_2$$

Applying Newton's second law of motion to each of the two masses, we then obtain

$$M_1 \frac{dv_1}{dt} + (k_1 + k_2)y_1 = k_2 y_2$$

$$M_2 \frac{dv_2}{dt} + k_2 y_2 = k_2 y_1$$

Rearrangement of these four equations yields the desired set of state-variable equations

$$\frac{dy_1}{dt} = v_1$$

$$\frac{dv_1}{dt} = -\left(\frac{k_1 + k_2}{M_1}\right) y_1 + \frac{k_2}{M_1} y_2$$

$$\frac{dy_2}{dt} = v_2 \qquad (17\text{-}13)$$

$$\frac{dv_2}{dt} = +\frac{k_2}{M_2} y_1 - \frac{k_2}{M_2} y_2$$

The two alternative solutions produce two different sets of state variables. In the first case they consist of the displacement of mass M_1 and its first three derivatives, whereas in the second case they are the displacements of both masses and their first derivatives (velocities). It should be noted that either set of state-variable equations is sufficient for describing the system; that is, the values of all four state variables of either set at a given instant of time will define the state of the system at that time.

For the preceding example the second set of state-variable equations (Eq. 17-13) would be more useful for most applications, since we would normally be more interested in knowing the velocity and displacement of the second mass than the second and third derivatives of the displacement of the first mass. In addition, it is easier to determine the initial conditions required in a practical problem of this type if they are velocities and displacements.

17-2 STATE-VARIABLE FORMULATION

This example points out that there is not always a single unique set of state variables that must be used to describe a given system; there may be two or more alternative sets that are satisfactory. The same total number of state variables must be used in any case, however.

Although our purpose in this chapter is to obtain state-variable equations, the reader will recognize the fact that such equations can be manipulated into single higher-order equations if desired. That is, if one were given a set of n first-order differential equations but would prefer a single nth-order differential equation, the modification can be readily accomplished. The approach is that illustrated throughout most of the book, namely, to introduce the p operator and algebraically eliminate response variables of no interest in favor of the one response variable of interest.

Example 17-4. Given the set of state-variable equations,

$$\frac{dx_1}{dt} = -Ax_1 + Bx_2$$

$$\frac{dx_2}{dt} = -Cx_2 - Dx_3$$

$$\frac{dx_3}{dt} = Ex_1 + Fx_2 - Gx_3 + H$$

develop the corresponding single differential equation in x_3 that describes the system.

Solution: The first two equations are written in operator form as

$$x_1 = \frac{B}{p+A} x_2$$

$$x_2 = -\frac{D}{p+C} x_3$$

The third equation, written in operator form with the above substitutions, becomes

$$(p+G)x_3 = \left(\frac{EB}{p+A}\right)x_2 - \left(\frac{FD}{p+C}\right)x_3 + H$$

$$= -\left(\frac{EB}{p+A}\right)\left(\frac{D}{p+C}\right)x_3 - \left(\frac{FD}{p+C}\right)x_3 + H$$

Clearing the denominator,

$$(p+G)(p+A)(p+C)x_3$$
$$= -EBDx_3 - FD(p+A)x_3 + (p+A)(p+C)H$$

With further reduction and changing back from operator to derivative form, we obtain the final result

$$\frac{d^3x_3}{dt^3} + (C + G + A)\frac{d^2x_3}{dt^2}$$
$$+ (CG + AC + AG + FD)\frac{dx_3}{dt}$$
$$+ (AGC + EBD + FDA)x_3$$
$$= \frac{d^2H}{dt^2} + (A + C)\frac{dH}{dt} + ACH \qquad (Ans.)$$

17-3 State-Space Trajectories

The term *state space* is often used in conjunction with dynamic system analysis based on the state-variable formulation, because the *state* of a system having n state variables can be represented as a point in n-dimensional space. The coordinates of the state space are the state variables. Since the state of a dynamic system will normally vary with time, the locus of the state points will trace a *trajectory* in the state space as time increases.

It should be noted that since time is not one of the coordinates of the state space, it is an implicit function; that is, there is a specific time associated with each point of the state-space trajectory.

The state of the system at any instant of time can also be represented as a *vector* from the origin of the state-space coordinates to the state point. The components of the vector are the values of the state variables at that instant of time. Note that the tip of the *state vector* will trace out the trajectory as time increases (as can be visualized in state space of three dimensions or less).

For a system with two state variables state space consists of a plane; with three state variables it is three-dimensional space. Trajectories can be plotted graphically for these cases, and visual observation of them may give insight into the system's performance. For systems with more than three state variables, however, the n-dimensional state space cannot be visualized, and therefore no plot of the trajectory can be made, even though the significance of the trajectory is the same.

As an example of a state-space trajectory, consider the system shown in Fig. 17-4. The set of state-variable equations describing the system is

$$\frac{dy}{dt} = v$$
$$\frac{dv}{dt} = -\frac{k}{M}y \qquad (17\text{-}14)$$

The corresponding single differential equation for this system is

$$M\frac{d^2y}{dt^2} + ky = 0 \qquad (17\text{-}15)$$

17-3 STATE-SPACE TRAJECTORIES

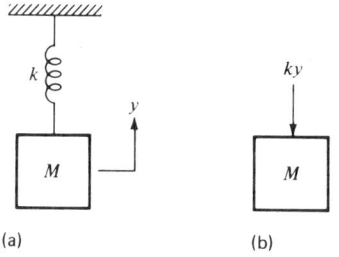

FIG 17-4. (a) Mass-spring system.
(b) Free-body diagram of the mass.

With the initial conditions

$$y\bigg|_{t=0} = y_0$$

$$\frac{dy}{dt}\bigg|_{t=0} = 0$$

the state-space trajectory will have the form illustrated in Fig. 17-5. Note that the coordinate scales have been chosen so that the trajectory appears as a circle. With other scales it would be an ellipse. It should be mentioned at this point that this particular type of state-space trajectory (a plot of velocity versus displacement, with time as an implicit function) is also called a *phase-plane* plot, or trajectory.

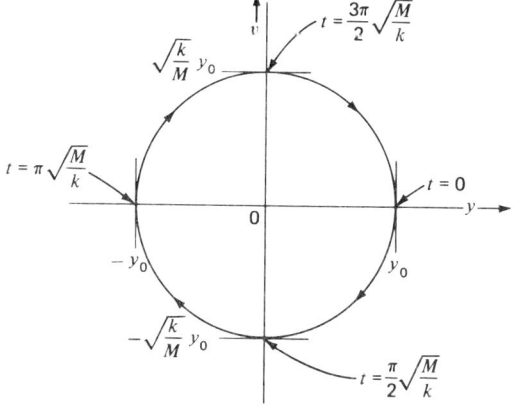

FIG 17-5. State-space trajectory of the system of Fig. 17-4.

17-4 Analog Computer Solutions

Three fundamental mathematical operations that can be performed by an analog computer are integration, summation, and multiplication by a constant. These operations are sufficient for the solution of linear differential equations. The general nature of analog computer solutions will be illustrated with the mechanical system shown in Fig. 17-1 and defined by Eqs. 17-5.

For simplicity, let Eqs. 17-5 be written as

$$\frac{dy}{dt} = v$$
$$\frac{dv}{dt} = -Ay - Bv + CF(t) \tag{17-16}$$

where A, B, and C are the overall coefficients. The first equation can be expressed as

$$y = \int_0^t v\, dt + y_0 \tag{17-17}$$

where y_0 is the initial condition. The solution of Eq. 17-17 can be implemented by an integrating device as illustrated in Fig. 17-6a; that is, displacement y can be obtained by integrating velocity v. The initial condition y_0 must, of course, be provided to the integrator.

A second integrator, Fig. 17-6b, is used to obtain velocity v through the integration of acceleration dv/dt. The initial condition v_0 must be supplied.

Multiplication by a constant is diagrammed in Fig. 17-6c. Here displacement y is multiplied by the constant $-A$ to produce $-Ay$.

The second equation of Eqs. 17-16 defines acceleration dv/dt as the

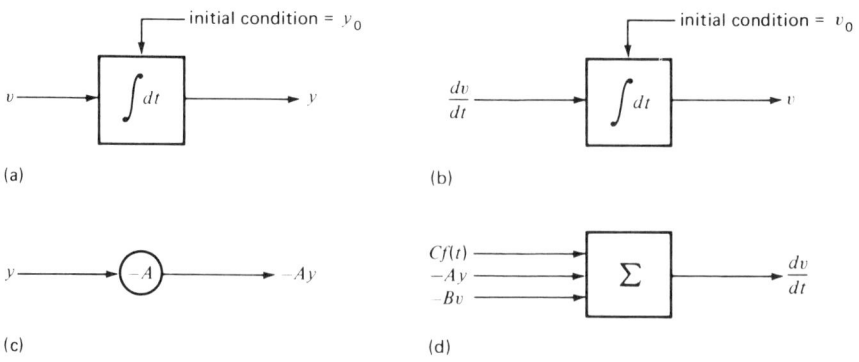

FIG 17-6. Analog computer operations.

17-4 ANALOG COMPUTER SOLUTIONS

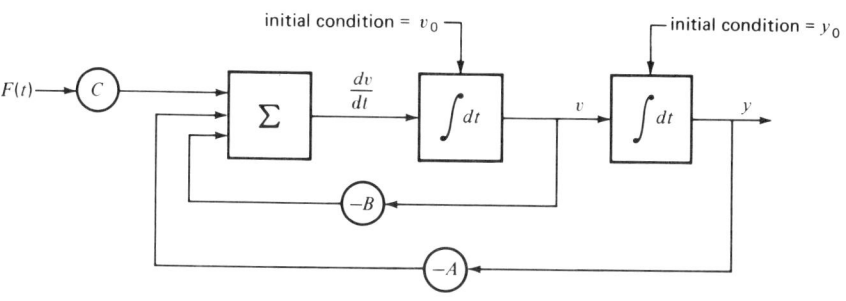

FIG 17-7. Analog computer diagram of the mass-spring-damper system of Fig. 17-1.

algebraic summation of three terms. The summation operation is provided by the analog computer and can be represented as shown in Fig. 17-6d.

The completed analog computer diagram is given in Fig. 17-7. It is obtained by combining the various elements shown in Fig. 17-6. To obtain a solution, one must be able to generate a voltage signal corresponding to the forcing function $F(t)$. The solution is in the form of the voltages in the computer circuit, which are proportional to the acceleration, velocity, and position. It is convenient to record these as functions of time on a strip chart recorder.

The important relationship between the state-variable formulation and analog computer solutions is that *the state variables identify the integrator outputs and the equations supply, in a convenient form, the signals required for integrator inputs.*

Example 17-5. Determine the set of state-variable equations that describe the electric circuit of Fig. 17-8. Using them, develop an analog computer diagram for simulation of the system.

Solution: The system equations are developed by use of the loop analysis technique presented in Section 2-4. The two loop currents, i_1

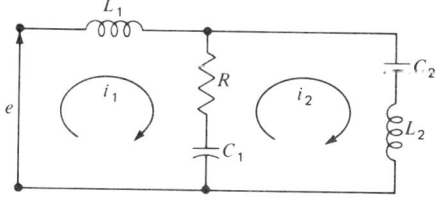

FIG 17-8. Electric circuit for Example 17-5.

and i_2, are defined in Fig. 17-8. The current through R and C_1 is $i_1 - i_2$ (flowing downward).

Applying Kirchhoff's first law (Eq. 2-34) to the left loop, in combination with the equations for voltage drop from Table 2-1 noting that, for capacitors,

$$e = \frac{1}{C}\int i\, dt = \frac{1}{C} q$$

yields

$$e - L_1 \frac{di_1}{dt} - R(i_1 - i_2) - \frac{1}{C_1}(q_1 - q_2) = 0$$

Application of Kirchhoff's first law to the loop on the right yields

$$-\frac{q_2}{C_2} - L_2 \frac{di_2}{dt} - \frac{1}{C_1}(q_2 - q_1) - R(i_2 - i_1) = 0$$

The relationships between the currents and charges are

$$i_1 = \frac{dq_1}{dt} \qquad i_2 = \frac{dq_2}{dt}$$

Rearrangement of the above four equations yields the desired set of state variable equations,

$$\frac{dq_1}{dt} = i_1$$

$$\frac{di_1}{dt} = -\frac{1}{L_1 C_1} q_1 - \frac{R}{L_1} i_1 + \frac{1}{L_1 C_1} q_2 + \frac{R}{L_1} i_2 + \frac{e}{L_1}$$

$$\frac{dq_2}{dt} = i_2$$

$$\frac{di_2}{dt} = \frac{1}{L_2 C_1} q_1 + \frac{R}{L_2} i_1 - \left(\frac{1}{L_2 C_1} + \frac{1}{L_2 C_2}\right) q_2 - \frac{R}{L_2} i_2$$

By applying the analog computer operations to the four state-space equations and combining the various elements as required, the desired diagram (Fig. 17-9) is obtained.

The discussion up to this point has been concerned with the solution of linear differential equations. However, analog computers can provide for the multiplication of varying signals and for various nonlinear functions. Thus, solutions for nonlinear differential equations can be generated, making the analog computer a powerful tool for the analysis of dynamic systems.

The basic principles of analog computer simulation have been presented in this section. The practical details of wiring up a circuit board and making the runs is not covered here, however. Additional techniques (e.g.,

17-5 DIGITAL COMPUTER SOLUTIONS

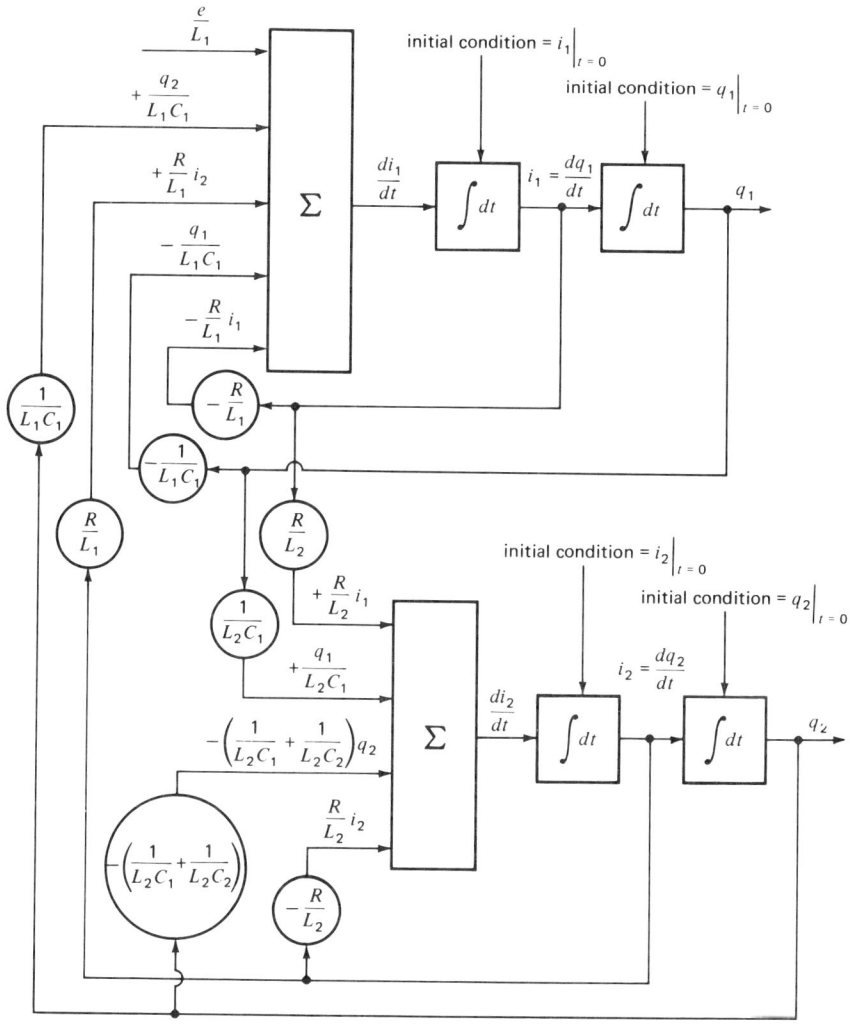

FIG 17-9. Analog computer diagram for the electric circuit of Fig. 17-8.

sign reversals, and magnitude and time scaling) are required for many analog computer problems [7, 16].

17-5 Digital Computer Solutions

The computational nature of the digital computer makes it convenient to obtain a numerical solution to an nth-order differential equation by working with the corresponding set of n first-order differential equations. For this reason, the state-variable formulation of system equations offers an appropriate starting point. For illustrative purposes, we will again utilize the mechanical system shown in Fig. 17-1 and defined by Eqs. 17-16, repeated here as

$$\frac{dy}{dt} = v$$
$$\frac{dv}{dt} = -Ay - Bv + CF(t) \qquad (17\text{-}18)$$

A graphical representation of the approach to be described is given by Fig. 17-10. The time axis is broken into equal time increments of size Δt. The strategy is to compute slopes for both displacement y and velocity v for a specific instant of time. With the assumption that the slopes remain constant for the time increment Δt, they are then used to calculate new values for y and v. The process is then repeated for as many times as are required to complete the solution. More detail follows.

To obtain a solution we must specify the forcing function $F(t)$ and the initial conditions y_0 and v_0. The first computation step is to compute the slopes at $t = 0$ using Eq. 17-18. The result is

$$\left.\frac{dy}{dt}\right|_{t=0} = v_0 \qquad (17\text{-}19)$$

$$\left.\frac{dv}{dt}\right|_{t=0} = -Ay_0 - Bv_0 + CF(0) \qquad (17\text{-}20)$$

With the slopes available, the next step is to calculate values for y and v at

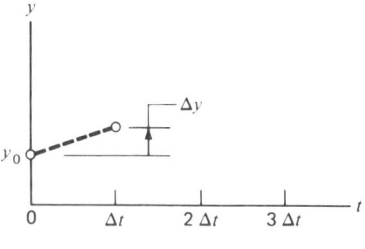

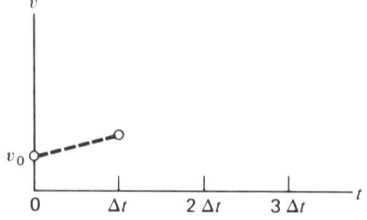

FIG 17-10. Graphs illustrating the first step in generating a digital computer solution.

17-5 DIGITAL COMPUTER SOLUTIONS

$t = \Delta t$. The new value for y is

$$y\bigg|_{t=\Delta t} = y_0 + \Delta y = y_0 + \frac{dy}{dt}\bigg|_{t=0} \Delta t \qquad (17\text{-}21)$$

In a similar manner,

$$v\bigg|_{t=\Delta t} = v_0 + \frac{dv}{dt}\bigg|_{t=0} \Delta t \qquad (17\text{-}22)$$

With the new values of y and v available, new slopes are calculated and the basic pattern is repeated over and over again.

Based on the computational scheme employed, the reader can appreciate the fact that nonlinear differential equations pose no problem for the digital computer. For example, if the second equation of Eq. 17-18 were

$$\frac{dv}{dt} = -Ay - Bv^2 + CF(t) \qquad (17\text{-}23)$$

the initial slope would be readily calculated as

$$\frac{dv}{dt}\bigg|_{t=0} = -Ay_0 - Bv_0^2 + CF(0) \qquad (17\text{-}24)$$

Although conceptually sound, the approach described is somewhat crude. Fortunately, refined techniques are readily available that will yield the desired degree of accuracy without undue use of computer time [11]. In addition, there are special digital computer programs in wide use that have been developed to facilitate the setting up and solution of dynamic system problems.

Example 17-6. The two-tank liquid-level system of Fig. 17-11 has outlet valves with nonlinear flow-head characteristics. Using the system parameters given in the figure, and assuming initial flow equilibrium, obtain a numerical solution for the change in head of tank 2 when a step disturbance flow of 500 in.^{3}/s is applied. Use 1-second time increments for the solution, and carry it out over a period of 5 seconds.

Solution: The basic equation for a single liquid-level tank system, based on the conservation of mass, is

$$A\frac{dh}{dt} = q_i - q_o$$

Applying this equation to each of the two tanks of Fig. 17-11 yields, in state-variable form,

$$\begin{aligned} \frac{dh_1}{dt} &= -\frac{200}{A_1}\sqrt{h_1} + \frac{q_i}{A_1} + \frac{q_d}{A_1} \\ \frac{dh_2}{dt} &= \frac{200}{A_2}\sqrt{h_1} - \frac{240}{A_2}\sqrt{h_2} \end{aligned} \qquad (17\text{-}25)$$

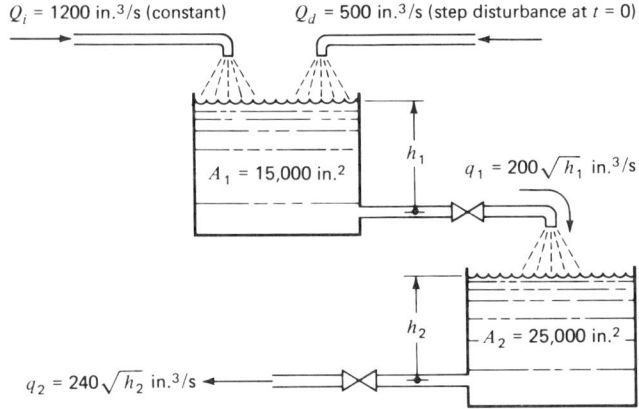

FIG 17-11. A two-tank liquid-level system with nonlinear outlet valves.

The system is initially at equilibrium with $q_i = q_1 = q_2 = 1200$ in.^{3}/s. Using the given head-flow characteristics of the valves, we can solve for the equilibrium liquid levels in the two tanks

$$\sqrt{h_1} = \frac{q_1}{200} = \frac{1200}{200} = 6$$

$$h_1 = 36 \text{ in.}$$

$$\sqrt{h_2} = \frac{q_2}{240} = \frac{1200}{240} = 5$$

$$h_2 = 25 \text{ in.}$$

At $t = 0$, when the step disturbance flow is applied,

$$h_1 = 36 \text{ in.}$$
$$h_2 = 25 \text{ in.}$$
$$q_d = Q_d = 500 \text{ in.}^3/\text{s}$$
$$q_i = Q_i = 1200 \text{ in.}^3/\text{s}$$

Solving Eqs. 17-25 with these values yields

$$\frac{dh_1}{dt} = -\frac{200}{15,000}\sqrt{36} + \frac{1200}{15,000} + \frac{500}{15,000} = 0.033333 \text{ in./s}$$

$$\frac{dh_2}{dt} = \frac{200}{25,000}\sqrt{36} - \frac{240}{25,000}\sqrt{25} = 0$$

The above values are assumed to remain constant over the first second of operation. At $t = 1$ second we reevaluate the liquid levels in

17-5 DIGITAL COMPUTER SOLUTIONS

the following manner:

$$h_1 = h_{1_\text{previous}} + \frac{dh_1}{dt} \Delta t$$

$$= 36 + (0.033333)(1) = 36.033 \text{ in.}$$

$$h_2 = h_{2_\text{previous}} + \frac{dh_2}{dt} \Delta t$$

$$= 25 + 0 = 25 \text{ in.}$$

We then solve Eqs. 17-25 again using the new values of h_1 and h_2,

$$\frac{dh_1}{dt} = -\frac{200}{15,000} \sqrt{36.033} + \frac{1200}{15,000} + \frac{500}{15,000} = 0.033296 \text{ in./s}$$

$$\frac{dh_2}{dt} = \frac{200}{25,000} \sqrt{36.033} - \frac{240}{25,000} \sqrt{25} = 2.2216 \times 10^{-5} \text{ in./s}$$

The process is continued, with the liquid levels and their derivatives reevaluated at the end of each one-second time increment. Table 17-1 shows the results of the numerical simulation for the first 5 seconds. If a digital computer were used, the time increments could be conveniently made smaller and a more sophisticated method of evaluating the system variables used (for the sake of accuracy), but the basic approach would remain the same.

Digital computer solutions make it unnecessary to use linearization techniques. Without linearization, however, superposition cannot be used; for example, if the input amplitude is doubled, the response cannot be obtained by proportion. With nonlinear models the system equations must be solved for each different input.

TABLE 17-1
Results of Numerical Simulation of Example 17-6

Time	h_1	h_2	$\dfrac{dh_1}{dt}$	$\dfrac{dh_2}{dt}$
0	36	25	0.033333	0
1	36.033	25	0.033296	2.2216×10^{-5}
2	36.067	25.00002	0.033259	4.4399×10^{-5}
3	36.100	25.00007	0.033222	6.6546×10^{-5}
4	36.133	25.00013	0.033185	8.8658×10^{-5}
5	36.166	25.00022		

17-6 Conclusion

The purpose of this chapter is to introduce the concept of the state-variable formulation of system equations and to point out the basic relationships between it and computer solutions. However, a comment about obtaining solutions "by hand" is in order at this point. Although linear state-variable equations are readily put into a matrix form and can be solved using a matrix approach, this is frequently the tedious way to go. In general, when seeking a solution by hand, it is preferable to work with the system defined by a single nth-order differential equation in the variable of primary interest.

Problems

17-1. For each of the following nth-order differential equations write an equivalent set of n state-space equations and draw an analog computer diagram for simulating the system.

(a) $$\frac{d^2 y}{dt^2} + 5 \frac{dy}{dt} + 4y = 3$$

(b) $$3 \frac{d^3 y}{dt^3} + \frac{d^2 y}{dt^2} + 2 \frac{dy}{dt} + y = 4t + 1$$

(c) $$\frac{d^4 y}{dt^4} + 2 \frac{d^3 y}{dt^3} + 3 \frac{d^2 y}{dt^2} + 4 \frac{dy}{dt} + 5y = 0$$

17-2. For each of the given sets of state-variable equations develop an equivalent single differential equation in x to describe the system.

(a) $$\frac{dx}{dt} = 2y$$
$$\frac{dy}{dt} = -5x - y + 3t + 4$$

(b) $$\frac{dx}{dt} = y$$
$$\frac{dy}{dt} = z$$
$$\frac{dz}{dt} = -x - 3y - 4z + 2t + 10$$

(c)
$$\frac{dx}{dt} = 4x + 3y$$
$$\frac{dy}{dt} = 2x + y + z$$
$$\frac{dz}{dt} = 2y + 3z - 45$$

17-3. Sketch the state-space trajectories of the mechanical systems with the given initial conditions. Put values of t on the trajectories.

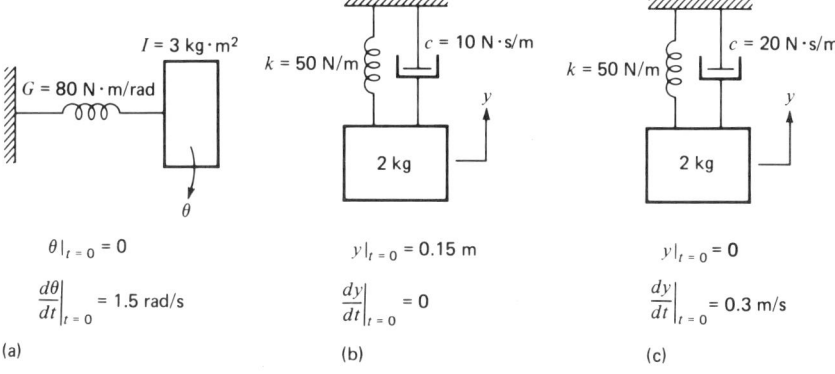

17-4. Write a set of state-variable equations to describe each of the given mechanical systems, and draw an analog computer diagram to simulate it.

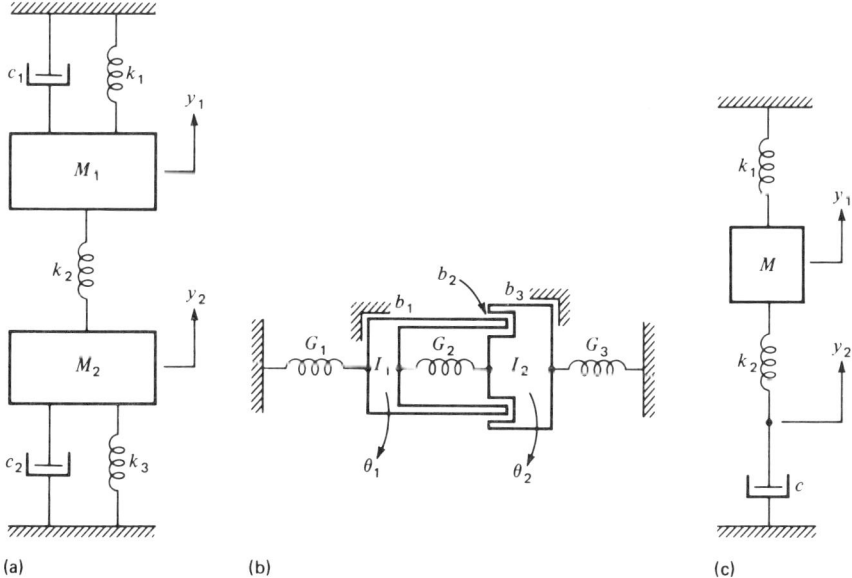

17-5. Represent each of the given electric systems by a set of state-variable equations.

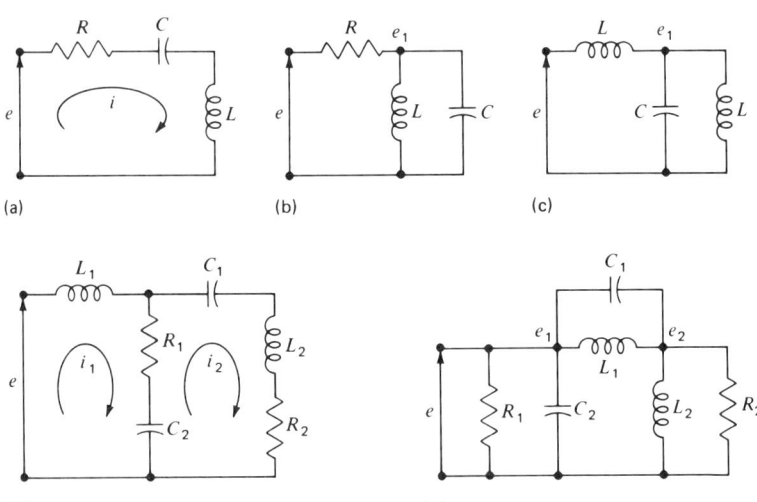

17-6. A pendulum with a mass of 1.5 kg and a 0.6-meter rod of negligible mass is swinging through an angle of $\pm 45°$.
  (a) Write the set of nonlinear state-variable equations that describe the pendulum and set up a method for obtaining a digital computer simulation.
  (b) Simulate the pendulum with a digital computer program and compare to the linearized solution.

17-7. Carry out the solution of Example 17-6 on a digital computer. Compare to a linearized solution.

17-8. For the system shown, the springs are linear but the center spring contacts the mass only when $y \geq 0.5$ in. Simulate the free vibration of the system on a digital computer for the initial conditions $y = -2$ and $dy/dt = 0$ at $t = 0$. Assume massless springs and the following system parameters: $k_1 = 20$ lb/in., $k_2 = 30$ lb/in.

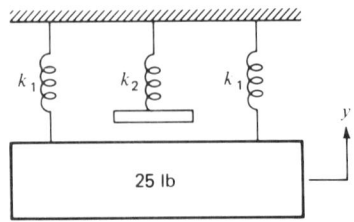

17-9. For the system shown, the empirical equation $F_s = 5z + 1.2|z|z$ was developed to represent accurately the nonlinear spring over its usable range of operation; z is the increase in length (inches) from the unloaded condition and F_s is the tensile load (pounds). The damper has a combination of Coulomb friction and friction proportional to the square of velocity, as represented by the equation

$$F_f = 4.8 \frac{dy/dt}{|dy/dt|} + 3.2 \left|\frac{dy}{dt}\right| \frac{dy}{dt}$$

with F_f in pounds and y in inches. Simulate the response of the system on a digital computer for a step input $x = +5$ in., if the system is initially at rest.

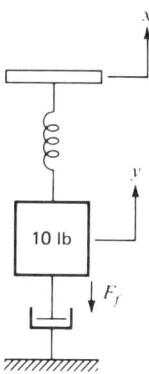

Experimental Determination of System Dynamic Characteristics

18-1 Introduction

When working with a real dynamic system, analytical techniques alone are seldom adequate for proper design and/or understanding of its characteristics. Successful study of an actual dynamic system invariably requires systematic experimental investigations. Therefore this book would not be complete without at least a brief consideration of some of the more important principles and techniques of experimental system study.

In order to plan and conduct experimental work satisfactorily, it is necessary to have a knowledge of the instruments that are available for this purpose, to know their range of application, their limitations, and proper calibration techniques. There will be no attempt in this chapter, however, to give specifications for any particular instrument. We will merely point out the approximate range of application for instruments based on differ-

18-1 INTRODUCTION

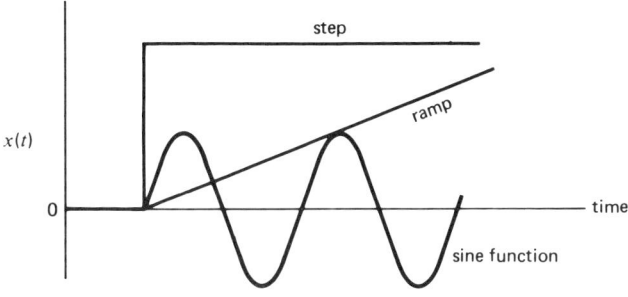

FIG 18-1. System inputs commonly used for experimentally determining a system's response characteristics.

ent fundamental principles, indicating the most important parameters that should be considered and investigated when determining the suitability of an instrument for a given task. The details of operation of specific instruments will not be considered. Instead we will concentrate on only the more important basic principles applying to a whole class of instruments. Operating details vary considerably among instruments made by different manufacturers. Probably the best place to learn the details of operation and the specifications of any specific model is from literature supplied by the manufacturer.

A dynamic system may be exposed to inputs of widely varying characteristics. Its response to all inputs is important, but it would obviously be an insurmountable task to test the system response to all conceivable types and magnitudes of inputs. There are, however, three generally accepted system inputs that are commonly used for experimentally determining a system's response characteristics. They are the step, the ramp, and the sinusoidal functions, as shown in Fig. 18-1. (These, incidentally, are the same inputs normally used for theoretical system analysis.) Knowledge of how the system responds to one or more of these three inputs will usually be adequate for deciding whether or not its performance would be satisfactory for a given application.

The problems at the end of the chapter are designed to help the student develop an appreciation for the basic factors that should be considered when setting up an experimental test of a system and to give some practice in the design of a test and the interpretation of its results. A good understanding of the material presented in this chapter will prepare the student for a more detailed study of experimental techniques for dynamic systems.

18-2 Test Equipment and Instrumentation

This section discusses some of the more important types of equipment and instrumentation commonly used for dynamic system testing. The principles of operation upon which each is based and the general range of applicability are covered. Specific operating techniques and detailed specifications are not considered; these should be obtained from the manufacturers' literature.

CATHODE RAY TUBE OSCILLOSCOPE

The cathode ray tube oscilloscope (commonly referred to as the CRT oscilloscope, or simply CRT), is extremely useful for the study of a wide range of dynamic systems, especially those that operate at high frequencies. A typical CRT oscilloscope is illustrated in Fig. 18-2.

The heart of the CRT oscilloscope is the cathode ray tube itself. We are all very familiar with this device since it is also used as the television picture tube. In the cathode ray tube, an electron gun produces a continuous stream of electrons aimed toward the center of the screen. The screen is coated with a fluorescent material so that the face of the tube will glow at the point where it is struck by the electron beam. The beam passes between two pairs of electrostatic plates. A voltage difference across each pair of plates will bend the electron beam, giving a displacement at the screen proportional to that voltage. The two pairs of plates therefore control the vertical and horizontal position of the trace produced by the electron beam.

Because of the small mass of an electron, the electron beam can be made to follow signals with frequencies of thousands or even millions of hertz. The signal to be displayed must be supplied to the CRT as a voltage. A transducer with an electric voltage output is therefore necessary for all system signals except those that occur naturally as voltages.

The most common way of using a CRT is to display a single signal of interest on the vertical scale, with the horizontal movement of the trace taking place at a speed chosen by the operator. With CRT models having dual channels (i.e., two electron beams, each controlled by its own set of plates) it is possible to display two signals simultaneously as functions of time. With the horizontal speed the same for both traces, this feature allows for easy comparison of amplitudes, shapes, frequencies, and the determination of phase relationships.

The CRT can also be used so that the horizontal position of the trace corresponds to some function of interest of the system, rather than moving at a fixed speed. The display of a system's output versus its input is very useful in many cases for study of frequency-response characteristics.

18-2 TEST EQUIPMENT AND INSTRUMENTATION

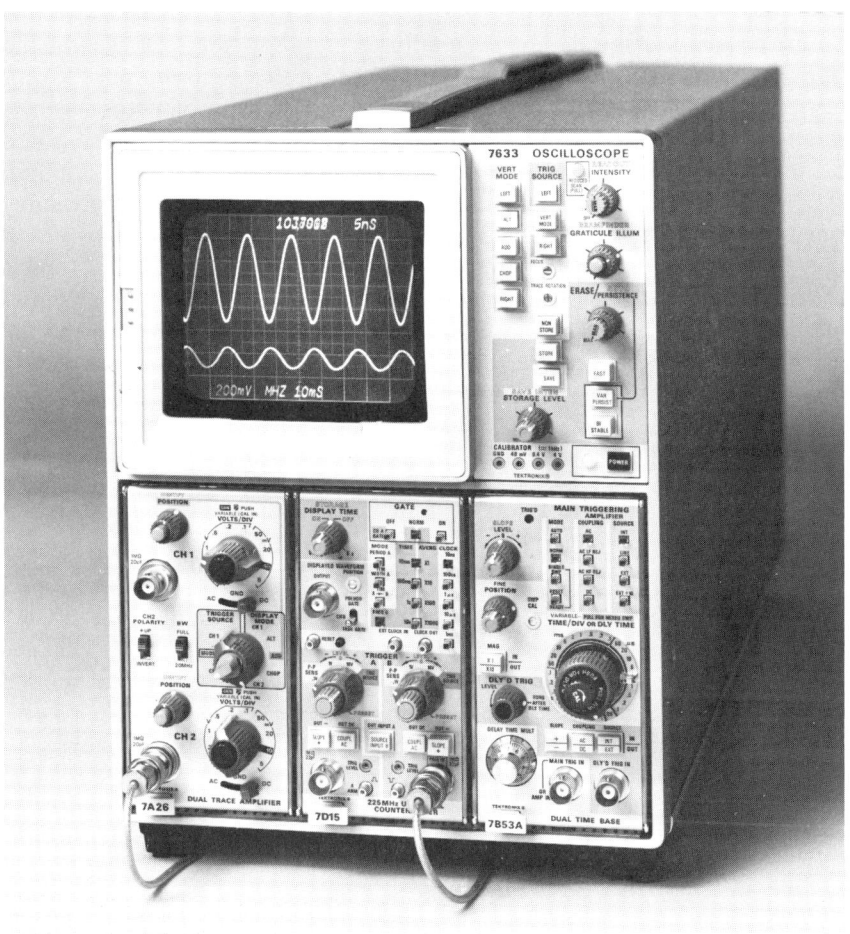

FIG 18-2. A cathode ray tube oscilloscope. (*Courtesy of Tektronix, Inc.*)

When a time base is used for the horizontal sweep, the CRT trace will move from the left edge of the screen to the right edge at the chosen speed. Normally, the control knobs will be set so that as the trace reaches the right edge and disappears there, it will reappear virtually instantaneously at the left edge, and the process then repeats itself. With a steady-state periodic signal the time factor of the horizontal sweep can be adjusted to produce a steady image on the screen; that is, the sweep time can be synchronized with the signal frequency.

With a transient signal (such as the response to a step input), however, a different approach is necessary. It is possible to set the CRT so that a single sweep occurs, from left to right, only when triggered by some voltage signal, such as a step. For a slow transient signal, a single sweep is normally best. Its speed should be adjusted to allow the total part of the response that is of interest to appear before the trace disappears. With a fast-responding system, such as one for which the response to a step is essentially completed in less than a half-second or so, a steady image can be achieved by having a series of transient responses, one after another, with each sweep triggered in the same manner so that the consecutive traces lie on top of one another.

Often, for a slow transient response, a more permanent picture is required than that given by a single sweep on the CRT screen. One method of achieving this result is to take a photograph of the response, leaving the camera shutter open for the total sweep. Special cameras and mounting apparatus are available to make this a simple procedure. Another approach is to use a special type of CRT, called a "memory 'scope," which is designed so that the total trace of a transient signal will remain visible until the operator erases it.

X-Y RECORDER

The X-Y recorder (Fig. 18-3) is in some respects quite similar to the CRT oscilloscope, except that it is useful only in recording signals that change somewhat slowly, since the recording is produced by mechanical motion rather than the electrostatic displacement of an electron beam. The recording pen is moved over a piece of paper in accordance with voltage signals for the vertical and horizontal axes. As with the CRT, time may be used for the horizontal signal if desired. One advantage the X-Y recorder has over the CRT is that it automatically supplies a permanent record of the test.

STRIP CHART RECORDER

There are a great number of recording instruments available, under the general category of strip chart recorder, that can record continuously one or more signals as functions of time on long strips or rolls of paper marked with two-dimensional coordinates. Many such recorders will handle up to eight or more signals at a time, recording them side by side on wide rolls of paper (Fig. 18-4). The availability of a permanent record of a number of system functions recorded simultaneously is convenient for determining amplitude and phase relations for varying types of operations. Such a record allows one to readily see the effect of a change in one variable (e.g., the system input) on the other recorded variables. Most strip chart recorders provide a mark at the edge of the paper once each second to produce an automatic record of the speed at which the paper was moved as the signals were recorded.

18-2 TEST EQUIPMENT AND INSTRUMENTATION 353

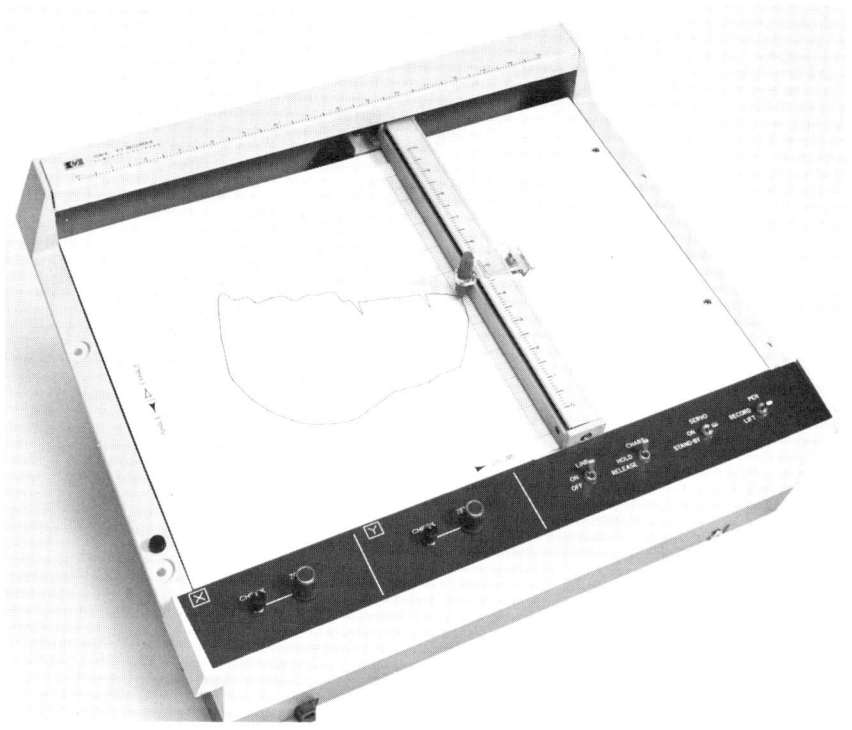

FIG 18-3. An *X-Y* recorder. (*Courtesy of Hewlett-Packard Company*.)

Different methods are used for marking the signal trace on the recorder paper. Some recorders use ink, others use a hot pen to record on heat-sensitive paper, and still others use a light beam on photosensitive paper. The frequency of signal that can be accurately recorded on a strip chart recorder will vary from less than 50 Hz to hundreds of hertz—or even thousands of hertz for models using the light beam principle.

ACCELEROMETERS AND VIBRATION METERS

An accelerometer is a device to measure the acceleration of a mechanical object to which it is attached. In general, an accelerometer will produce a voltage that is proportional to its instantaneous value of acceleration. Accelerometers are normally unidirectional, so they must be mounted with the proper orientation to give a valid signal.

The acceleration signal can be integrated electronically to produce a velocity signal, that is, a voltage proportional to the instantaneous value of velocity. A second integration will produce a displacement signal.

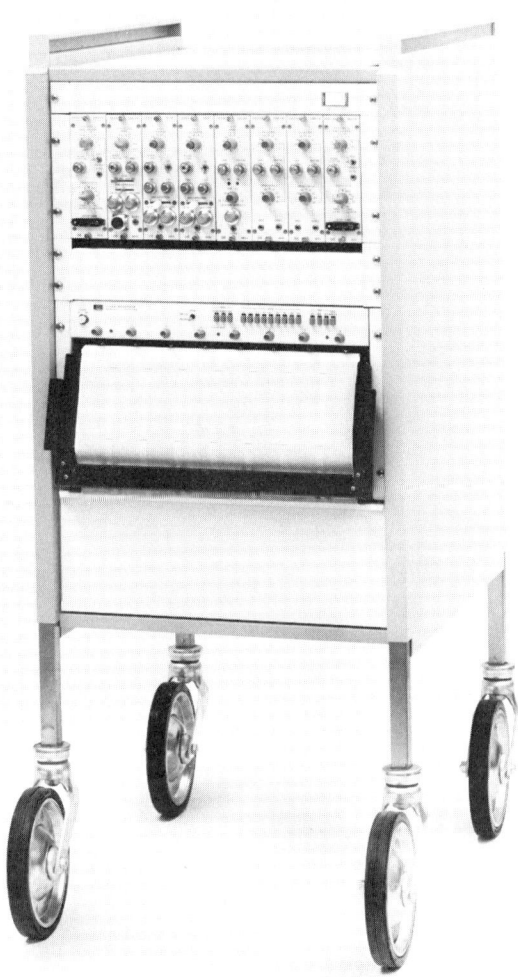

FIG 18-4. An eight-channel strip chart recorder. (*Courtesy of Hewlett-Packard Company*.)

Although these signals may be displayed or recorded with a CRT oscilloscope or strip chart recorder, they are more commonly read by an ac voltmeter to obtain acceleration, velocity, and displacement values for steady-state vibration. Instruments called *vibration meters* are designed to perform the above functions when used with an accelerometer as the signal pickup device.

POSITION PICKUPS

Although an accelerometer can be used to produce a displacement signal by means of a double integration, better accuracy can be obtained by using a transducer that reads position directly. For relatively slow motion, a probe attached to an electric potentiometer may give a satisfactory signal. Another device that is commonly used is the linear variable differential transformer (LVDT). The LVDT is basically an ac transformer, with a magnetic core that changes the mutual inductance of a primary and two secondary coils as it is moved axially (Fig. 18-5). The position of the coils is normally fixed, and the core is attached to the object whose motion is to be measured by a probe. The LVDT produces a voltage signal that is proportional to displacement, and some versions are designed to give good signals for very low movements—of the order of 0.001 inch or less. The ac output signal can be rectified if desired to a dc signal.

For some tests a probe or other device placed on the vibrating mass will have a significant effect on its dynamic characteristics and therefore give misleading data. Devices based on electric capacitance may be useful in such cases. A capacitive pickup, fixed with respect to ground, is placed close to the vibrating piece. The capacitance between the pickup and the vibrating part (a function of the distance separating them) is used to produce a displacement signal.

In addition to the above, there are many other types of position transducers available on the market to handle the special requirements of a wide variety of applications.

MECHANICAL SHAKERS

Many mechanical vibrating systems are self-excited; that is, the system includes an unbalanced rotating element, or some other source, that causes vibration to occur. In other cases the system is passive and will vibrate only when acted upon by an external excitation.

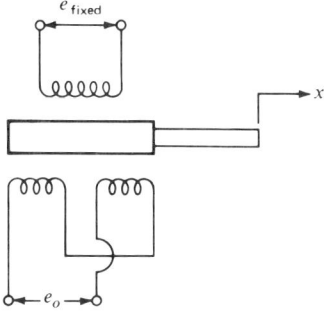

FIG 18-5. The linear variable differential transformer (LVDT).

Mechanical shakers are devices used to provide a periodic input to a passive mechanical system so that its response characteristics may be experimentally determined. The inputs are usually sinusoidal, or very nearly so, and can be varied over a wide range of frequencies.

There are various types of shakers available. Some work on the electromagnetic principle, some on positive mechanical displacement, and others with a unidirectional force produced by two counterrotating unbalanced shafts. The shakers normally apply translational inputs, but versions giving rotary inputs are available. Shakers come in a wide variety of sizes, from small ones up to those weighing many tons.

In terms of vibration input, the various types of shakers fall into two basic categories:

1. Those producing a specific *position input amplitude*, with the resultant force on the system depending upon this amplitude, the frequency, and the system mass. A good example of this type of shaker is a platform that is given sinusoidal motion by a piston-crank or scotch-yoke mechanism. Figure 18-6 shows a shaker of this type.

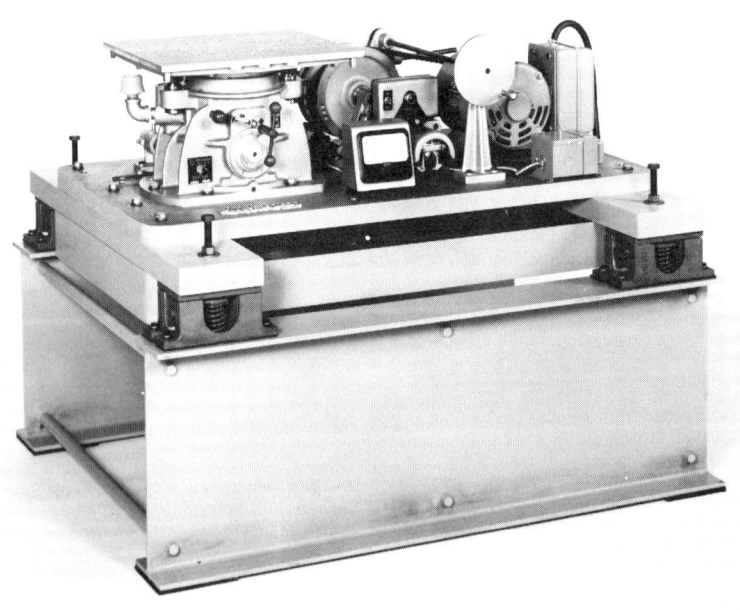

FIG 18-6. A mechanical shaker that provides sinusoidal position excitation. (*Courtesy of All American Tool & Manufacturing Co., Inc.*)

18-2 TEST EQUIPMENT AND INSTRUMENTATION 357

2. Those producing a specific *force input amplitude*. The electromagnetic shaker (Fig. 18-7) fits in this category. The displacement it produces on a given system under test depends not only on the force input but on the system mass and other dynamic characteristics as well.

Force-producing shakers can again be divided into two main types: those fixed to ground having a movable (vibrating) table or platform to which the test item or system is attached and those that are attached to the test system and move with it as vibration is induced. In the latter case the mass of the shaker must be taken into account when evaluating the test results.

THE STROBOSCOPE

The stroboscope (Fig. 18-8) is a very useful instrument for investigating the vibration characteristics of many mechanical systems in which vibration amplitudes are not extremely small. (It can be used to advantage when amplitudes are even of the order of 0.005 to 0.010 inch.)

FIG 18-7. An electromagnetic shaker that provides sinusoidal force excitation. (*Courtesy of Ling Electronics, A Division of Altec Corporation.*)

FIG 18-8. A stroboscope. (*Courtesy of GenRad, Inc.*)

The stroboscope produces pulses of light of very high intensity but very short duration. The frequency at which these pulses occur is readily varied by a simple dial. The motion of a vibrating object may be "frozen" by illuminating it with a stroboscope whose frequency is the same as the vibration frequency. If steady-state vibrating conditions are present, the stroboscope frequency may be set to a slightly different value to produce a slow-motion picture of the actual vibration, allowing amplitudes, nodes, phase relationships, and other vibration characteristics to be clearly observed. In some cases use of the stroboscope will give insight into many aspects of a vibration problem that would be difficult to obtain by any other method.

SPECIAL-PURPOSE EQUIPMENT

Due to the importance in many engineering projects of obtaining good frequency-response data in a quick and reliable manner, a great deal of special equipment has been developed for this purpose. By inputting to such instruments voltages proportional to the input and output of a dynamic system undergoing frequency-response tests, it is possible to read directly the frequency, amplitude ratio, and phase angle.

Since such special-purpose instruments do most of the work for us, their use will not be emphasized in this chapter.

OTHER TYPES OF INSTRUMENTATION

Much of the equipment discussed up to this point can be used only with mechanical and/or electric dynamic systems. Although more experimental work is probably done in these areas than in any others, this in no way implies the lack of importance of experimentally studying other types of dynamic systems.

For slow-moving dynamic systems, satisfactory instrumentation equipment is usually easy to find. For systems that respond quickly, equipment adequate for obtaining the required data may be more sophisticated and expensive. Temperature probes, flow meters, pressure transducers, and so on that will respond to changes occurring hundreds of times per second are often required.

18-3 Applying System Inputs

Step, ramp, and sinusoidal inputs to dynamic systems must be properly applied if accurate data are to be obtained. There are a few special techniques that are useful to know when setting up some types of experimental tests based on the use of one or more of these inputs.

APPLYING A STEP INPUT

A perfect step input would be an instantaneous magnitude change. Practically speaking, an instantaneous change is impossible, but it should be quick enough that the response is not perceptibly different than it would be with a perfect step. In a real mechanical system a "step" occurring quickly enough is sometimes difficult to achieve. Consider the system of Fig. 18-9. The input (as well as the output) is a mechanical position. The straightforward way of applying a step input would be to move the top end of the spring to a new value of x "instantaneously." In practice, it may be difficult to move a piece of machinery such as this quickly enough to

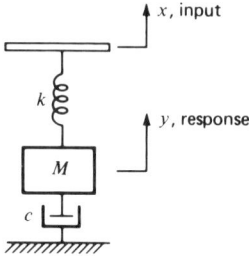

FIG 18-9. A simple mass-spring-damper system.

obtain a valid response. It is quite likely that, without some sophisticated equipment to perform the operation, the "step" movement itself would take a period of time that was a fairly high percentage of the period of free vibration.

For systems similar to that of Fig. 18-9, however, there are alternative ways of obtaining valid step response data. For example, one may hold y fixed while moving x the distance of the desired step. A quick release of the mass will then produce the required step response. Another approach is to allow x to remain fixed but to move the mass downward the step distance. When the mass is quickly released, the same step response is obtained. As discussed earlier in Section 6-2, proper initial conditions can produce system action equivalent to a step response if the parameters are properly chosen and defined.

Figure 18-10 illustrates a simple position-control system, designed so that the output angular position θ_o will follow the input θ_i. Although the input and output are mechanical positions, electrical components are employed in the system. Again, a real step input for this system would involve turning the input shaft instantaneously (or at least very quickly) to a new angular position. This operation may be impractical or at least inconvenient for the real system, but there are other techniques that can be employed to produce the same effect, that is, cause the same response. Since the error signal is converted into an electric voltage by the potentiometer, a bias voltage suddenly applied as e_{bias} in Fig. 18-10 will cause the control system to respond as if it has received a true mechanical step input. A change in voltage that is abrupt enough to be considered a true step is easy to obtain.

APPLYING A RAMP INPUT

A ramp input is defined as a system input whose magnitude is changing at a fixed rate. Since there is no "instantaneous" movement necessary, this

18-3 APPLYING SYSTEM INPUTS

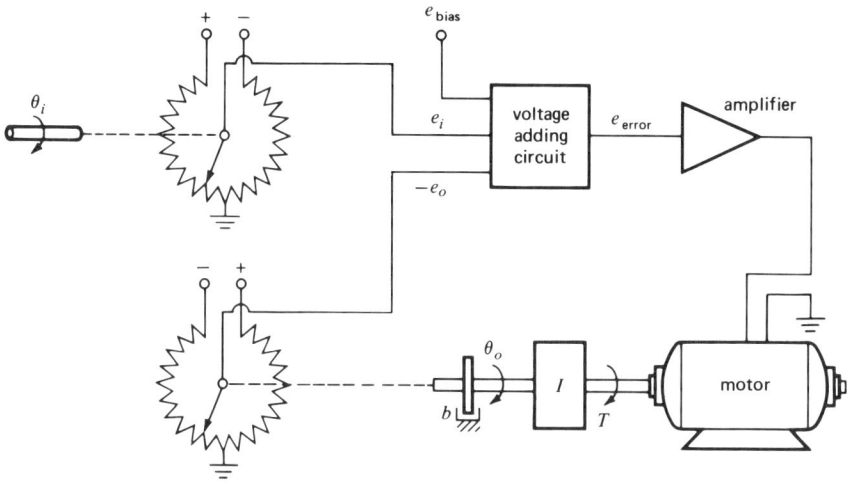

FIG 18-10. An electromechanical feedback position-control system.

input is not normally very difficult to achieve, even where a mechanical ramp is required. For systems that combine mechanical and electrical components (e.g., Fig. 18-10), an electrical ramp can often be used as an alternative to a true mechanical ramp.

When a ramp is first applied, transient effects occur. These will vary depending upon how the ramp is begun. An abrupt change in input velocity to the ramp value will, for example, produce different transients than by starting with a constant acceleration. Since steady-state response to a ramp is the information usually required, however, the question of how the ramp should be started seldom arises. The ramp is simply applied for a period of time long enough that the transients will die out.

APPLYING A SINUSOIDAL INPUT

A sinusoidal input, like the ramp, is often easier to apply than a step input since there is no requirement for an "instantaneous" change in amplitude. A sinusoidal force input for a mechanical system may be achieved by an electromagnetic shaker, and a sinusoidal position input (if translation) can be applied by a vibration table. A sinusoidal torsional input can be accomplished by torsional versions of the shaker and vibrating table.

For a control system in which the position input is converted to an electric voltage signal (such as the system of Fig. 18-10), the system may be given the equivalent of a sinusoidal position input by applying a sinusoidal voltage signal.

18-4 Recording and Interpreting System Response

DISPLAY OF INPUT AND OUTPUT SIGNALS ON A COMMON TIME SCALE

The most straightforward method of displaying or recording system response characteristics is to have the input and the output (response) side by side or superimposed on a common time basis. Figure 18-11 illustrates such recordings from tests to determine step response, response to a ramp input, and frequency response. Although the common time scale is essential if the traces are to be useful (and in most cases it would be difficult *not* to have the common time base), it is not essential that the amplitude scales be the same as long as the scale factors are accurately known.

A two-channel cathode-ray tube oscilloscope is very useful for displaying such response data, particularly for very fast-responding systems where the range of alternative instruments may be inadequate. The CRT may be adjusted so that the input and output signals appear stationary and the system response characteristics are obtained from measurements made directly on the face of the tube. For step and ramp responses, it is often advantageous to take a picture or use a memory scope.

Recording input and response data side by side on a strip chart recorder is quite straightforward once one becomes familiar with the calibration procedures for any particular recorder; the recorder will automatically produce a permanent record on the recording paper. With the automatic 1-second time interval marker on many recorders, frequencies are easy to determine and/or check.

Many strip chart recorders have more than two channels (eight being a common number), so response at several locations within a complex system may be simultaneously recorded if desired, as may other dynamic

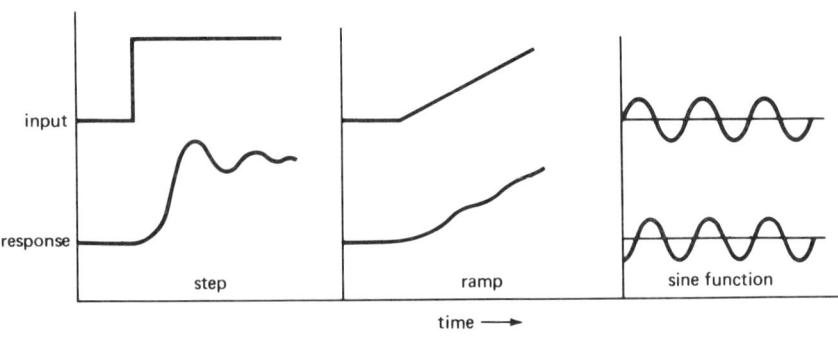

FIG 18-11. System input and response recorded side by side on a common time scale.

18-4 RECORDING AND INTERPRETING SYSTEM RESPONSE 363

data. A direct recording of the instantaneous error (difference between input and output amplitudes) is often very useful for control systems study.

For input and output amplitudes plotted side by side, the measurement of system characteristics is generally quite straightforward. It is important to remember, however, that the real system being investigated may be nonlinear and/or have other complexities that make it useful to consider time constants, natural frequencies, and other characteristics of linear models that approximate the real system with good accuracy.

STEP RESPONSE

Consider Fig. 18-12, which shows experimentally obtained step response traces for several different systems. The step response of Fig. 18-12a can be seen to have essentially the characteristics of a first-order system, although there is a slight deviation from the ideal curve. Based on the experimental trace, the first-order model is found to give a reasonably accurate representation if the time constant is chosen on the basis of the point where the amplitude is 63.2 percent of the final value.

In Fig. 18-12b the response is oscillatory and can be seen to be, within a small range of error, that of an underdamped second-order system. The damped natural frequency ω_d can be obtained by measuring the time between successive oscillation peaks. The damping ratio ζ can be determined by comparing the characteristics of the curve to the family of curves in Fig. 14-7 or by using the log decrement method. A first-order model would obviously be an inappropriate approximation for any system having this oscillatory type of step response.

In Fig. 18-12c is illustrated a commonly seen type of experimental step response. There is some question here as to whether a first-order model or a second-order model with large damping ratio would be more appropriate. The concave upward shape of the curve at the origin is a characteristic of a second-order system, but in this particular case the concave upward portion is a very small part of the range of interest and causes the curve to deviate very little from the corresponding first-order approximation. Since the second-order approximation can also be made to represent the experimental curve accurately, this is an example where a choice can be made between the two types of models. An example of a system that would have a step response of this type is a simple mass-spring-damper system (Fig. 18-13) in which the mass M is quite small in comparison to the spring constant k and the damping coefficient c.

The step response of Fig. 18-12d shows resonance at two different frequencies, and therefore was most likely produced by a fourth- or higher-order system. Representation of a system with such a complex response cannot be accurately given by a simple first- or second-order model.

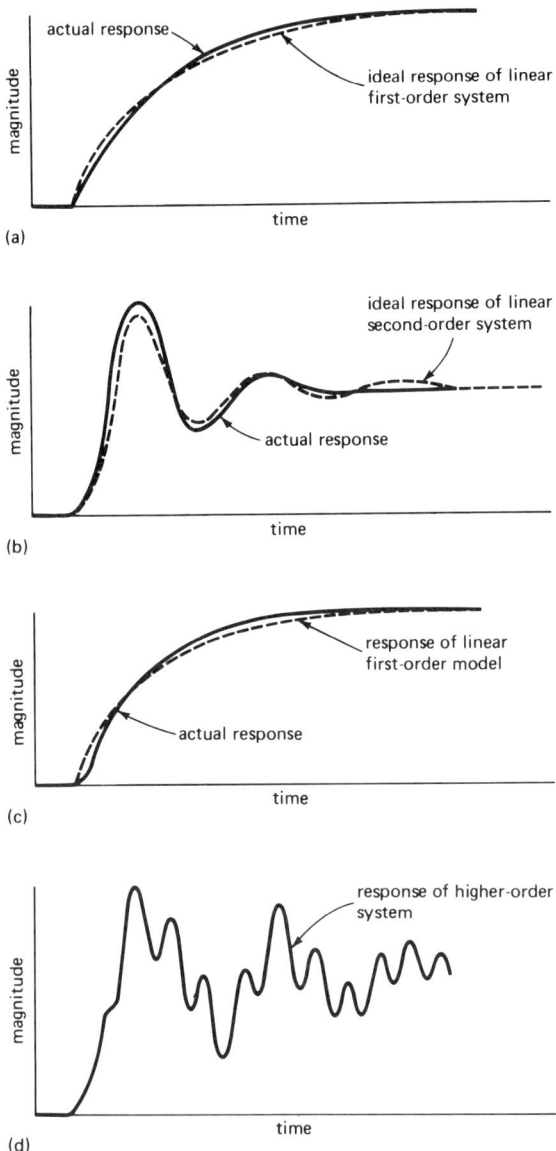

FIG 18-12. Experimental step response curves, with models based on ideal linear systems.

18-4 RECORDING AND INTERPRETING SYSTEM RESPONSE

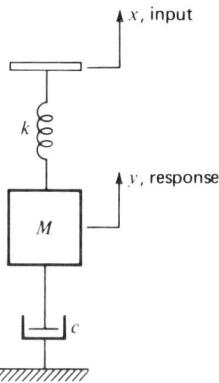

FIG 18-13. A simple mass-spring-damper system, which may be fairly accurately modeled as a first-order system if mass M is very small.

RAMP RESPONSE

Figure 18-14 shows various types of ramp response characteristics obtained with typical dynamic systems. The input and response traces are superimposed in this case, rather than being merely side by side.

In all the cases illustrated, the ramp starts with an abrupt change in input velocity (this is the commonly used manner of applying a ramp), which will produce oscillations in underdamped systems that are of second or higher order. Such oscillations are apparent in Fig. 18-14b and c, whereas Fig. 18-14a is more typical of what will occur with a first-order system or a higher-order system with high damping. The oscillations will eventually die out except in the case of systems with zero or negative damping. In general the steady-state response characteristics present after the transient oscillations have disappeared are normally of most interest with ramp response tests, since the system damping characteristics are better studied with either step or frequency response tests. The significant factor in the steady-state part of the response is the amount by which the output lags the input. This is normally measured as the difference in amplitude at a given instant of time, such as x_i-x_o for a mechanical translational system, θ_i-θ_o for a mechanical rotational system, T_i-T_o for a temperature control system, and so on. It is also possible to speak in terms of a time lag—that is, the difference in time for the points of equal input and output amplitudes.

The difference between Fig. 18-14b and c should be carefully noted. There is an appreciable lag between input and output in Fig. 18-14b but none in Fig. 18-14c. Most dynamic systems will have a lag when responding to a ramp input, but there are control system compensation techniques

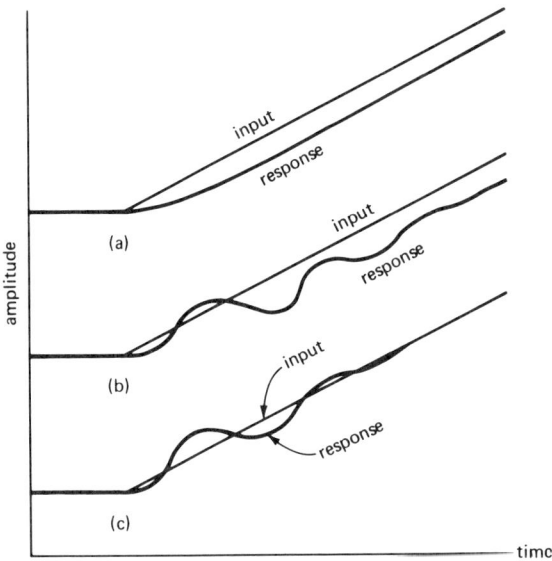

FIG 18-14. Experimental ramp response characteristics of some typical dynamic systems.
  (a) First-order system or overdamped second-order system.
  (b) Underdamped second-order system.
  (c) Underdamped second-order system with compensation to give zero steady-state error to ramp input.

that can be employed to reduce the steady-state lag to zero (e.g., the *proportional plus integral* control covered in Section 15-4).

Since the difference between input and output amplitudes is usually the major item desired in a ramp test, it may be more important to record this directly if it is conveniently possible to obtain such a signal.

FREQUENCY RESPONSE

Some typical frequency-response test results, with input and output plotted side by side on the same time scale, are illustrated in Fig. 18-15. In order to get the full-range frequency-response characteristics of the system and to plot a frequency-response curve such as that of Fig. 14-8, it is necessary to have a set of recordings for a number of frequencies to cover the range of interest of the system. In general, one should have recordings for at least five different frequencies, properly chosen, and preferably considerably more.

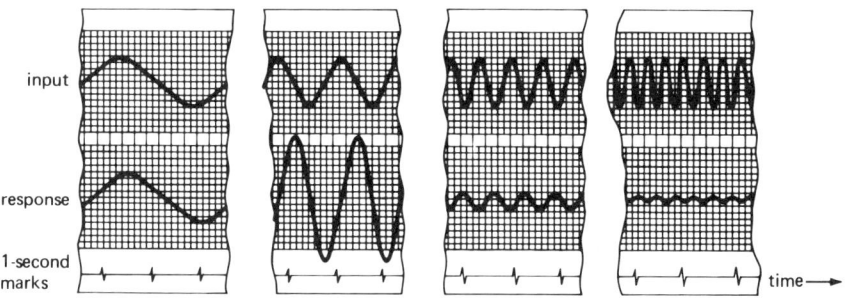

FIG 18-15. A frequency-response test as recorded on a strip-chart recorder.

One way sometimes used to obtain the strip chart data necessary for plotting the frequency-response curve is to start the input at a very low frequency and then slowly and steadily increase this frequency while maintaining its amplitude fixed until the required range of frequencies has been covered. In reducing the data for plotting, then, one will merely choose a number of points (i.e., frequencies) along the recording and measure the amplitude ratio and phase angle at each. In some instances this approach is satisfactory, but in many cases the results are not accurate because the transients caused by the changing of the frequency are continually present and thus the output signal does not exhibit the true steady-state response. It is better to run the test at a number of discrete frequencies, with the system kept at each frequency for a long enough period of time for the transients to die out. In general the effect of transients on the experimental frequency-response data is more significant for lightly damped systems.

Amplitude ratio and phase angle are the two significant parameters to be obtained from the frequency-response data. Amplitude ratio is basically determined by measuring the input and output amplitudes and taking the ratio of the two, taking into account any difference in the scale factors of the two channels.

For cases where the input and output have different units (e.g., a vibrating system with a sinusoidal force as input and position as output), the ratio of amplitudes should be normalized. For most systems, in which the frequency response curves are flat at frequencies well below the natural frequency (or break frequency), as illustrated in Figs. 14-6 and 14-8, this means normalizing the amplitude ratio to a value of 1 at the low frequencies. One commonly used way of accomplishing this during an experimental test is to make the first recording at a frequency that is a few orders of magnitude below the system's natural (or break) frequency and adjusting the gain controls of the two channels so that the amplitudes of the input and output traces are equal.

The phase angle is obtained by measuring the time displacement of the output trace with respect to the input, and comparing this with the period of the harmonic curve. (See the example given in Fig. 18-16.)

DISPLAY OF OUTPUT AMPLITUDE VERSUS INPUT AMPLITUDE

One commonly used method of obtaining amplitude ratio and phase angle data for a dynamic system is to display output amplitude on the vertical axis vs. input amplitude on the horizontal axis. This type of display can be done with a cathode ray tube oscilloscope or an X-Y recorder, although most X-Y recorders can handle only low-frequency signals. This type of display is useful only for steady-state harmonic inputs and has no value for step or ramp functions.

The types of pattern obtained, often referred to as *Lissajous figures*, are illustrated in Fig. 18-17. With equal vertical and horizontal scales, a 45° line will be obtained when the two signals are of equal amplitude with 0° phase angle. In general, a closed figure will occur, ranging from a circle to ellipses that are inclined at various angles.

The amplitude ratio is determined by simply measuring the relative magnitudes of the vertical and horizontal components of the trace, taking into account the two scales.

The figure will be a straight line in quadrants 1 and 3 (i.e., inclined at an angle less than 90°) when the phase angle is 0°, 360°, or multiples

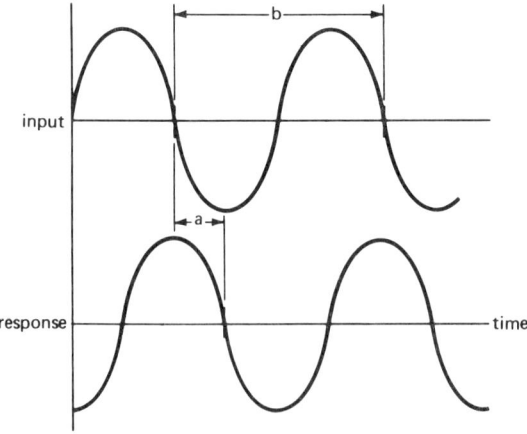

FIG 18-16. Method of obtaining phase angle between two sinusoidal curves recorded side by side on a common time scale. $\phi = -(a/b)360°$, where ϕ is the phase angle of the output with respect to the input.

18-4 RECORDING AND INTERPRETING SYSTEM RESPONSE

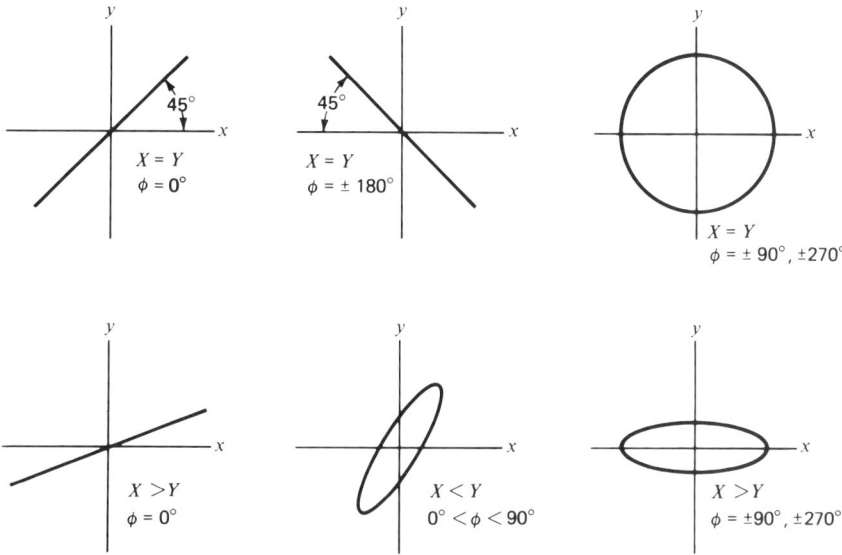

FIG 18-17. Some representative Lissajous figures obtained by recording response versus input for sinusoidal signals, with equal vertical and horizontal scale factors.

thereof. It is straight line in quadrants 2 and 4 when the phase angle is 180° or multiples thereof. At phase angles of 90°, 270°, and so on the figure is an ellipse with the major axis either vertical or horizontal (a circle if the amplitude ratio is 1.0).

At phase angles other than those mentioned in the paragraph above, the figure will be an ellipse with major axis inclined at some angle other than 0° or 90°.

The use of the Lissajous figure is most useful for determining the exact frequencies at which significant phase angles (e.g., $-90°$, $-180°$) occur. (Note that the frequency cannot be measured from the figure itself, but must be determined in some other manner.) The phase angle can be accurately estimated at all values by comparing the Lissajous figure shape with Fig. 18-18 if the scales are adjusted in each case to have equal horizontal and vertical displacements.

STROBOSCOPIC TECHNIQUES

Probably the most important use of the stroboscope is to accurately determine the frequency of vibration or of rotation of a mechanical system. There are a few pitfalls that must be avoided, however, if one is to be sure that the frequency found is valid.

370 EXPERIMENTAL DETERMINATION OF SYSTEM DYNAMIC CHARACTERISTICS

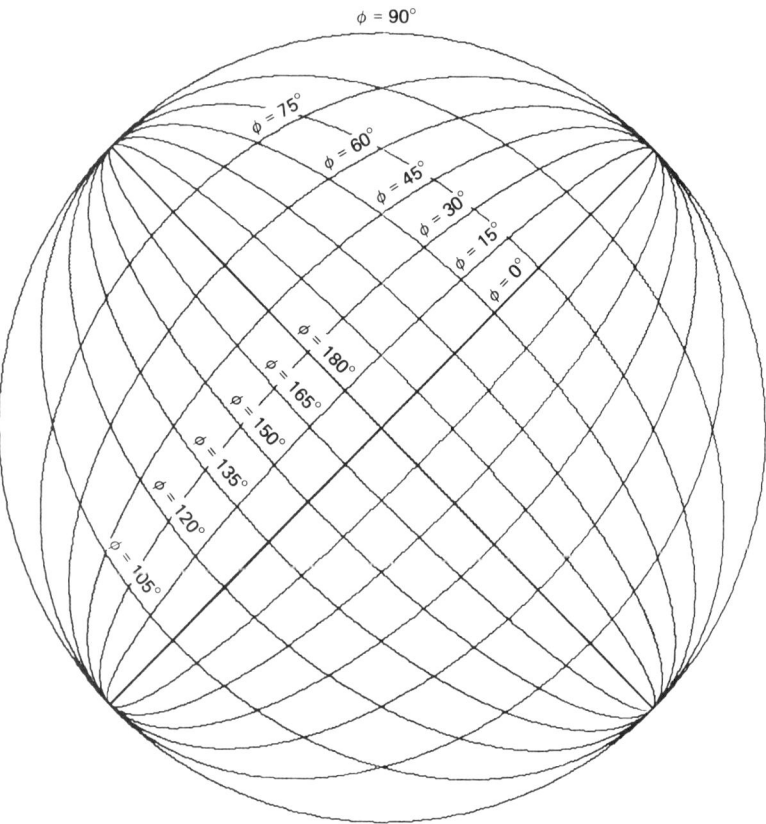

FIG 18-18. Lissajous figures obtained for various phase angles if scale factors are adjusted so that the vertical and horizontal amplitudes are equal.

Assuming the stroboscope is properly calibrated (and this is normally a very simple operation, covered by instructions included with the instrument), the main problem is that it is possible to set the stroboscope frequency at a value that will cause the system motion to "freeze" even though the frequency is not the correct one. This will occur, first of all, if the stroboscope light does not flash once for each cycle of the system being observed, but every other cycle, every third cycle, every fourth cycle, or the like. With this situation the frozen image is clear and sharp, and there is no indication of a problem unless one senses that the stroboscope frequency is too low. The only way to be absolutely sure the stroboscope frequency is correct is simply to find the highest frequency at which a clear image is formed. This is facilitated if one keeps in mind the principles involved. Assume, for example, that a clear "frozen" image is produced at a stroboscope frequency of 1000 Hz and there is some doubt that the true system frequency is that low. If the frequency is in fact higher, it must be at even multiples of 1000, so that one should try 2000, 3000, 4000, 5000, and so on.

Another type of problem can occur when the stroboscope frequency is some multiple of the actual system frequency. In that case, the frozen image, although not perfect, may appear to be so if not carefully inspected, or if it has certain geometrical characteristics. For example, a rotating gear with an even number of teeth may form an apparently perfect image if the stroboscope frequency is set at twice its speed of rotation, since the gear may look virtually the same when rotated 180°. This problem can be avoided if a single identifying mark, such as a spot of paint, is put on the side of the gear. The correct stroboscope frequency will show a single mark as it actually appears, while doubling the stroboscope frequency will show two marks, with each of them a little less distinct; tripling the frequency will produce three marks, and so forth.

If, instead of a rotating object, a vibrating beam is being studied, setting the stroboscope frequency to match that of the beam, except for a small difference, will cause the image to appear as a single beam that is vibrating slowly. Running the stroboscope at twice this frequency will cause an image of two beams moving 180° out of phase to appear, the two merging as a single beam at the midpoint of the vibratory motion.

The stroboscope can also be used to check phase relationships, although only rough approximate values can be determined unless auxiliary equipment is employed. If two vibratory motions (either translational or rotational) at the same frequency are viewed together, while illuminated by a stroboscope set at a frequency just below theirs, the phase relationships can be seen. A precise phase angle is difficult to determine, but it can be seen if the movements are approximately in phase ($\phi = 0°$), out of phase ($\phi = 180°$), or near 90°. One word of caution is in order, however. If the stroboscope frequency is set slightly higher than that of the system, rather than lower, the phase relationships will be reversed; for example, a true phase angle of $-90°$ will appear as $+90°$.

The stroboscope is useful only for studying mechanical systems operating at steady-state frequencies. It is difficult if not impossible to use it to obtain useful data for responses to step and ramp inputs.

18-5 Conclusion

The results of a dynamic system test reflect the response characteristics of all measuring and recording equipment used as well as the characteristics of the system under study. Although it is possible to correct experimental data for instrumentation characteristics, it is better to use equipment with "flat" response characteristics (i.e., with a fixed amplitude ratio and negligible phase angle) over the frequency and amplitude range used.

A knowledge of how to conduct valid and accurate experimental tests on dynamic systems is very important for today's practicing engineer. Some basic fundamental principles have been introduced in this chapter. Much additional information must be learned, however, before one can be truly proficient in dynamic system testing. This information may be obtained from further formal study or from actual testing experience with help from manuals and other information furnished by the instrumentation manufacturers.

Problems

18-1. The input x and output y of a dynamic system were recorded side by side at the same gain level on a strip-chart recorder, as shown. If the input can be represented by the equation

$$x = 10 \sin 7.85t$$

and we want to represent the output by the equation

$$y = A \sin(\omega t + \phi)$$

determine
(a) The proper values of A, ω, and ϕ.
(b) The recording speed, in inches per second.

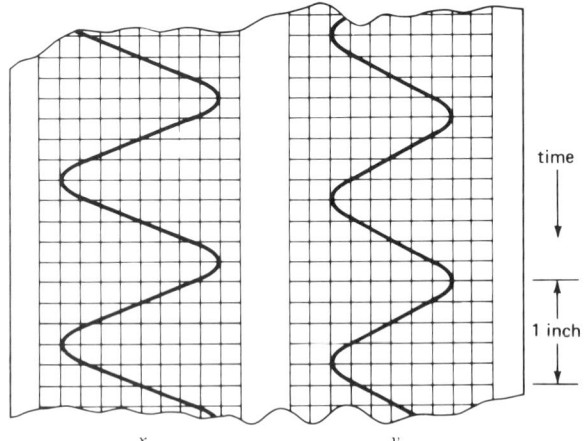

PROBLEMS

18-2. A two-channel strip chart recorder is used to record the response of an angular position control system. The system output (θ_o) is recorded on one channel, and error ($\theta_i - \theta_o$) on the other, as illustrated. Determine the system amplitude ratio and phase angle from the two traces recorded with equal gains.

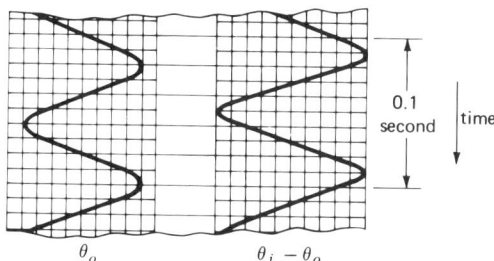

θ_o $\theta_i - \theta_o$

18-3. From the given set of strip chart recordings, plot the system frequency response curves (amplitude ratio and ϕ vs. ω). Note that different recording speeds were used, and that the input amplitudes (on the left in each case) were maintained constant for the total test. The gains of the two channels were not changed. Each division is 1 mm. Determine ω_n and ζ from a comparison with the theoretical curves of Fig. 14-8.

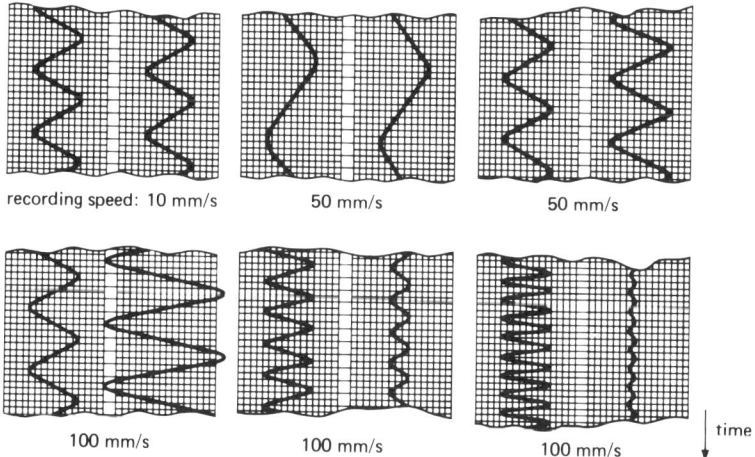

recording speed: 10 mm/s   50 mm/s   50 mm/s

100 mm/s   100 mm/s   100 mm/s

18-4. A system has the given step response characteristics.
(a) If it is to be modeled as a first-order system, determine a good approximate value of τ.
(b) Determine approximate values of ω_n and ζ for a second-order model.

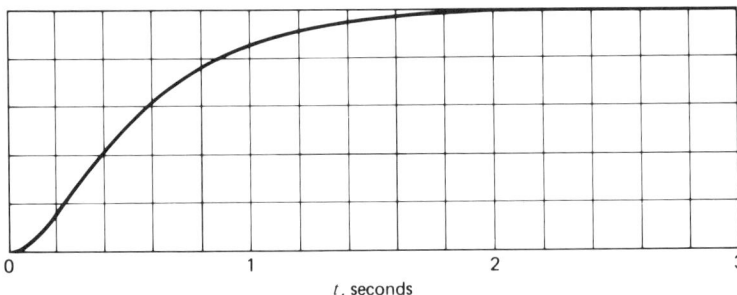
t, seconds

18-5. The system with the given parameters is approximated as a first-order system with no mass.
(a) Determine the error in step response characteristics (e.g., error in percent of steady-state value reached at $t = \tau$ and $t = 4\tau$) with the first-order model.
(b) Determine the error in frequency-response characteristics (by comparing amplitude ratio and phase angle at about four frequency values) with the first-order model.

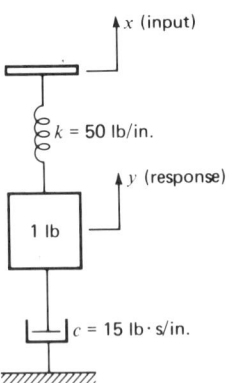

PROBLEMS

18-6. A system has the given step response characteristics. If it is to be modeled as a second-order system, determine good approximate values of ω_n and ζ.

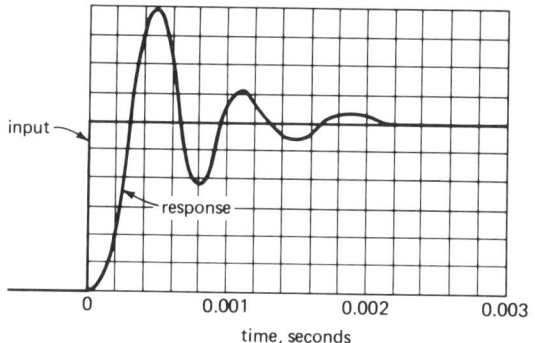

18-7. The frequency response of a system is studied by using a CRT oscilloscope to display output on the vertical axis versus input on the horizontal axis. The scale factors have been adjusted in each case to give equal vertical and horizontal amplitudes on the CRT screen. Determine for each display shown the approximate phase angle of the output with respect to the input.

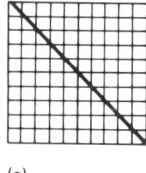 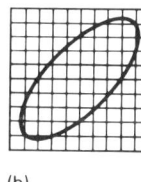 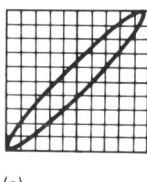 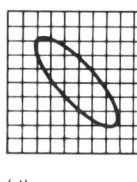
(a)   (b)   (c)   (d)

18-8. Frequency response is studied as described in Prob. 18-7 except that vertical and horizontal scale factors are kept equal. Determine for each display shown the amplitude ratio and phase angle of the output with respect to the input.

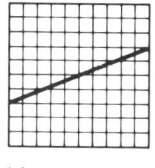 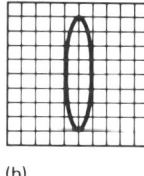 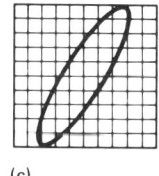 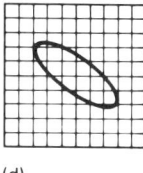
(a)   (b)   (c)   (d)

18-9. (a) A stroboscope is used to "freeze" the motion of a 12-tooth gear with no distinguishing marks, rotating at 1000 r/min. List the speed settings of the stroboscope (above the true speed) that may give an apparently perfect image and therefore a faulty reading of revolutions per minute.
(b) Repeat for a 15-tooth gear.

Appendix A

Deflection of Beams

The compliant elements (i.e., the "springs") of mechanical vibrating systems are often structures that may be closely approximated as simple beams. Beam deflection equations are very useful for determining effective spring constants for such systems.

The equations presented here are for beams of *uniform cross section* and give the deflection y due to an applied load F. The deflection is measured at the point of the applied load, since this is the deflection that is needed for determining the spring characteristics of the beam.

For beams of nonuniform cross section (or those that do not fit the equations given here for other reasons) the force-deflection characteristics can be determined by the application of fundamental beam theory [19].

Definitions

1. E = *modulus of elasticity*
   Typical values;
   (a) $E = 30 \times 10^6$ lb/in.^{2} for steel
   (b) $E = 10 \times 10^6$ lb/in.^{2} for aluminum
2. I_a = *area moment of inertia of cross section*
   (a) Rectangular cross section:

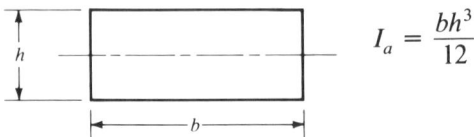

DEFLECTION OF BEAMS

(b) Circular cross section:

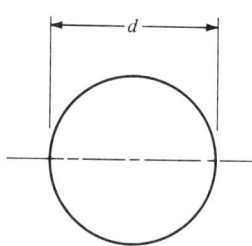

$$I_a = \frac{\pi d^4}{64}$$

(c) Other cross sections:

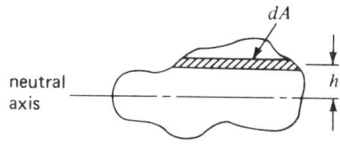

$$I_a = \int h^2 \, dA$$

3. *y = deflection at point of applied load F, in direction of applied load*

Equations

1. Cantilever beam:

$$y = \frac{Fl^3}{3EI_a}$$

2. Simply supported beam; load at center:

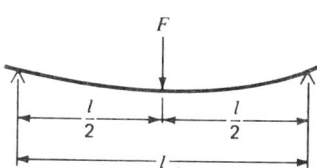

$$y = \frac{Fl^3}{48EI_a}$$

3. Simply supported beam; load at any point:

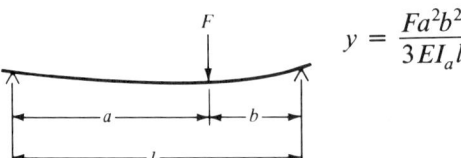

$$y = \frac{Fa^2b^2}{3EI_a l}$$

4. Simply supported beam; overhanging load:

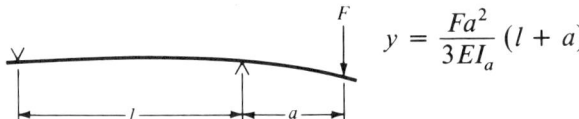

$$y = \frac{Fa^2}{3EI_a}(l + a)$$

5. Beam with one end fixed and the other simply supported; load at center:

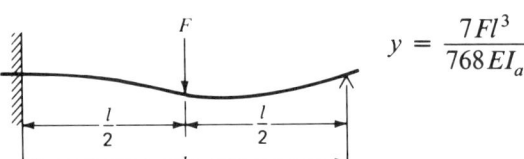

$$y = \frac{7Fl^3}{768EI_a}$$

6. Beam with one end fixed and the other simply supported; load at any point:

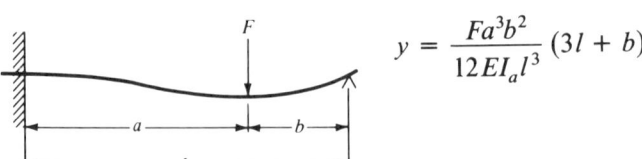

$$y = \frac{Fa^3b^2}{12EI_a l^3}(3l + b)$$

DEFLECTION OF BEAMS

7. Beam fixed at both ends; load at center:

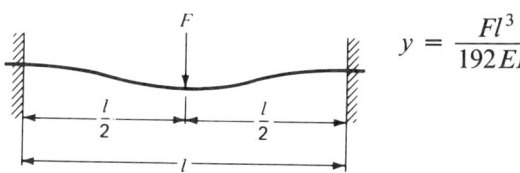

$$y = \frac{Fl^3}{192EI_a}$$

8. Beam fixed at both ends; load at any point:

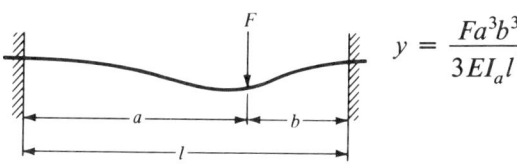

$$y = \frac{Fa^3b^3}{3EI_a l^3}$$

Appendix B

Alternative Mathematical Expressions for Certain Harmonic Functions

When the roots of the characteristic equation of a second-order system are complex conjugates, the solution is

$$y = A_1 e^{(a+jb)t} + A_2 e^{(a-jb)t}$$
$$= e^{at}\left(A_1 e^{jbt} + A_2 e^{-jbt}\right) \quad \text{(B-1)}$$

To determine A_1 and A_2, initial conditions must be considered (see Section 3-3). It is possible to write

$$y_{t=0} = y_0 \quad \text{(B-2)}$$

$$\left.\frac{dy}{dt}\right|_{t=0} = \dot{y}_0 \quad \text{(B-3)}$$

to describe any combination of initial conditions. Differentiation of the solution (Eq. B-1) yields

$$\frac{dy}{dt} = ae^{at}\left(A_1 e^{jbt} + A_2 e^{-jbt}\right) + e^{at}\left(A_1 jb e^{jbt} - A_2 jb e^{-jbt}\right) \quad \text{(B-4)}$$

With $t = 0$, substitution of Eq. B-2 into Eq. B-1 yields

$$A_1 + A_2 = y_0 \quad \text{(B-5)}$$

and substitution of Eq. B-3 into Eq. B-4 yields

$$a(A_1 + A_2) + jb(A_1 - A_2) = \dot{y}_0 \quad \text{(B-6)}$$

ALTERNATIVE MATHEMATICAL EXPRESSIONS FOR CERTAIN HARMONIC FUNCTIONS

Simultaneous solution of Eqs. B-5 and B-6 produces the expressions

$$A_1 = \frac{y_0}{2} - \frac{j(\dot{y}_0 - ay_0)}{2b} \quad \text{(B-7)}$$

$$A_2 = \frac{y_0}{2} + \frac{j(\dot{y}_0 - ay_0)}{2b} \quad \text{(B-8)}$$

Note that A_1 and A_2 are therefore complex conjugates for any set of initial conditions (i.e., for any values of y_0 and $\dot{y}_0$).

The solution (Eq. B-1) may be manipulated to an alternative form by use of the Euler formula

$$e^{jbt} = \cos bt + j \sin bt \quad \text{(B-9)}$$

Substitution of Eq. B-9 into Eq. B-1 yields

$$y = e^{at}\left[(A_1 + A_2)\cos bt + (A_1 - A_2)j \sin bt\right] \quad \text{(B-10)}$$

By writing Eqs. B-7 and B-8 in the simplified forms

$$A_1 = \alpha + j\beta \quad \text{(B-11)}$$

$$A_2 = \alpha - j\beta \quad \text{(B-12)}$$

and substituting them into Eq. B-10, we obtain

$$y = e^{at}\left[(\alpha + j\beta + \alpha - j\beta)\cos bt + (\alpha + j\beta - \alpha + j\beta)j \sin bt\right]$$
$$= e^{at}\left[2\alpha \cos bt + 2j\beta(j \sin bt)\right] \quad \text{(B-13)}$$

or

$$y = e^{at}(C_1 \sin bt + C_2 \cos bt) \quad \text{(B-14)}$$

where $C_1 = -2\beta$ and $C_2 = 2\alpha$. Note that constants C_1 and C_2 are both real numbers.

Equation B-14 shows that Eq. B-1 may alternatively be written as a damped harmonic function. It will now be shown that the expression in parentheses in Eq. B-14, containing a sine and a cosine term, can be replaced by the single harmonic term $C_3 \sin(bt + \phi)$. This will be done by the use of vectors.

In Fig. B-1a the two vectors of lengths C_1 and C_2 are 90° apart. With the C_1 vector positioned at the angle bt from the horizontal axis as shown, its projection on the vertical axis is $C_1 \sin bt$, and the projection of the C_2 vector on the vertical axis is $C_2 \cos bt$. The expression in parentheses in Eq. B-14 can therefore be represented by the sum of these two vertical projections.

If the vectors C_1 and C_2 are added, as in Fig. B-1b, the vertical projection of their vector sum C_3 is seen to be given by two equivalent

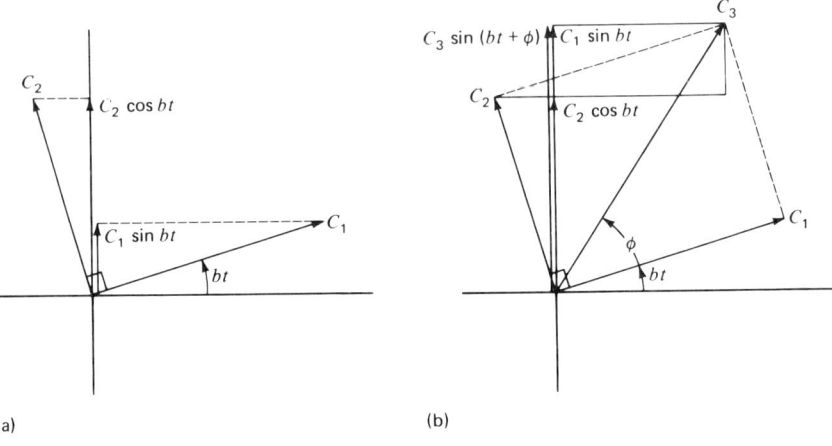

FIG B-1

expressions: (1) $C_1 \sin bt + C_2 \cos bt$ and (2) $C_3 \sin(bt + \phi)$. Substitution of the second expression allows Eq. B-14 to be written

$$y = C_3 e^{at} \sin(bt + \phi) \tag{B-15}$$

The length of C_3 is, by the Pythagorean theorem,

$$C_3 = \left(C_1^2 + C_2^2\right)^{1/2} \tag{B-16}$$

The angle ϕ is seen, by basic trigonometry, to be

$$\phi = \tan^{-1} \frac{C_2}{C_1} \tag{B-17}$$

Appendix C

Solution of Equations

In control work it is sometimes necessary to determine the roots of an equation or to factor a polynomial. Since the approach is the same in either case, only root solving will be discussed here.

Consider the general equation

$$a_n p^n + a_{n-1} p^{n-1} + \cdots + a_1 p + a_0 = 0 \tag{C-1}$$

in which the coefficients are real numbers. The coefficient of p^n can be made unity by division:

$$p^n + \frac{a_{n-1}}{a_n} p^{n-1} + \cdots + \frac{a_1}{a_n} p + \frac{a_0}{a_n} = 0$$

or

$$p^n + b_{n-1} p^{n-1} + \cdots + b_1 p + b_0 = 0 \tag{C-2}$$

Several observations can be made regarding the roots of Eq. C-2:

1. The number of roots is equal to the degree of the equation n.
2. The roots may be real or complex. However, complex roots occur in pairs. This means that if n is odd, there must be at least one real root.
3. The roots r are related to certain coefficients by the expressions

$$r_1 + r_2 + \cdots + r_n = -b_{n-1} \tag{C-3}$$

and

$$r_1 r_2 \cdots r_n = (-1)^n b_0 \tag{C-4}$$

There are literally dozens of methods for root solving. One popular way of seeking real roots is by using synthetic division. This approach will be illustrated by an example. Suppose that the roots of the equation

$$f(p) = p^3 + 5p^2 + 11p + 15 = 0$$

are required. The first step is to assume a real root, say -1 (an integer for simplicity). Then the coefficients of the equation are written down. The steps are as follows: Multiply the first coefficient by -1, add the result to the second coefficient, multiply the sum by -1 and add to the third coefficient, and so forth. Thus, we obtain

```
   1      +5      +11     +15   | -1
          -1      - 4     - 7
   1      +4      + 7     + 8
```

The value of $f(-1)$ is 8; therefore, -1 is not a root of the equation. Try -4.

```
   1      +5      +11     +15   | -4
          -4      - 4     -28
   1      +1      + 7     -13
```

Thus, $f(-4) = -13$. Not only is -4 not a root, but the sign change [remember that $f(-1) = +8$] indicates the presence of a root between -1 and -4. Next, try -3.

```
   1      +5      +11     +15   | -3
          -3      - 6     -15
   1      +2      + 5      0
```

The value of $f(-3)$ is zero, and so -3 is a root. Once a real root has been found by synthetic division, the complete solution can easily be obtained. Division of $f(p)$ by the factor $(p + 3)$ yields $p^2 + 2p + 5$, from which $p = -1 \pm j2$. Therefore, the three roots are $r_1 = -3$, $r_2 = -1 + j2$, $r_3 = -1 - j2$. Equations C-3 and C-4 can be used as a check. From Eq. C-3,

$$-3 - 1 + j2 - 1 - j2 = -5$$
$$-5 = -5$$

From Eq. C-4,

$$-3(-1 + j2)(-1 - j2) = (-1)^3 15$$
$$-15 = -15$$

A popular approach to solving equations in which n is even and greater than 2 is known as Lin's method [10].[1] The method involves

---

[1] It should be pointed out that Lin's method is not restricted to equations in which n is even. However, in this discussion it is assumed that at least one real root has been determined by synthetic division if n is odd.

SOLUTION OF EQUATIONS

repeated division by assumed quadratic factors until the remainder is sufficiently small. In this manner a polynomial can be factored into quadratic factors from which the roots of the equation can easily be obtained. Lin's method is best illustrated by an example. Suppose that the roots of the equation

$$f(p) = p^4 + 6p^3 + 18p^2 + 30p + 25 = 0$$

are to be determined. In using Lin's method, the polynomial $f(p)$ is divided by an assumed quadratic factor in an attempt to express $f(p)$ as a product of two quadratic factors. The last three terms of $f(p)$ are used to obtain the first trial divisor.

$$p^2 + \frac{30}{18}p + \frac{25}{18} = p^2 + 1.7p + 1.4$$

Division yields

$$\begin{array}{r}
p^2 + 4.3p + 9.3 \\
p^2 + 1.7p + 1.4 \overline{\smash{\big)}\, p^4 + 6.0p^3 + 18.0p^2 + 30.0p + 25.0}\\
p^4 + 1.7p^3 + 1.4p^2 \\
\hline
4.3p^3 + 16.6p^2 + 30.0p \\
4.3p^3 + 7.3p^2 + 6.0p \\
\hline
(9.3p^2 + 24.0p + 25.0)\\
9.3p^2 + 15.8p + 13.0\\
\hline
8.2p + 12.0
\end{array}$$

The remainder is somewhat large, so a second trial divisor is obtained from the expression enclosed in parentheses in the foregoing step. It is

$$p^2 + \frac{24.0}{9.3}p + \frac{25.0}{9.3} = p^2 + 2.6p + 2.7$$

Division with this assumed factor produces a remainder $3.9p + 7.4$, which is smaller than the previous remainder. The process can be continued by using a third trial divisor determined in the same manner. The remainder is then $2.8p + 5.2$. After eight trials, the remainder is $-0.12p - 0.10$, which is reasonably small. (In later trials, after the remainder became small, a second decimal place was carried. The procedure of carrying progressively more decimal places as the remainder gets smaller results in less computational labor without sacrificing accuracy.) The resulting factors are $(p^2 + 4.01p + 4.98)(p^2 + 1.99p + 5.04)$, which are close to the actual factors $(p^2 + 4p + 5)(p^2 + 2p + 5)$. The roots of the equation can easily be obtained, since each factor can be equated to zero.

Appendix D

Steady-State Solutions by Rotating-Vector and Complex-Number Techniques

Introduction

The main text has been based on a single unified approach for obtaining solutions. The differential equations that represent the various dynamic systems are solved in each case by applying the single classical method of solution presented in Chapter 3. The use of a single method of solution gives the advantage of coherency and procedural simplicity. The student does not have to consider which method of solution is proper for which type of problem and thus is able to spend more time on the various types of dynamic systems and their characteristics.

For certain types of problems, however, the classical approach of Chapter 3 can become somewhat laborious even though its application is straightforward. For the study of steady-state system response to harmonic inputs another method, the *method of complex numbers*, allows solutions to be obtained with somewhat less labor. Because of this advantage, the method is widely employed, particularly in the study of mechanical vibrating systems.

The complex number method does have limitations. It is usable only for steady-state response and only where the input is a harmonic signal. But since this case is so common and important, the method is quite valuable as a means of saving computation labor.

This appendix is devoted to the presentation of the complex-number method for obtaining steady-state response. As part of the presentation, the similar *rotating-vector method* is used as an aid in explaining the vector techniques involved. The appendix is designed to be used as a supplement to Chapter 7 and can be omitted if desired without affecting the student's ability to understand the basic text.

Rotating-Vector Technique

The basic principle of representing harmonic motion by a rotating vector was presented in Section 5-2. Included in that presentation was the use of two vectors rotating at the same speed while separated by a fixed phase angle ϕ to represent two harmonic functions of a common frequency with the phase difference ϕ.

The rotating-vector concept can be extended to serve as an aid in the solution of a forced vibration (or other steady-state response) problem. As an example we will present here an alternative solution, based on rotating vectors, to the general second-order equation (Eq. 7-5), which is

$$\frac{1}{\omega_n^2}\frac{d^2y}{dt^2} + \frac{2\zeta}{\omega_n}\frac{dy}{dt} + y = K_1 \sin \omega t \tag{D-1}$$

For this method, we assume the solution

$$y_{ss} = Y \sin(\omega t + \phi) \tag{D-2}$$

which has the derivatives

$$\frac{dy_{ss}}{dt} = Y\omega \cos(\omega t + \phi) \tag{D-3}$$

$$\frac{d^2y_{ss}}{dt^2} = -Y\omega^2 \sin(\omega t + \phi) \tag{D-4}$$

Substitution into Eq. D-1 yields

$$Y\left[-\frac{\omega^2}{\omega_n^2}\sin(\omega t + \phi) + 2\zeta\frac{\omega}{\omega_n}\cos(\omega t + \phi) + \sin(\omega t + \phi)\right] = K_1 \sin \omega t \tag{D-5}$$

In order for each term in Eq. D-5 to be equal to the *horizontal projection* of its rotating vector, it is necessary to change all the sine terms to cosine terms. To do so, we use the basic trigonometric relationship

$$\sin \alpha = \cos\left(\alpha - \frac{\pi}{2}\right) \tag{D-6}$$

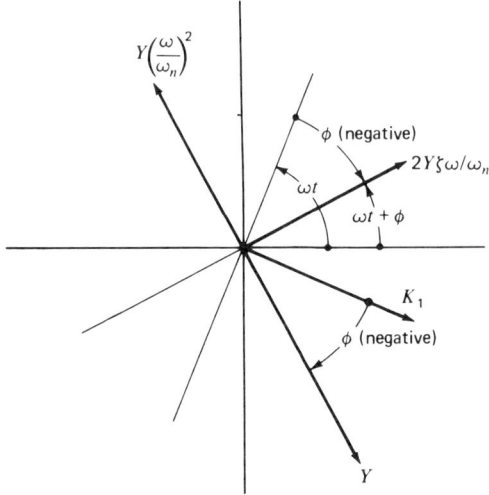

FIG D-1. The set of rotating vectors that represents Eq. D-7.

Application of Eq. D-6 to Eq. D-5 results in

$$Y\left[-\left(\frac{\omega}{\omega_n}\right)^2 \cos\left(\omega t + \phi - \frac{\pi}{2}\right) + 2\zeta\left(\frac{\omega}{\omega_n}\right)\cos(\omega t + \phi)\right.$$

$$\left. + \cos\left(\omega t + \phi - \frac{\pi}{2}\right)\right] = K_1 \cos\left(\omega t - \frac{\pi}{2}\right) \quad \text{(D-7)}$$

Equation D-7 is, of course, satisfied when Y and ϕ are such that the left-hand side is equal to the right-hand side. If we represent each term in Eq. D-7 by a rotating vector, as illustrated in Fig. D-1,[1] then we can also say that the equation is satisfied when the algebraic sum of the horizontal projections of the vectors for the terms on the left-hand side of Eq. D-7 is equal to the horizontal projection of the vector for the right-hand term, $K_1 \cos(\omega t - \pi/2)$. Going one step further, we know that the above horizontal projection requirement can be achieved *for all values of ωt* only if the *vector sum* of the vectors representing the left-hand terms is equal to the vector for $K_1 \cos(\omega t - \pi/2)$.

---

[1] Figure D-1 has been drawn with a negative value of ϕ to correspond to reality. The phase angle ϕ will be found to be negative whenever the damping ratio ζ is positive, which is always true for a mechanical vibrating system. For other types of second-order systems ζ may be negative (see Chapter 14), but this results in an unstable system that has no steady-state solution.

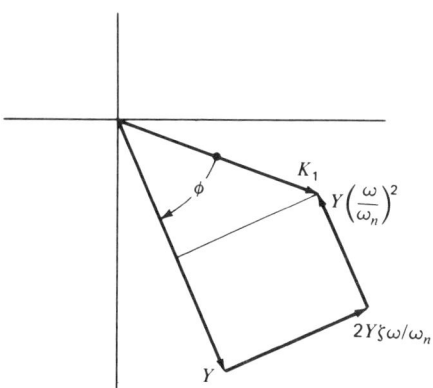

FIG D-2. Vector polygon drawn to aid in the solution of Eq. D-7 by the rotating-vector method.

The solution to Eq. D-7 therefore can be accomplished through the solution of a vector polygon problem. To aid in the development of the solution, the vectors of Fig. D-1 are redrawn as a vector polygon in Fig. D-2. In order for Eq. D-7 to be satisfied, the polygon must be closed. With this constraint the resulting trigonometric equations allow us to solve for the two unknowns, Y and ϕ.

In Fig. D-2, the Pythagorean theorem applied to the right triangle yields

$$K_1^2 = \left[Y - Y\left(\frac{\omega}{\omega_n}\right)^2\right]^2 + \left(2Y\zeta\frac{\omega}{\omega_n}\right)^2$$

Solving for the amplitude ratio,

$$\frac{Y}{K_1} = \frac{1}{\sqrt{\left[1 - (\omega/\omega_n)^2\right]^2 + 4\zeta^2(\omega/\omega_n)^2}} \qquad \text{(D-8)}$$

The phase angle is found by the basic trigonometric relationship

$$\phi = -\tan^{-1}\frac{2\zeta\omega/\omega_n}{1 - (\omega/\omega_n)^2} \qquad \text{(D-9)}$$

with the negative sign occurring because ϕ is a negative angle.

Equations D-8 and D-9 are identical to those derived in Chapter 7 (Eqs. 7-7 and 7-8) by direct mathematical solution of the differential equations.

The solution of this problem has been simplified somewhat by the use of the vector representation, and this tends to be true in general for the

Complex-Number Method

The complex-number method of solving for the steady-state response to a sinusoidal input is based on the same fundamental concept as the rotating-vector method; that is, a rotating vector is also used to represent each harmonic function, and the horizontal projection gives the actual value of the function. The main difference is that the complex-number method uses a vector that rotates in a *complex plane* (Fig. D-3). A complex plane has one real coordinate axis (the horizontal axis) and one imaginary axis (the vertical one). Any point on the plane represents a complex number $\alpha + j\beta$ (where $j = \sqrt{-1}$) with α the horizontal distance and β the vertical distance from the origin.

The vectors and horizontal projections appear the same whether the plane is complex or not. The reason for using the complex plane is that the unique characteristics of complex numbers allow solutions to be obtained by relatively simple mathematical techniques without the necessity of drawing and analyzing vector polygons.

A cosine function

$$y = Y \cos \omega t \tag{D-10}$$

is represented by a vector of length Y rotating counterclockwise at ω radians per second on the complex plane and positioned at the angle ωt

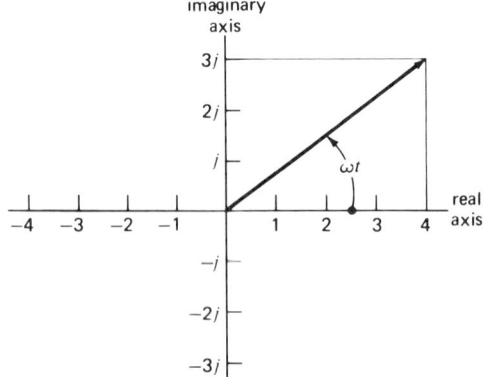

FIG D-3. Representation of the harmonic function $y = 5 \cos \omega t$ by the vector $\bar{y} = Y(\cos \omega t + j \sin \omega t) = Y e^{j\omega t}$ rotating counterclockwise in the complex plane.

COMPLEX-NUMBER METHOD 391

from the real (horizontal) axis (see Fig. D-3). The vector may in turn be represented by the complex number corresponding to its vertical and horizontal components,

$$\bar{y} = Y(\cos \omega t + j \sin \omega t) \tag{D-11}$$

A comparison of Eqs. D-10 and D-11 shows that the true harmonic function in this case is the *real component* of the vector (i.e., the real part of Eq. D-11), which corresponds to the horizontal projection of the vector.

All harmonic functions should be written as cosine functions (e.g., $\sin \omega t$ should be changed to $\cos(\omega t - \pi/2)$ so that the real part of the complex number representing its vector will be the true function.

The work involved in the solution is also simplified by use of the Euler formula [9]

$$e^{j\theta} = \cos \theta + j \sin \theta \tag{D-12}$$

which allows each rotating vector to be represented by a single exponential term.

The technique used to obtain the solution to a particular problem can be summarized as follows: Each term of the differential equation is assumed to be a vector rotating counterclockwise on the complex plane and is represented by an exponential term (e.g., $x = X \sin \omega t$ would be represented by the vector $Xe^{j(\omega t - \pi/2)}$). The resultant complex-number equation is solved by a combination of algebraic and vectorial methods. The real part of that solution is the true solution of the differential equation.

The student should not attempt to attach any physical significance to the complex-number representation. It is simply a mathematical tool that has been found to produce a solution with a minimum of labor.

In order to illustrate the details of the complex-number method, it will now be used to obtain the solution to the general second-order equation

$$\frac{1}{\omega_n^2} \frac{d^2 y}{dt^2} + \frac{2\zeta}{\omega_n} \frac{dy}{dt} + y = K_1 \sin \omega t \tag{D-13}$$

Equation D-13 is identical to Eqs. 7-5 and D-1, and its solution by the complex-number method can be directly compared to the two other solutions already accomplished (i.e., the direct solution of the differential equation and solution by the ordinary rotating-vector method).

As the first step we assume the solution

$$y_{ss} = Y \sin(\omega t + \phi) \tag{D-14}$$

This is then rewritten as the equivalent cosine function

$$y_{ss} = Y \cos\left(\omega t + \phi - \frac{\pi}{2}\right)$$

which is represented as a rotating vector in the complex plane by the equation

$$\bar{y} = Y\left[\cos\left(\omega t + \phi - \frac{\pi}{2}\right) + j\sin\left(\omega t + \phi - \frac{\pi}{2}\right)\right]$$

Application of the Euler formula then yields

$$y = Ye^{j(\omega t + \phi - \pi/2)} \tag{D-15}$$

Equation D-15 is differentiated twice to obtain

$$\frac{dy}{dt} = j\omega Ye^{j(\omega t + \phi - \pi/2)} \tag{D-16}$$

$$\frac{d^2y}{dt^2} = -\omega^2 Ye^{j(\omega t + \phi - \pi/2)} \tag{D-17}$$

Substitution of the above into Eq. D-13, with the forcing function also changed to vector representation, yields

$$\left[-\left(\frac{\omega}{\omega_n}\right)^2 + j2\zeta\frac{\omega}{\omega_n} + 1\right]Ye^{j(\omega t + \phi - \pi/2)} = K_1 e^{j(\omega t - \pi/2)}$$

which reduces to

$$\left[-\left(\frac{\omega}{\omega_n}\right)^2 + j2\zeta\frac{\omega}{\omega_n} + 1\right]Ye^{j(\omega t + \phi)} = K_1 e^{j\omega t} \tag{D-18}$$

Equation D-18 must still be considered a vector equation. The term in brackets, with real and imaginary components, is a vector on the complex plane. It can be represented by the alternative expression

$$\left[1 - \left(\frac{\omega}{\omega_n}\right)^2 + j2\zeta\frac{\omega}{\omega_n}\right] = Ce^{j\psi} \tag{D-19}$$

with

$$C = \sqrt{\left[1 - \left(\frac{\omega}{\omega_n}\right)^2\right]^2 + \left(2\zeta\frac{\omega}{\omega_n}\right)^2} \tag{D-20}$$

by the Pythagorean theorem, and

$$\psi = \tan^{-1}\frac{2\zeta\omega/\omega_n}{1 - (\omega/\omega_n)^2} \tag{D-21}$$

Combining Eqs. D-18 and D-19 yields

$$YCe^{j(\omega t + \phi)}e^{j\psi} = K_1 e^{j\omega t}$$

or

$$Ye^{j(\phi + \psi)} = \frac{K_1}{C} \tag{D-22}$$

CONCLUSION

Since the right-hand side of Eq. D-22 is real, the left-hand side must also be real, which will be true only if

$$\phi = -\psi \qquad (D\text{-}23)$$

and therefore

$$Y = \frac{K_1}{C} \qquad (D\text{-}24)$$

The final solution is now obtained by combining Eqs. D-20, D-21, D-23, and D-24. The resulting amplitude ratio is

$$\frac{Y}{K_1} = \frac{1}{\sqrt{\left[1 - (\omega/\omega_n)^2\right]^2 + 4\zeta^2(\omega/\omega_n)^2}} \qquad (D\text{-}25)$$

and the phase angle is

$$\phi = -\tan^{-1} \frac{2\zeta\omega/\omega_n}{1 - (\omega/\omega_n)^2} \qquad (D\text{-}26)$$

As expected, Eqs. D-25 and D-26 are the same as the solutions obtained previously by the other two methods of solution (Eqs. 7-7, 7-8, D-8, and D-9).

It should again be pointed out that even though the complex-number method is based on rotating vectors, it is not necessary to draw a vector polygon since the complex-number representation gives the vertical and horizontal components in each case.

Conclusion

The general complex-number approach illustrated by the example above can be used to obtain the steady-state response of a wide range of linear dynamic systems with harmonic forcing functions. Its use can often save considerable labor over the classical method of obtaining the particular integral, especially for more complex systems.

Appendix E

Rayleigh's Energy Method

Rayleigh's energy method can be used to determine the fundamental natural frequency of transverse vibration (or the first critical speed) of a lumped parameter beam or rotor with two or more masses. The method is based on the principle that in a conservative vibrating system (one with no damping) the maximum kinetic energy that occurs each cycle (when all masses are simultaneously at their maximum velocity with zero displacement) is equal to the maximum potential energy that occurs each cycle (when all the masses simultaneously have zero velocity but maximum displacement).

With the beam acting as a linear spring, so that the force at each mass is proportional to its displacement, the maximum potential energy is

$$PE = \frac{F_1 Y_1}{2} + \frac{F_2 Y_2}{2} + \cdots \qquad (\text{E-1})$$

where

F_1 = transverse force on M_1 required to hold it at displacement Y_1

F_2 = transverse force on M_2 required to hold it at displacement Y_2

With the assumption that all masses move with harmonic motion at the fundamental natural frequency ω_n, we can write $y_1 = Y_1 \sin \omega_n t$, $y_2 = Y_2 \sin \omega_n t$, and so on. Since the corresponding maximum velocities are $Y_1 \omega_n$, $Y_2 \omega_n$, and so on, the maximum kinetic energy of the system is

$$KE = \frac{\omega_n^2}{2} \left[M_1 Y_1^2 + M_2 Y_2^2 + \cdots \right] \qquad (\text{E-2})$$

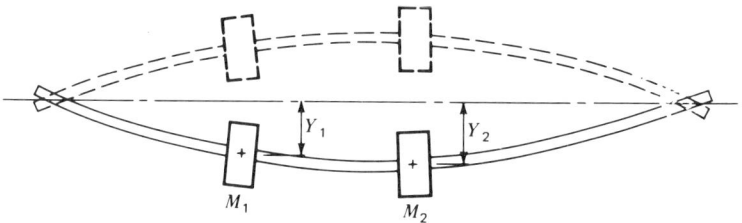

FIG E-1. Rotor in transverse vibration at lowest natural frequency, shown at point of maximum amplitude.

In order to make use of these two equations, however, we must use the assumption (which is in actuality not precisely correct) that the maximum displacements $Y_1, Y_2, \ldots$ that occur with free undamped lateral vibration are proportional to the static deflections that would occur from gravity if the beam were mounted horizontally; that is,

$$\frac{Y_1}{\delta_1} = \frac{Y_2}{\delta_2} = \cdots \tag{E-3}$$

with the δ's being static deflections of the masses due to gravity (see Figs. E-1 and E-2).

With the beam acting as a linear spring, we may write

$$\frac{F_1}{Y_1} = \frac{M_1 g}{\delta_1}, \quad \frac{F_2}{Y_2} = \frac{M_2 g}{\delta_2}, \quad \cdots \tag{E-4}$$

By equating the expression for maximum potential energy (Eq. E-1) to the expression for maximum kinetic energy (Eq. E-2) and using the relationship of Eq. E-4, we obtain

$$\omega_n^2 = g \left[\frac{M_1 \delta_1 + M_2 \delta_2 + \cdots}{M_1 \delta_1^2 + M_2 \delta_2^2 + \cdots} \right] \tag{E-5}$$

In order to apply this equation, however, it is necessary to determine the static deflections δ_1, δ_2, and so on. To explain how this is done, we will use the example of a rotor having two masses mounted on a shaft of uniform

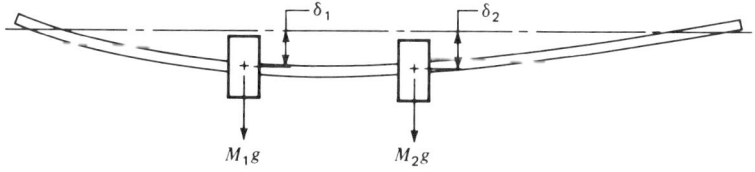

FIG E-2. Rotor mounted horizontally, with deflection from force of gravity alone.

cross section. The deflection δ_1 is due not only to the weight of M_1 but also to that of M_2, so it must be calculated in two parts, using the principle of superposition.

A simply supported beam with a load F at any point will have a deflection δ at that point (see Fig. E-3 for nomenclature)

$$\delta = \frac{Fa^2b^2}{3EI_a l} \qquad (E\text{-}6)$$

At a position x (with $x < a$), the deflection due to the load F is

$$\delta = \frac{Fbx}{6EI_a l}(l^2 - x^2 - b^2) \qquad (E\text{-}7)$$

With $x > a$, the corresponding equation for the deflection at x is

$$\delta = \frac{Faz}{6EI_a l}(l^2 - z^2 - a^2) \qquad (E\text{-}8)$$

To get the total deflection δ_1 at M_1, we apply Eq. E-6 to get the component of deflection due to the weight of M_1, apply Eq. E-7 to get the component due to the weight of M_2, and add the two. After obtaining the δ's for the other masses by similar sets of calculations, we have sufficient information to solve Eq. E-5 for ω_n, the lowest natural frequency of the system.

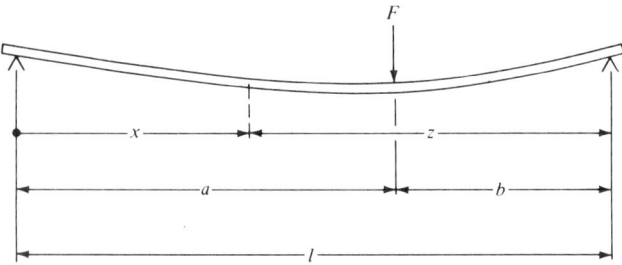

FIG E-3. Simply supported beam of uniform cross-section.

Deflection at F:
$$\delta = \frac{Fa^2b^2}{3EI_a l}$$

Deflection at x, with $x < a$:
$$\delta = \frac{Fbx}{6EI_a l}(l^2 - x^2 - b^2)$$

Deflection at x, with $x > a$:
$$\delta = \frac{Faz}{6EI_a l}(l^2 - z^2 - a^2)$$

Appendix F

Decibel Conversion Table ($m = 20 \log_{10} M$)

M	0	1	2	3	4	5	6	7	8	9
0.0	$m =$	−40.00	−33.98	−30.46	−27.96	−26.02	−24.44	−23.10	−21.94	−20.92
0.1	−20.00	−19.17	−18.42	−17.72	−17.08	−16.48	−15.92	−15.39	−14.89	−14.42
0.2	−13.98	−13.56	−13.15	−12.77	−12.40	−12.04	−11.70	−11.37	−11.06	−10.75
0.3	−10.46	−10.17	−9.90	−9.63	−9.37	−9.12	−8.87	−8.64	−8.40	−8.18
0.4	−7.96	−7.74	−7.54	−7.33	−7.13	−6.94	−6.74	−6.56	−6.38	−6.20
0.5	−6.02	−5.85	−5.68	−5.51	−5.35	−5.19	−5.04	−4.88	−4.73	−4.58
0.6	−4.44	−4.29	−4.15	−4.01	−3.88	−3.74	−3.61	−3.48	−3.35	−3.22
0.7	−3.10	−2.97	−2.85	−2.73	−2.62	−2.50	−2.38	−2.27	−2.16	−2.05
0.8	−1.94	−1.83	−1.72	−1.62	−1.51	−1.41	−1.31	−1.21	−1.11	−1.01
0.9	−0.92	−0.82	−0.72	−0.63	−0.54	−0.45	−0.35	−0.26	−0.18	−0.09
1.0	0.00	0.09	0.17	0.26	0.34	0.42	0.51	0.59	0.67	0.75
1.1	0.83	0.91	0.98	1.06	1.14	1.21	1.29	1.36	1.44	1.51
1.2	1.58	1.66	1.73	1.80	1.87	1.94	2.01	2.08	2.14	2.21
1.3	2.28	2.35	2.41	2.48	2.54	2.61	2.67	2.73	2.80	2.86
1.4	2.92	2.98	3.05	3.11	3.17	3.23	3.29	3.35	3.41	3.46
1.5	3.52	3.58	3.64	3.69	3.75	3.81	3.86	3.92	3.97	4.03
1.6	4.08	4.14	4.19	4.24	4.30	4.35	4.40	4.45	4.51	4.56
1.7	4.61	4.66	4.71	4.76	4.81	4.86	4.91	4.96	5.01	5.06
1.8	5.11	5.15	5.20	5.25	5.30	5.34	5.39	5.44	5.48	5.53
1.9	5.58	5.62	5.67	5.71	5.76	5.80	5.85	5.89	5.93	5.98
2.	6.02	6.44	6.85	7.23	7.60	7.96	8.30	8.63	8.94	9.25
3.	9.54	9.83	10.10	10.37	10.63	10.88	11.13	11.36	11.60	11.82
4.	12.04	12.26	12.46	12.67	12.87	13.06	13.26	13.44	13.62	13.80
5.	13.98	14.15	14.32	14.49	14.65	14.81	14.96	15.12	15.27	15.42
6.	15.56	15.71	15.85	15.99	16.12	16.26	16.39	16.52	16.65	16.78
7.	16.90	17.03	17.15	17.27	17.38	17.50	17.62	17.73	17.84	17.95
8.	18.06	18.17	18.28	18.38	18.49	18.59	18.69	18.79	18.89	18.99
9.	19.08	19.18	19.28	19.37	19.46	19.55	19.65	19.74	19.82	19.91
	0.	1.	2.	3.	4.	5.	6.	7.	8.	9.

NOTE: Table is easily extended. If $M' = M(10)^n$, $m' = m + 20n$. As an example: If $M' = 80 = 8(10)$, $m' = 18.06 + 20 = 38.06$.

Appendix G

Routh's Criterion

In 1877 E. J. Routh developed a method for determining whether or not an equation has roots with positive real parts without actually solving for the roots. When used with the characteristic equation of a control system, Routh's criterion offers a simple means for detecting system instability, because roots with positive real parts indicate transients that increase, rather than decay, with time.

The usual characteristic equation can be written in general form as

$$a_0 p^n + a_1 p^{n-1} + \cdots + a_{n-1} p + a_n = 0$$

in which the coefficients typically are positive numbers. An unstable system, or one with limited stability (roots with zero real parts), is immediately indicated if any coefficient is either zero or negative.

The first step in applying Routh's criterion is to form the Routh array. For simplicity of illustration, a characteristic equation with $n = 5$ will be used.

$$a_0 p^5 + a_1 p^4 + a_2 p^3 + a_3 p^2 + a_4 p + a_5 = 0$$

The array is

$$
\begin{array}{ccc}
a_0 & a_2 & a_4 \\
a_1 & a_3 & a_5 \\
b_1 & b_3 & \\
c_1 & c_3 & \\
d_1 & & \\
e_1 & &
\end{array}
$$

ROUTH'S CRITERION

The first two rows are composed of the coefficients of the characteristic equation. The remaining entries are obtained by computation with the following pattern:

$$b_1 = \frac{\begin{vmatrix} a_0 & a_2 \\ a_1 & a_3 \end{vmatrix}}{a_1} = \frac{a_1 a_2 - a_0 a_3}{a_1}$$

$$b_3 = \frac{\begin{vmatrix} a_0 & a_4 \\ a_1 & a_5 \end{vmatrix}}{a_1} = \frac{a_1 a_4 - a_0 a_5}{a_1}$$

Similarly,

$$c_1 = \frac{b_1 a_3 - a_1 b_3}{b_1} \qquad c_3 = \frac{b_1 a_5 - a_1 (0)}{b_1}$$

Entries d_1 and e_1 are determined in a similar manner.

Routh's criterion states that the number of sign changes in the first column of the array is equal to the number of roots with positive real parts.

Consider, as an example, the characteristic equation

$$p^5 + 3p^4 + 7p^3 + 20p^2 + 6p + 15 = 0$$

The Routh array is

1	7	6
3	20	15
1/3	1	
11	15	
6/11		
15		

The system in question is stable. There are no sign changes in the first column of the array, so there are no roots with positive real parts. However, the Routh criterion gives no indication of the degree of stability.

Next, consider the characteristic equation

$$p^4 + 2p^3 + 3p^2 + 8p + 2 = 0$$

The Routh array is

1	3	2
2	8	
−1	2	
12		
2		

There are two sign changes (plus to minus and minus to plus) in the first column, showing that there are two roots with positive real parts. Therefore, the system is unstable.

It can be shown that all the entries in any row can be multiplied or divided by a constant without affecting the sign changes in the first column. This procedure can be used to reduce the mathematical labor involved in forming the array.

Two special cases are worthy of note. Should the first entry in a row be zero while the other entries are not zero, the procedure is to replace the zero with a small positive number ϵ. The array can then be continued. Sign changes in the first column can be ascertained by letting ϵ approach zero. If a row of zeros occurs, the system is either unstable or possesses limited stability; it is *not* stable.

Bibliography

1. Andrewartha, H. G., and Birch, L. C. *Distribution and Abundance of Animals*. Chicago: University of Chicago Press, 1954.
2. Bollinger, J. G., and Bonesho, J. "How to Design a Self-Optimizing Vibration Damper." *Machine Design*, Feb. 1968. Also, U.S. Patent No. 3,483,951.
3. Burington, R. S. *Handbook of Mathematical Tables and Formulas*, 4th ed. New York: McGraw-Hill, 1965.
4. Den Hartog, J. P. *Mechanical Vibrations*, 4th ed. New York: McGraw-Hill, 1956.
5. Eckman, D. P. *Automatic Process Control*, New York: Wiley, 1958.
6. Gazis, D. C. "Traffic Flow and Control: Theory and Applications." *American Scientist*, vol. 60, no. 4, July–Aug. 1972, pp. 414–424.
7. Harrison, H. L., and Bollinger, J. G., *Introduction to Automatic Controls*, 2nd ed. Scranton, Pa.: International Textbook, 1969.
8. Harrison, H. L., Loucks, O. L., Mitchell, J. W., Parkhurst, D. F., Tracy, C. R., Watts, D. G., and Yannacone, V. J., Jr. "Systems Studies of D.D.T. Transport." *Science*, vol. 170, Oct. 30, 1970, pp. 503–508.
9. Kreysig, E. *Advanced Engineering Mathematics*, 2nd ed. New York: Wiley, 1967.
10. Lin, Shih-Nge. "A Method of Successive Approximations of Evaluating the Real and Complex Roots of Cubic and Higher Order Equations," *Journal of Mathematics and Physics*, vol. 20, 1941, pp. 231–242.
11. McCormick, J. M., and Salvadori, M. G. *Numerical Methods in FORTRAN*. Englewood Cliffs, N.J.: Prentice-Hall, 1965.
12. Merritt, H. E. *Hydraulic Control Systems*. New York: Wiley, 1967.

13. Myklestad, N. O. "A New Method of Calculating Natural Modes of Uncoupled Bending Vibration of Airplane Wings and Other Types of Beams." *Journal of the Aeronautical Sciences*, vol. II, no. 2, Apr. 1944, pp. 153–162.
14. Otten, D. D. "Attitude Control for an Orbiting Observatory: OGO." *Control Engineering*, Dec. 1963, pp. 81–85.
15. Prohl, M. A. "A General Method for Calculating Critical Speeds of Flexible Rotors." *Journal of Applied Mechanics*, Sept. 1945, pp. A142–A148.
16. Rekoff, M. G., Jr. *Analog Computer Programming*. Columbus, Ohio: Charles E. Merrill, 1967.
17. Richardson, L. F. *Arms and Insecurity*. London: Stevens & Sons, 1960.
18. Thomson, W. T. *Vibration Theory and Applications*. Englewood Cliffs, N.J.: Prentice-Hall, 1965.
19. Timoshenko, S. P., and Gere, J. M. *Mechanics of Materials*. New York: Van Nostrand Reinhold, 1972.
20. Timoshenko, S. P., Young, D. H., and Weaver, W., Jr. *Vibration Problems in Engineering*, 4th ed. New York: Wiley, 1974.
21. Tondl, A. *Some Problems of Rotor Dynamics*. London: Chapman & Hall, 1965.
22. Wilcox, J. B. *Dynamic Balancing of Rotating Machinery*. London: Sir Isaac Pitman and Sons, 1967.

Index

Absorber, vibration, 198
Accelerometers, 353–354
Amplitude ratio, 122, 268
Analog computer
  mathematical operations, 336–337
  schematic diagrams, 337–339
  solutions, 336–339

Balance
  dynamic, 245–251
  machines for, 251
  static, 243–245
Balancing machines, 251
Bang-bang control, 293–297
Beam deflection equations, 376–379
Beams, lateral vibration of, 210–219
Bearing compliance, effect on critical speeds, 226–229
Block diagram algebra, 308–314
Block diagrams, 303–314
  block diagram algebra, 308–314
  cascaded blocks, 308–309
  manipulation, 311–314
  summing point, 307, 311–313
  take-off point, 307, 313
  terminology, 307–309

Bode diagrams, 268–269, 272–274
  experimental determination, 366–368
  straight-line approximation, 269–270, 274
Boundary conditions, 213, 218
Break frequency, 270

Capacitance
  electrical, 21–22
  heat, 26
Cascade, blocks in, 308–309
Cathode ray tube oscilloscope, 350–352
  two-channel, 362
Characteristic equation, definition, 43
Closed-loop systems, 259
Combining simultaneous differential equations, 53
Compensation, 296, 321–323
Complementary function (solution), 49, 52, 194
Complex number method, 390–393
Complex plane, 390
Conservative vibrating system, definition, 394
Control actions
  bang-bang, 293–297

Control actions (continued)
  derivative, 297
  integral, 288–291, 314
  proportional, 285–288
  proportional plus derivative, 297
  proportional plus integral, 291–293
  proportional plus integral plus derivative, 297
Control systems
  closed-loop, 259
  feedback, 2, 256, 258
  home heating system, 25, 257
  liquid level, 285–293, 314–321
  open-loop, 259
Corner frequency, 270
Coulomb friction, 12, 235
Critical damping, 105, 107
Critical speeds, 221–234
Curvilinear translation, 16

D'Alembert, principle of, 223, 245
Damped natural frequency, 107, 200, 274
Damping
  Coulomb, 12, 235
  critical, 105, 107
  equivalent, 176–177
  nonlinear, 61
  viscous, 11–13
Damping coefficient
  rotational, 15
  translational, 11–12
Damping ratio, 105–106, 108, 200
Dashpot
  rotational, 15
  translational, 12
DC motor with load, 23–25
Decibels
  conversion table, 397
  definition, 268
Degrees of freedom, 86–88, 166–169, 209
Derivative control, 297
Differential equations
  degree of, 41
  first-order, 43
  higher-order, 46
  homogeneous, 43
  linear, 42
  nonhomogeneous, 49
  nonlinear, 42
  order of, 41
  ordinary, 41
  partial, 41
  second-order, 44
Digital computer solutions, 339–343
Distributed parameter systems
  definition, 209
  rigorous analysis of, 211–216
Disturbance, 257
Driving function, 42
Dynamic balance, 245–251
Dynamic system, definition, 1

Eccentricity, 222, 243
Ecological systems, 2–3, 26–29
Electric elements, table of, 22
Electric systems, 21–25
Equivalent damping, 177
Equivalent inertia, 176–177
Equivalent spring rate, 177
Euler formula, 381, 391

Feedback compensation, 321–323
Feedback control systems, 2, 256, 258
Feedback loops, 309–311
Fluid power system, 20–21
Forced vibration. *See* Vibration
Forcing function, 42
Free vibration. *See* Vibration
Frequency, 82
  circular, 84
  damped natural, 107
  natural, 85, 215
  resonance, 85
  undamped natural, 96, 98
Frequency response
  analytical determination of, 125–127, 141–143, 266–270
  curves, 128–129, 138–139, 144–145, 268, 269, 272–273
  definition, 122
  experimental determination of, 366–368

Friction
  Coulomb, 12, 235
  viscous, 11–13

Gravity constant g, 89

Half-speed whirl, 235
Harmonic motion, 82
Home heating system, 25–26, 257–258
Hydraulic equipment
  actuator (cylinder), 20, 288
  valve, 20, 288
Hydraulic systems, 18–21
Hysteresis whirl, 235

Impulse, 290
Inductance, electrical, 21–22
Inertia, equivalent, 176–177
Initial conditions, 43, 47, 314
Inputs
  displacement, 122–123, 141–143
  force, 122
  ramp, 349, 360–361
  rotating unbalance, 124–125, 137–141
  simultaneous, 149–153
  sinusoidal, 121, 261, 349, 361
  step, 99, 261, 349, 359–360
Integral control, 288–291, 314

Journal bearing whirl, 234

Kirchhoff's first and second laws, 21

Laplace transforms, 324
Limit cycle, 296
Limited stability, 276–277
Linearization techniques and principles, 57, 62–63
Linear variable differential transformer (LVDT), 355
Lin's method, 384–385
Liquid-level systems, 19, 59, 64, 135, 262, 285–293, 314, 329, 341
Lissajous figures, 368, 370
Logarithmic decrement
  definition, 111
  method, 110, 274

modified, 112
Loop analysis of electric systems, 21–23
Lumped parameter systems, 209–211
LVDT. *See* Linear variable differential transformer

Magnification factor, 122
Mechanical shakers, 355–357
Moment of inertia
  area, 90, 376–377
  mass, 90

Newton's second law of motion
  rotation, 15
  translation, 4
Node analysis of electric systems, 23
Nonlinear damper, 61
Nonlinearities, examples of, 58–62
Nonlinear liquid-level systems, 59–60, 64–68, 341
Nonlinear springs, 59, 63–64

Oil whip, 234
Open-loop systems, 259
Operator p, 42

Particular integral, 49, 52, 194
  table of assumed values, 49
Pendulum, 16–18, 61–62, 86
Period, 82
Phase angle, definition, 84–85
Phase-plane plot, 335
Political-military system, 29
p operator, 42
Position pickups, 355
Principal modes of vibration, 193
Proportional control, 285–288
Proportional plus derivative control, 297
Proportional plus integral control, 291–293
Proportional plus integral plus derivative control, 297

Ramp input, 349, 360–361
Ramp response, 365–366
Rayleigh's energy method, 394–396

Resistance, electrical, 21–22
Resonance, 85, 130, 197
Response
  frequency. *See* Frequency response
  ramp, 365–366
  step. *See* Step response
Right-hand rule, 247
Root solving, 383–385
Rotating unbalance, vibration caused by, 124–125, 137–141
Rotating vector method, 387–390
Rotor, definition, 221
Routh's criterion, 277, 398–400

Satellite, 87, 294
Shape function, 216
Slug, 89
Spacecraft, 294
Spring-mass-damper systems, 13
Spring-mass systems, 5
Spring rate
  equivalent, 177
  linearized, 63–64
  torsional, 15
  translational, 6
Springs
  beam, 10
  parallel, 7, 8, 16
  series, 7, 9–10, 16
Stability, 259, 275–277, 398–400
  limited, 276–277
  Routh's criterion for, 398–400
State space
  definition, 334
  trajectories, 334–335
State variable formulation, 327–328
State variables, 327
State vector, 334
Static balance, 243–245
Steady-state response, 52
Step input, 99, 261, 349, 359–360
Step response
  analytical determination of, 99–100, 264–266, 271, 274
  curves, 266, 271

experimental determination of, 363–365
Straight-line approximation for frequency response curves, 130–131, 269–270, 274
Strip chart recorder, 352–354, 362
Stroboscope, 357–358, 369–371
Stroboscopic techniques, 369–371
Summing point, 307
  relocation of, 311–313
Superposition, principle of, 42, 150, 320, 343
Synthetic division, 384

Tachometer feedback, 322–323
Takeoff point, 307–308
  relocation of, 313
Thermal system, 25–26
Threshold speed, 234–235
Time constant, 264, 270
Transfer function, 303–306
Transient response, 52
Transmissibility, 146–149

Undamped natural frequency, 96, 98, 200
Units
  SI, 88–90
  U.S., 88–90

Vibration
  definition, 80
  forced, 85, 121, 194, 200–201
  free, 81, 85, 97, 104, 192, 199
  principal modes of, 193
Vibration absorber, 198
Vibration meters, 353–354
Viscous friction, 11–13

Whipping, 221
Whirling, 221
Whirl modes, 230–233

X-Y recorder, 352–353

80 81 82 9 8 7 6 5 4 3 2

$1000
M. Jackson
Billy Jean
30th caller
570-101-033
Before 7 pm